QUANTUM SEMICONDUCTOR STRUCTURES

Fundamentals and Applications

QUANTUM SEMICONDUCTOR STRUCTURES

Fundamentals and Applications

CLAUDE WEISBUCH
BORGE VINTER

Thomson-CSF
France

ACADEMIC PRESS, INC.
Harcourt Brace Jovanovich, Publishers

Boston San Diego New York
London Sydney Tokyo Toronto

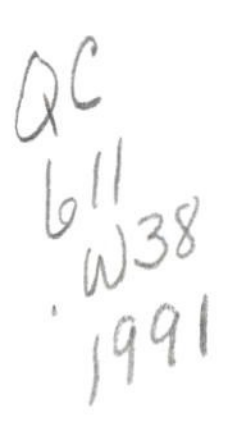

This book is printed on acid-free paper. ∞

ACADEMIC PRESS, INC.
1250 Sixth Avenue, San Diego, CA 92101

United Kingdom Edition published by
ACADEMIC PRESS LIMITED
24–28 Oval Road, London NW1 7DX

Library of Congress Cataloging-in-Publication Data:

Weisbuch, C. (Claude), date.
Quantum semiconductor structures : fundamentals and applications / Claude Weisbuch, Borge Vinter.
p. cm.
Includes bibliographical references and index.
ISBN 0-12-742680-9 (acid-free paper)
1. Semiconductors. 2. Quantum Hall effect. I. Vinter, Borge.
II. Title
QC611.W38 1991 90-46358
621.381′ 52—dc20 CIP

Printed in the United States of America

91 92 93 94 9 8 7 6 5 4 3 2 1

Contents

Foreword

Once in a while a topic comes along that has universal appeal. It is both understandable to the undergraduate and challenging to the most sophisticated researcher, no matter what language he or she speaks. Such is the field of quantum semiconductor structures.

The fundamental, theory—that of a particle in a one-dimensional box—has been known since the earliest days of quantum mechanics. Even so, it is hard to believe that the field as we know it today was founded almost 20 years ago. At that time, groups at AT&T Bell Laboratories and IBM T. J. Watson Research Laboratories began studying very thin layers of GaAs and $A1_X$ Ga_{1-X} As heterostructures in which layers were less than 250Å in thickness; the first successful results were reported in 1973.

Both groups benefited greatly from early fundamental work in molecular beam epitaxy, although the materials, the physics, and the device engineering were quite distinct and separate fields. Devices, namely quantum well lasers, field effect transistors, and high-frequency oscillators, were proposed or actually discovered very early by the original research groups and have been evolving steadily ever since. Many basic discoveries—some readily understood by existing theory while others confounded the theories and required new models and concepts, have also been made. Although new and fascinating developments continue, the field is beginning to mature. The stage is set for a major publication to bring together the many elements of diverse achievement.

This book, with its encyclopaedic clarity, should be ranked with the best! When I read it for the first time, thoughts of Linus Pauling's *Nature of the Chemical Bond* and other great teaching/research texts passed through my mind. This book is of the "right stuff" and could achieve equivalent status in the years to come.

The wide range of achievement that is presented here with clarity, and liberally illustrated from the original literature, is all the more astonishing when one recalls that 15 years ago it was not possible to have a paper on this topic accepted in the mainstream conferences of the day. Now perhaps 50% of meetings on semiconductor and optoelectronic physics, materials, and devices are dominated by talks describing two-dimensional, one-dimensional, and zero-dimensional electronic systems in III–V semiconducting materials.

Dr. Claude Weisbuch is an excellent choice as the senior author of this book. His 20 years of experience in leading physics and semiconductor electronics research in major institutions on both sides of the Atlantic have developed the background required to speak with authority. Dr. Borge Vinter has added his own flavor, especially through theoretical derivations that are instructive and clear.

For those like myself, who have been waiting for a comprehensive exposé of this quantum semiconductor electronic field to be available, they need wait no longer—and they will not be disappointed. This book will be an authoritative reference for decades to come and will stimulate both newcomers and experienced researchers to greater achievement in this fascinating field.

Raymond Dingle
Sydney, Australia

Preface

This book is the outgrowth of a previous work published as an introductory chapter in *Applications of Multiquantum Wells, Selective Doping, and Superlattices* (Volume 24 of the series *Semiconductors and Semimetals*, R. K. Willardson and A. C. Beer series editors, R. Dingle volume editor). Due to both the demand for an introductory text for student use and rapid progress in the field, a new, expanded student edition was needed.

The teaching demand is, of course, dictated by the extraordinary successes of low-dimensionality semiconductor heterostructures, both in fundamental and applied fields. After a slow start in the Seventies, pioneered by Esaki and Tsu for the transport properties, and by Dingle for optical properties, the field got a sudden impulse at the end of the decade due to improved growth methods, and breakthroughs in fundamental science (the Quantum Hall effect by von Klitzing) and applications (modulation doping by Störmer and Dingle and high-quality injection quantum well lasers by Tsang). The situation in the mid-Eighties is well reviewed in the aforementioned volume, where outstanding fundamental properties and applications were described: high-yield optical properties, electron mobilities up to 2.10^6 cm^2 V^{-1} s^{-1}, the normal and "anomalous" Quantum Hall effect, new material pairs from strained-layer epitaxy, low-threshold quantum well lasers, ultrahigh speed microwave and digital integrated circuits, and electro-optical and nonlinear materials with unexpected efficiencies.

Since then, the rate of progress has not slowed down: on the fundamental side electron mobilities now reach 12.10^6 cm^2 V^{-1} s^{-1}, the anomalous Quantum Hall effect has even fractional numbers, and lower dimensionality electrical systems exhibit numerous new effects (quantum point resistance, electron focusing, ballistic motion up to 100 μm, nonlocal propagation etc.). In the applications side, we are witnessing the ubiquity of the quantum well laser, which wins on all performance segments (threshold, power, linewidth, temperature coefficient, etc.) with the recent advent of strained-active layers, vertical-emitting lasers, and phased-arrays; electro-optical processing arrays reach large IC size with more than 2000 active elements. Devices based on both horizontal and vertical transport still progress at a fast pace, in speed and integration level. Finally, quantized semiconductor structures are an industrial reality as millions of low noise transistors (mainly for direct-broadcast-satellite receivers) and quantum-well lasers (mainly for compact-disc players) are produced every month.

The aim of the present book is to provide a short introduction to this many fascinating applications of quantized semiconductor structures and convey some of the bewilderment that one should encounter at the extraordinary field, which developed in a decade, and the pace at which it still expands. These applications are based on a theoretical background, which should be acquired not only by those who intend to study fundamental properties of quantized systems but also by those who devote their efforts to inventing and fabricating devices with ever-increasing performance. The number of fundamental topics relevant to the general understanding of the field of quantized semiconductor structures is quite large, and some choice had to be made to keep the size of this introductory book within reasonable bounds: we selected only those calculations that we feel every student should go through once, and we give only sketchy descriptions of all other theories, referring to more specialized texts for details.

The material in this book comes from various interactions with many individuals. Claude Weisbuch wishes especially to thank R. Dingle, who introduced him to the field back in 1979 and has since been not just a colleague but a friend and a source of major inspiration as well. At Thomson, since 1983, E. Spitz has provided unrestricted support, both in the company and as a friend. H. Störmer was an especially close colleague and friend during the Bell Labs years. Bell Labs was an outstandingly welcoming institution; very fruitful collaborations occurred, principally with A. Gossard, W. Wiegmann, W. Tsang, A. Cho, J. Hegarty, A. Sturge, R. Miller, P. Petroff, C. Shank, R. Fork, B. Greene, A. Pinczuk, and V. Narayanamurti. At Thomson-CSF, both of us would like to thank J. P. Harrang, J. Nagle, N. Vodjdani, P. Bois, E. Costard, S. Delaître, T. Weil, F. Chevoir, E. Rosencher, and A. Tardella, who at various times and in various capacities provided a stimulating environment and useful discussions. During three years, an especially fruitful and friendly collaboration was established with M. J. Kelly and his colleagues at GEC. More recently, another excellent teaming with J. Kotthaus, J. Williamson, S. Beaumont, C. Sotomayor-Torres, P. Van Daele, R. Baets, F. Briones, C. Harmans, E. Böckenhoff and H. Benisty provided many important insights. Colleagues have kindly supplied us with preprints and photographs for which we are most thankful. Jane Ellis of Academic Press has been a most supportive editor. Brigitte Marchalot made the physical production of this new edition possible through her talent and cooperation.

Claude Weisbuch
Borge Vinter

CHAPTER I

Introduction

1. The Advent of Ultrathin, Well-Controlled Semiconductor Heterostructures

Although the search for ultrathin materials can be traced quite far back,[1,2] the motivation for their production went up sharply when new types of devices[3,4] were predicted, such as the Bloch oscillator. At the same time, the advent of a new growth technique, molecular beam epitaxy (MBE),[5-12] opened the way to the growth of semiconductors atomic layer upon atomic layer. In 1974 two basic experiments were carried out: Esaki and Chang reported the oscillatory behavior of the perpendicular differential conductance due to resonant electron tunneling across potential barriers,[13] and the optical measurements of Dingle[14] showed directly the quantization of energy levels in quantum wells, the well-known elementary example of quantization in quantum mechanics textbooks.[14a] Studies of ultrathin semiconductor layers have since then proliferated at an explosive rate.

Owing to progress in crystal availability and control, basic understanding of low-dimensional systems, and applicability of heterostructure concepts, the recent years have also seen the emergence of a wide family of structures and devices, which can be classified into four main (overlapping) families, as shown in Table I. At this point it seems worthwhile to emphasize the various structures that will be described or mentioned in this review, as their abundance can sometimes be confusing. They are depicted in Fig. 1 by means of their band diagrams. In many of these structures, we will be

TABLE I

THE FOUR MAIN FAMILIES OF DEVICES ORIGINATING FROM ULTRATHIN, WELL-CONTROLLED SEMICONDUCTOR HOMO- AND HETEROSTRUCTURES [a]

TWO-DIMENSIONAL SYSTEMS

SDHT-TEGFET-HEMT-MODFET
NIPI
Quantum Wells
Quantum Hall Devices

CHARGE TRANSFER SYSTEMS	ONE-DIMENSIONAL SYSTEMS
SDHT-TEGFET-HEMT-MODFET	Tunneling Structures
NIPI	Superlattices, NIPI (in perpendicular transport)
Real SpaceTransfer Devices	Quantum-Well Wires

BANDGAP ENGINEERED STRUCTURES

all of the above plus non-quantized-motion structures:

Double-Heterostructure Lasers
Graded-Gap APD
Heterostructure Bipolar Transistors (graded base or not)
Separate Absorption- Multiplication APD
Staircase Solid State Photomultiplier

[a] Note that the same structures can belong to several of the families and that, using the term *bandgap engineering* in its most general description of engineered structures with desired properties obtained by a tailoring of the band structure, all of the structures can be considered "bandgap-engineered."

interested in *quasi-two-dimensional* properties; the free motion of the carriers occurs in only two directions perpendicular to the growth direction, the motion in the third direction z being restricted to a well-defined portion of space by momentum, energy, and wave-function quantizations. Compared to "classical" heterostructures like double-heterostructure (DH) lasers,[15,16] the "quasi-2D" term means that the z motion is defined by one or a few quantum numbers, which is only the case in ultrathin structures and/or at low enough temperatures. We use here the word *quasi* to mark the difference with *exact* 2D systems in which the wave function is exactly confined in a plane, with no extension outside of that plane. In the

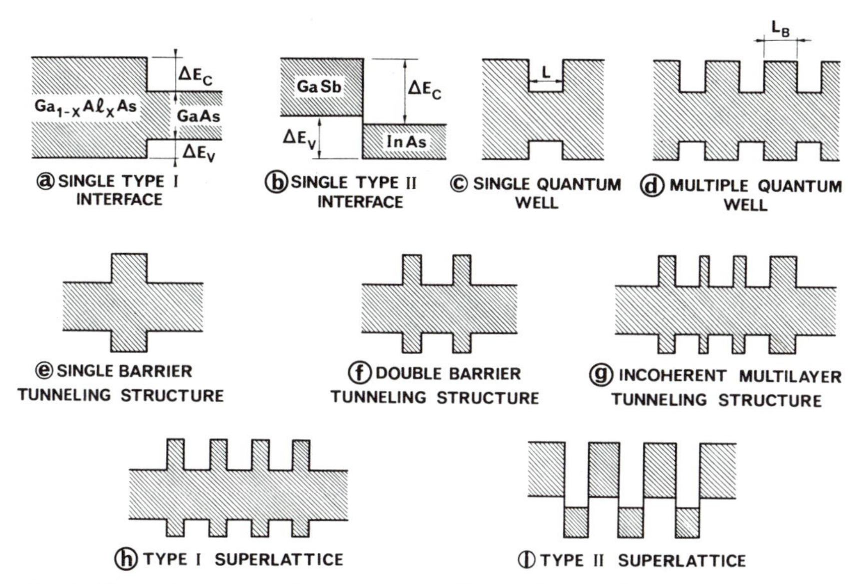

FIG. 1. The various types of heterostructures discussed or mentioned in this chapter. The widely used type-I heterostructure is shown in (a), with the band discontinuities such that both band edges of the smaller gap material are below those of the wide-bandgap material. In the type-II interface (b), the band structure is such that the top of the valence band of one of the compounds lies above the bottom of the conduction band of the other compound. Charge transfer occurs, leading to a conducting heterostructure. The type-I quantum well is shown in (c). The multiple-quantum-well structure [MQW, (d)] is such that L_B is large enough to prevent tunneling. Conversely, L_B in the single barrier (e), double barrier (f), type-I incoherent tunneling (g) and superlattice (h) structures is small enough to allow carrier tunneling across the barrier material. The difference between these two latter structures, (g) and (h), is that in the superlattice structure disorder and scattering are low enough to allow the *coherent* superlattice band states to build up, whereas in the incoherent tunneling structure scattering by disorder (here disordered interface fluctuations) destroys the phase coherence between the tunneling states. As charge transfer occurs in type-II multiple-quantum-well structures (i), these are considered as semimetallic superlattices, with the exception of ultrathin structures where energy quantization is so large that energy levels are raised enough in the respective bands to prevent any charge transfer.

rest of this chapter we shall refer to our quasi-2D systems merely as 2D systems.

The most widely known devices exploiting 2D motion are the quantum-well lasers[17,18] and the SDHT–TEGFET–HEMT–MODFET heterostructure transistors.[19–21] The parallel transport properties of $n-i-p-i$ structures[22] might prove useful in some devices like the heterojunction modulation superlattice. Due to the extraordinary properties of the quantum Hall effect,[23] some applications might be found in high-performance gyromagnetic devices. Quantum Hall structures are already being widely used as standard resistors in numerous national standards laboratories.

In some cases, we will be interested in the *one-dimensional* phenomena occurring along the z direction, either due to our search for perpendicular properties (i.e., perpendicular transport) or due to the unconfined extension of the wave functions in the z direction (superlattices or type-II multiple wells). Devices using these one-dimensional properties rely on tunneling or superlattice transport. Whereas tunneling devices such as the tunneling transistors[24] or negative differential resistance (NDR) tunneling diodes[25] have been demonstrated, clear superlattice effects have so far remained elusive. Great efforts are being devoted to the fabrication and understanding of true quantum one-dimensional systems best viewed as *quantum well wires.*[26,27]

The third family of devices shown in Table I relies on charge transfer, either *static* or *dynamic.* In the *static* case the charge transfer occurs between heterodoping and/or heterocomposition structures, leading to the appearance of electrostatic confining potentials due to depleted charges. Some of the 2D systems discussed above rely on this charge-transfer effect. *Dynamic* charge transfer occurs when electric-field-heated carriers can overcome potential barriers in heterostructures, leading to diminishing conductance and thus to NDR.[28-30]

Bandgap engineering[31] consists of the tailoring of an association of materials in order to custom design the structure for some desired properties unattainable in homostructures. A very good prototype of such structures is the double-heterostructure laser,[15,16] where one increases both the carrier confinement and optical wave confinement by using a heterostructure. It is clear that all the devices described above can be viewed as being due to bandgap-engineered structures. A number of other structures have been recently developed that do not involve space quantization in ultrasmall structures. These are shown in the lower part of Table I.

As can be seen in Table I, the variety of devices which have now been demonstrated is quite overwhelming, although the first devices (quantum-well lasers and modulation-doped structures) only appeared in the late 1970s.

The present book aims at presenting the basic physical phenomena encountered in these devices. The field is already so large, however, that we have concentrated on the basic phenomena encountered in the simplest and most widely used semiconductor pairs, the so-called type-I quantum wells and interfaces, where the small-bandgap material has both its electron and hole levels confined by the wider-bandgap materials. The other configurations (type-II quantum wells) have been thoroughly reviewed.[32-35] More details on strained-layer superlattices and their applications can be found in reviews of this young but rapidly developing field,[36–38] with applications in lasers[38] and infrared detectors.[39] The new field of amorphous semiconductors[40–43] is too far afield and will not be considered here, although many of the tools developed here can be applied to that subject. As new promising materials systems, not covered, one should point out SiGe/Si[44] and II-VI systems, the former for Si-compatible heterostructure and quantum devices (as evidenced by HBTs and quasi-direct gap optoelectronic material), the latter for optoelectronic material from middle infrared to blue and for new magnetic superlattice properties.

2. A Prerequisite: The Mastering of Semiconductor Purity and Interfaces

The mastery of layer growth is a prerequisite to all the structures which will be discussed in this book. We therefore wish to give an overview of the achievements in that field, referring the reader to more specialized texts for details. Quite different techniques have been used to grow quantized structures such as MBE,[5–12] metal–organic chemical vapor deposition (MOCVD),[45,46] hydride vapor transport,[47,48] hot-wall epitaxy[49] (HWE), or even liquid- phase epitaxy[50] (LPE). One can even trace through time how progress brought about by such a near-perfect growth technique as MBE has induced parallel spectacular progress in other growth techniques by demonstrating new and attainable goals.

The highly detailed control of crystal growth in MBE has been crucial to its progress and is due to the UHV environment, which allows for the implementation of powerful *in situ* analytical techniques. The growth sequence in an MBE chamber uses specific characterizations to ensure that each growth step is correctly carried out: before growth has started, mass analysis of residual molecules in the chamber detects any unwanted molecular species. Molecular beam intensities are precisely controlled by ion gauges. Substrate cleaning is checked by Auger electron spectrometry, which analyzes the chemical nature of the outer atomic layer. Reflection high-energy electron diffraction (RHEED) patterns monitor surface reconstruction after ion cleaning, annealing, and also during atomic layer growth. Studies of atomic layer growth through desorption measurements and RHEED analysis have provided a detailed understanding of MBE growth mechanisms.[51–54] RHEED oscillations due to recurrent atomic patterns in the layer-after-layer growth mode provide a very useful means of measuring layer thickness and are being more and more widely used.[55,56] TEM measurements of grown films have evidenced the smoothing effect of MBE growth on the starting substrate's roughness[57] (Fig. 2).

Although the growth kinetics of the MOCVD process is not as well monitored as that of MBE, recent progress leads to believe that MOCVD

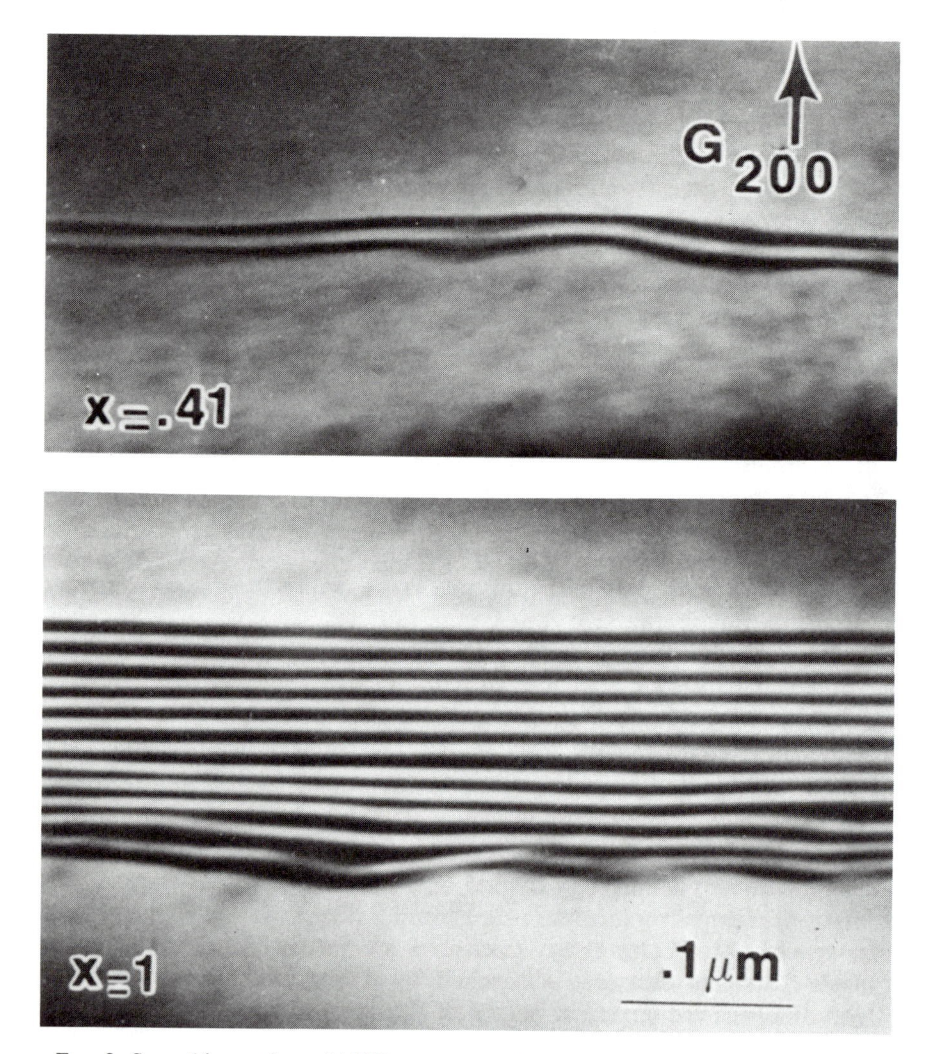

FIG. 2. Smoothing action of MBE quantum-well growth on interface roughness as observed in a dark-field transmission electron micrograph. The roughness of the starting GaAs surface is smoothed out by the growth of 3 to 5 quantum wells (courtesy of P.M. Petroff, AT&T Bell Laboratories).

growth leads to similar control of impurity content and interface abruptness[58–61a] (Fig. 3).

A vast amount of effort has also been devoted to characterization of interfaces, using various *ex situ* techniques such as chemical etching,[62] beveling,[62] SIMS,[63] Auger,[64] TEM,[65] and x rays.[66,67] The latter two tech-

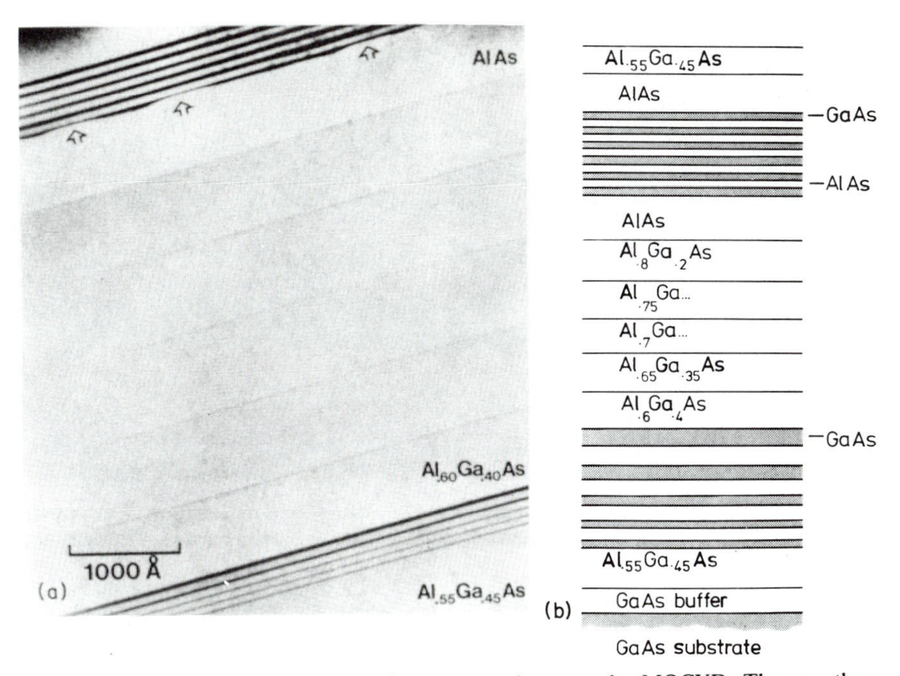

FIG. 3. (a) TEM characterization of a test sample grown by MOCVD. The growth sequence and the structure are shown in (b). The remarkable features are the sharpness of the very narrow GaAs layers (minimum ≈ 25 Å) appearing at the lower right-hand side corner, the interface roughness showing up at the uppermost interface of the AlAs layer, and the subsequent smoothing of this roughness by the multilayer growth (upper left-hand side corner) (after Leys *et al.*[61]).

niques have been shown to yield extremely precise information on a microscopic scale (Fig. 4). It has been thus shown that the preferred growth techniques, MBE and MOCVD, which are far from equilibrium growth processes, allow very low growth rates and thus good control for desired abrupt changes. Hot-wall epitaxy, an evaporation method, also leads to good interface control but has been used much less, due to the required high-purity bulk material. VPE and LPE are near-equilibrium methods with large growth rates and instabilities in the regime where redissolution (LPE) or etching (chlorine VPE) could diminish the total deposition rate. Stringfellow[68] also involved Cl absorption in the Cl-VPE method as a limitation to atomic in-plane motion and hindered coalescence of islands during atomic layer formation. Frijlink *et al.*[58] pointed out the strong reactivity of aluminum chloride with reactor material, forbidding growth of Al-containing structures with the Cl method.

FIG. 4. High-resolution electron micrograph of a GaAs–GaAlAs interface. The arrows point to the interface plane, which appears very smooth on the atomic scale (courtesy P. M. Petroff, AT&T Bell Laboratories).

In terms of purity, the two techniques have now emerged as those yielding the best bulk material ever grown by any technique; high mobilities and sharp luminescence intrinsic peaks attest the high quality of MOCVD[69,70] and MBE-grown GaAs.[71] Recently, MOCVD-grown InP[72] has given mobilities of 195,000 $cm^2 V^{-1} s^{-1}$ at 77 K. The steady progress in recent years can be traced to the availability of purer source materials and to a better control of the growth environment. The latter point is especially well documented for MBE, where introduction of better pumping systems, liquid-nitrogen shrouding of the growth space, and vacuum interlock transfer of substrates each brought significantly improved material properties. The amount of effort still being made in the basic understanding of growth methods, the better quality of starting substrates, and the availability of ever-purer starting materials should lead to still increasing material quality.

The range of materials now grown in ultrathin layers is extremely wide, and we shall not attempt to list them, as the rate of appearance of new ultrathin materials is still high. It has been widely believed that high-quality material could only be grown with layers perfectly lattice matched to the substrate, although it was remarked very early[73,74] that no misfit dislocation generated by the mismatch would occur if the epitaxial layers were sufficiently thin, allowing the mismatch to be fully accomodated by elastic strain. The realization of this effect led to the consideration of ultrathin multilayer structures with a much wider set of materials than with lattice-matched combinations.[75] Within the allowed range, the choice of layer thickness allows one to select a strain value which offers an additional parameter for the tailoring of electronic properties.[44] The most promising recent systems are at present HgTe/CdTe,[75a,b] where the superlattice growth should allow an easier control of bandgap than in LPE-grown alloys[76]; InAsSb/InAsSb, where the lattice strain should permit one to *decrease* the bandgap in the 10 μm range[77]; CdMnTe/CdTe,[78,79,79a] which has fascinating magnetic properties; and Ge_xSi_{1-x}/Si,[80,80a] where the strain could allow one to reach the 1.77 μm range for photodetectors and might also lead to direct-gap material on a Si substrate.

As will be described in Section 23, quantum well lasers incorporating strained layers as active regions evidence superior performance, thanks to the modification of valence band levels introduced by the strain.

CHAPTER II

The Electronic Properties of Thin Semiconductor Heterostructures

3. Quantum Well Energy Levels

a. Conduction Electron Energy Levels

The simplest quantum situation to be dealt with consists of a single layer of material A embedded between two thick (thickness much greater than the penetration length of the confined wave function) layers of material B, where B has a bandgap larger than A and where the band discontinuities[81] are such that both types of carriers are confined in the A material (Fig. 1c). This is the situation exemplified by the pairs of materials GaAs/GaAlAs, GaInAsP/InP, GaInAs/AlInAs, GaSb/AlSb, etc. The energy levels in the conduction band can be calculated quite easily in the *approximation of the envelope wave function,*[82-82d] using a Kane model[83] for describing the electron and hole states of the parent A and B materials.[84] The approximation assumes (1) an interface potential strongly localized at the $A-B$ interface, which means that on the scale of variation of the envelope wave function the interface potential is well localized at the geometrical interface, and (2) an interface potential which does not mix the band-edge wave functions but only shifts them, which is plausible due to the very different symmetries of the conduction and valence bands. It can then be shown[82] that the electron wave function takes approximately the form

$$\psi = \sum_{A,B} e^{i\mathbf{k}_\perp \cdot \mathbf{r}} u_{c\mathbf{k}}^{A,B}(\mathbf{r}) \chi_n(z)$$

where z is the growth direction, $\mathbf{k}_\perp$ is the transverse electron wave vector, $u_{c\mathbf{k}}(\mathbf{r})$ is the Bloch wave function in the A or B material, and $\chi_n(z)$ is the envelope wavefunction, determined to a good approximation by the Schrödinger-like equation[82–82f]

$$\left(-\frac{\hbar^2}{2m^*(z)}\frac{\partial^2}{\partial z^2}+V_c(z)\right)\chi_n(z)=\varepsilon_n\chi_n(z) \tag{1}$$

where $m^*(z)$ is the electron effective mass of the A or B material, $V_c(z)$ represents the energy level of the bottom of the conduction bands, and ε_n is the so-called confinement energy of the carriers. Therefore, the early description of the energy level scheme by simple confinment in a quantum well due to energy-band discontinuities in Dingle's work[1,14] can be well justified.

The continuity conditions at the interfaces are that $\chi_n(z)$ and $[1/m^*(z)][\partial\chi_n(z)/\partial z]$ should be continuous. Rather than the continuity condition of the derivative of the wave function as derived in quantum mechanics textbooks[14a] it is necessary for conservation of particle current to use the continuity of $(1/m^*)(\partial\chi/\partial z)$.

In the *infinitely deep well approximation,* the solution to Eq. (1) is very simple, as the wave function must be zero in the confining layer B, and therefore also at the interface because of the continuity equations. Taking the z abscissa origin at one interface (Fig. 5), the solution of Eq. (1) can evidently only be $\sim\sin(n\pi z/L)$, n being an odd or even integer. The confining energy ε_n is then simply $n^2(\pi^2\hbar^2/2m^*L^2)$ from Eq. (1).

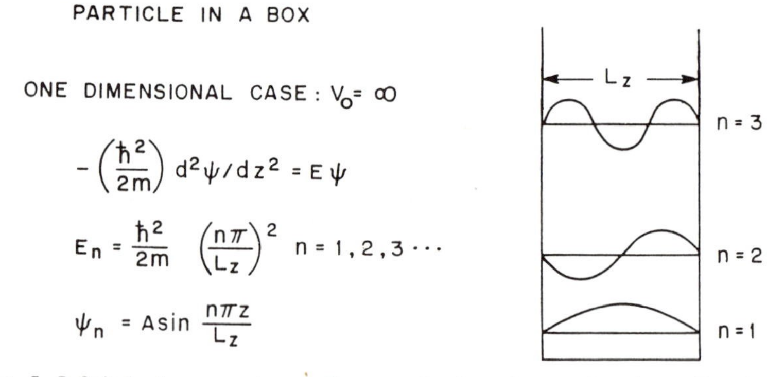

FIG. 5. Infinitely deep quantum-well energy levels and wave functions. (Reprinted with permission from Friedr. Vieweg & Sohn Verlagsgesellschaft mbH, R. Dingle, *Festkoerperprobleme* **15**, 21 (1975)).

The Schrödinger-like equation (1) in the *finite-well* case with the aforementioned boundary conditions can be exactly solved to yield the wave functions and energies.

Noting that the problem has an inversion symmetry around the center of the well now taken as the center of coordinates (Fig. 6), the solution wave functions of (Eq. 1) can only be even or odd. Therefore, they can be written as

$$\begin{aligned} \chi_n(z) &= A\cos kz, && \text{for} \quad |z| < L/2 \\ &= B\exp[-\kappa(z - L/2)], && \text{for} \quad z > L/2 \\ &= B\exp[+\kappa(z + L/2)], && \text{for} \quad z < -L/2 \end{aligned} \tag{2}$$

or

$$\begin{aligned} \chi_n(z) &= A\sin kz, && \text{for} \quad |z| < L/2, \\ &= B\exp[-\kappa(z - L/2)], && \text{for} \quad z > L/2 \\ &= B\exp[+\kappa(z + L/2)], && \text{for} \quad z < -L/2 \end{aligned} \tag{3}$$

where

$$\varepsilon_n = \frac{\hbar^2 k^2}{2m_A^*} - V_0, \qquad \varepsilon_n = -\frac{\hbar^2 \kappa^2}{2m_B^*}, \qquad -V_0 < \varepsilon < 0 \tag{4}$$

For the solution of Eq. (2), the continuity conditions at $z = \pm L/2$ yield

$$A\cos(kL/2) = B$$

$$(k/m_A^*)\sin(kL/2) = \kappa B/m_B^*$$

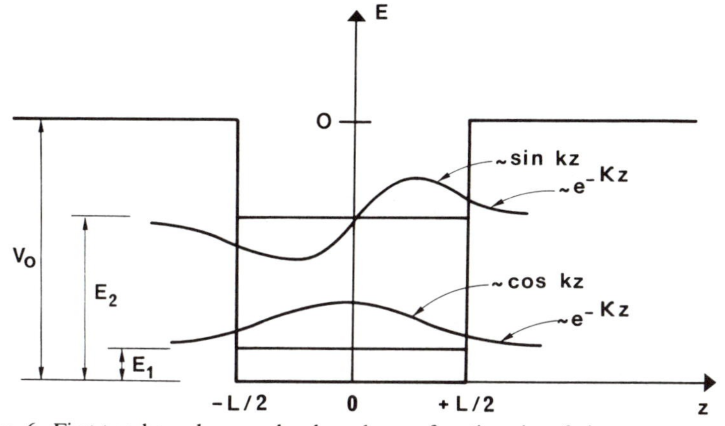

FIG. 6. First two bound energy levels and wave functions in a finite quantum well.

Therefore

$$(k/m_A^*)\tan(kL/2) = \kappa/m_B^* \tag{5}$$

Similarly, Eq. (3) yields

$$k/m_A^* \operatorname{cotan}(kL/2) = -\kappa/\mathrm{m}_B^* \tag{6}$$

The equations can be solved numerically or graphically. A very simple graphical type of solution can be developed if $m_A^* \approx m_B^*$. Then, using Eq. (4), Eqs. (5) and (6) can be transformed into implicit equations in k alone:

$$\cos(kL/2) = k/k_0, \qquad \text{for} \quad \tan kL/2 > 0 \tag{7}$$

$$\sin(kL/2) = k/k_0, \qquad \text{for} \quad \tan kL/2 < 0 \tag{8}$$

where

$$k_0 = 2m^* V_0/\hbar^2 \tag{9}$$

These equations can be visualized graphically (Fig. 7). There is always one bound state. The number of bound states is

$$1 + \mathrm{Int}\left[\left(\frac{2m_A^* V_0 L^2}{\pi^2\hbar^2}\right)^{1/2}\right] \tag{10}$$

where Int[x] indicates the integer part of x.

The important limiting case of the *infinitely high barriers* (Fig. 5) can be found again by putting $k_0 = \infty$ in Fig. 7. There is then an infinity of bound states with $k = n\pi/L$. Even solutions are $\chi_n \sim \cos kz$, with $kL = (2n+1)\pi$; odd solutions are $\chi_n \sim \sin kz$, with $kL = 2n\pi$. χ_n even and χ_n odd are the

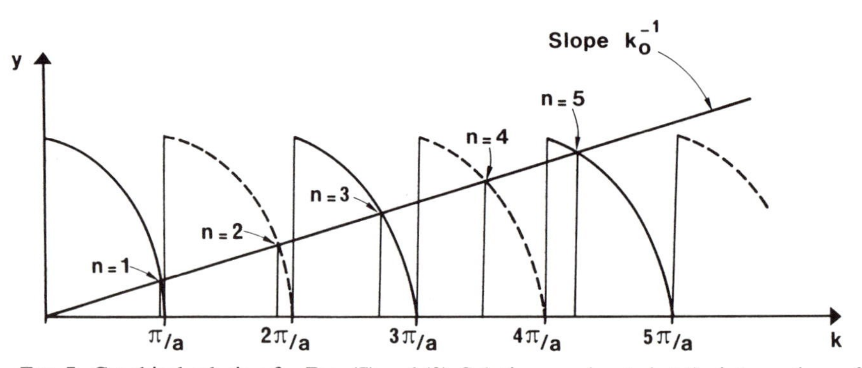

FIG. 7. Graphical solution for Eqs. (7) and (8). Solutions are located at the intersections of the straight line with slope k_o^{-1} with curves $y = \cos kL/2$ (with $\tan kL/2 > 0$; ——; even wave functions) or $y = \sin kL/2$ (with $\tan kL/2 < 0$; ---; odd solutions).

usual solution of the infinitely deep well $\chi_n = \sin(n\pi z/L)$, n integer even or odd, in the natural choice of coordinate origin $z = 0$ at one interface.

b. Hole Energy Levels

Turning to the hole quantization problem, the situation is much more complicated in usual semiconductor materials. The bulk hole bands are described in the Kane model by basis functions with angular momentum $J = \frac{3}{2}$ symmetry, i.e., 4-fold degeneracy at $k = 0$ (neglecting the spin-orbit split-off valence band).

The dispersion *near* $k = 0$ can be described by the Luttinger Hamiltonian[84a]:

$$H = \frac{\hbar^2}{2m_0}[(\gamma_1 + \tfrac{5}{2}\gamma_2)k^2 - 2\gamma_2(k_x^2 J_x^2 + k_y^2 J_y^2 + k_z^2 J_z^2) \\ - 4\gamma_3(\{k_x \cdot k_y\}\{J_x \cdot J_y + \cdots\}] \tag{11}$$

where γ_1, γ_2, γ_3 are the Luttinger parameters of the valence band and the symbol $\{\cdot\}$ represents the anticommutation

$$\{k_x \cdot k_y\} = k_x k_y + k_y k_x$$

In the bulk, propagation in a given direction can be described in terms of heavy- and light-hole propagation. Taking as a quantization axis z for the angular momentum the direction of propagation of the hole, the levels $J_z = \pm\frac{3}{2}$ and $J = \pm\frac{1}{2}$ give a simple dispersion relation from Eq. (11). Taking for example k_z in a [100] direction, the kinetic energy of holes is

$$\begin{aligned} E &= \frac{\hbar^2 k_z^2}{2m_0}(\gamma_1 - 2\gamma_2), \qquad \text{for} \quad J_z = \pm\tfrac{3}{2} \\ &= \frac{\hbar k_z^2}{2m_0}(\gamma_1 + 2\gamma_2), \qquad \text{for} \quad J_z = \pm\tfrac{1}{2} \end{aligned} \tag{12}$$

One obtains the usual [100] heavy-hole mass $m_0/(\gamma_1 - 2\gamma_2)$ and light-hole mass $m_0/(\gamma_1 + 2\gamma_2)$.

For hole levels in a quantum well, in a successive perturbation approach, one first treats the quantum-well potential as a perturbation to the $k = 0$ unperturbed states, then adds the Luttinger interaction as a new perturbation to the quantum-well levels.[85] As a first perturbation, the quantum-well potential lifts the degeneracy between the $J_z = \pm\frac{3}{2}$ and $\pm\frac{1}{2}$ bands as they correspond to different masses. According to the Luttinger equation, Eq. (11), inserting the values $k_x = k_\perp$, $k_y = k_z = 0$, the k dispersion in a [100] direction *perpendicular* to z is then given by

$$E = \frac{\hbar k_\perp^2}{2m_0}(\gamma_1 + \gamma_2), \qquad \text{for} \quad J_z = \pm\tfrac{3}{2}$$
$$E = \frac{\hbar k_\perp^2}{2m_0}(\gamma_1 - \gamma_2), \qquad \text{for} \quad J_z = \pm\tfrac{1}{2} \tag{12a}$$

The transverse dispersion equation corresponding to $J_z = \pm\frac{3}{2}$ (*heavy-hole band along the z direction*), now has a *light mass* $(m_0/\gamma_1 + \gamma_2)$, whereas the $J_z = \pm\frac{1}{2}$ level now has a *heavy mass* (Fig. 8). This situation is quite similar to that developed under a uniaxial compressive stress in the [100] direction.[86] The difference here is that the $\frac{3}{2}$ band is the higher-lying one. Due to the lighter mass of the $\frac{3}{2}$ band, one initially expects a crossing of the two bands. However, higher-order $\mathbf{k} \cdot \mathbf{p}$ perturbation terms lead to an anticrossing behavior, which increases the "heavy-hole" band mass and decreases the "light-hole" band mass.

Actually, the above procedure, which describes qualitatively the complicated valence-band effects, is not correct. One has to treat on equal footing the $\mathbf{k} \cdot \mathbf{p}$ perturbation, which yields the dispersion, and the dimensional

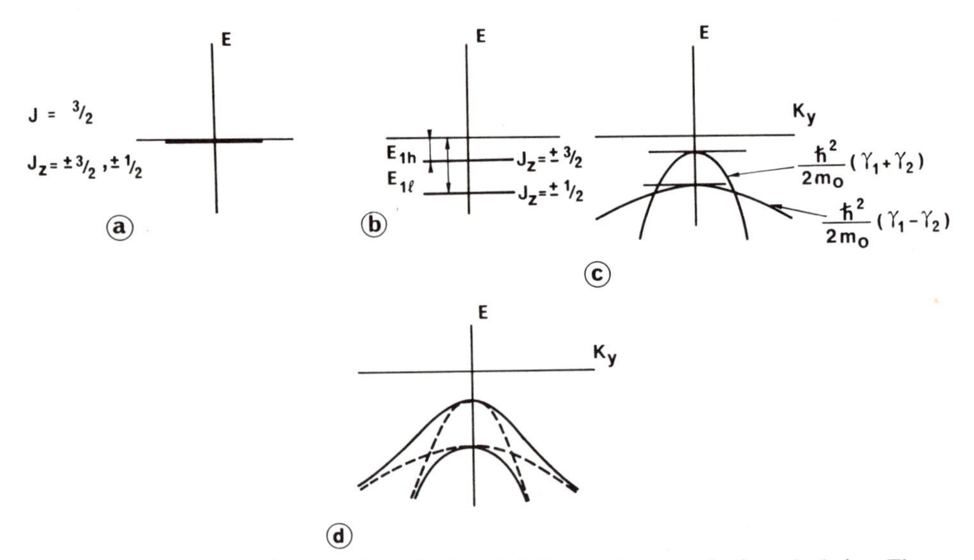

FIG. 8. Hole dispersion curve in a simple-minded successive perturbative calculation. The quantum-well potential lifts the 4-fold degeneracy of holes [in 3D, (a)] at $k = 0$. (b). The $\mathbf{k} \cdot \mathbf{p}$ interaction term as described by the Luttinger Hamiltonian then yields the dispersion in the y direction (for example) (c); finally, higher-order terms lead to an anticrossing behavior, (d).

perturbation introduced by the quantum well. In the degenerate set of valence-band levels at $k = 0$, one has to diagonalize a perturbative Hamiltonian:

$$H = H_{\mathbf{k}\cdot\mathbf{p}} + H_{\mathrm{QW}} \tag{13}$$

As is well known, the first-order solutions are linear combinations of the $k = 0$ valence-band wave functions [when $H_{\mathrm{QW}} = 0$, they are the $J_z = \pm\frac{3}{2}$ and $\pm\frac{1}{2}$ functions with the dispersion given by Eq. (12)]. Complications arise here because of the boundary conditions which have to be simultaneously satisfied for the quantum well. The set of functions which diagonalize $H_{\mathbf{k}\cdot\mathbf{p}}$ is not a basis set for H_{QW}, and strong mixing of the $J_z = \pm\frac{3}{2}$ and $\pm\frac{1}{2}$ bands is required to satisfy the boundary conditions, as was recognized as early as 1970 for the infinitely deep well.[87] More recent works have dealt with various band situations,[88] finite wells,[89] and the additional influence of magnetic fields.[89,90] One should note in addition that this effect strongly influences the value of the exciton Rydberg. Also, the strong nonparabolicity of the valence bands should influence dramatically the valence-band density of states from an exact steplike shape to more complicated shapes, which require a detailed knowledge of the valence-band levels.

The case of an infinitely deep well has been treated analytically.[87,89] Neglecting band warping in the spherical approximation (equality of the Luttinger parameters $\gamma_2 = \gamma_3 = \bar{\gamma}$), the energy levels at $k = 0$ are given by the usual uncoupled levels series:

$$E_{(\mathrm{l,h})h} = n^2 \frac{\pi^2\hbar^2}{2m_0L^2}(\gamma_1 \pm 2\bar{\gamma}) \tag{14}$$

The dispersion for $k_\perp = k_y \neq 0$ is given by the dispersion equation

$$\begin{aligned} &4[k_{\mathrm{l}z}^2 k_{\mathrm{h}z}^2 + k_y^2(k_{\mathrm{h}z}^2 + k_{\mathrm{l}z}^2) + 4k_y^4] \sin k_{\mathrm{h}z}L \sin k_{\mathrm{l}z}L \\ &\quad + 6k_y^2 k_{\mathrm{l}z} k_{\mathrm{h}z}(1 - \cos k_{\mathrm{h}z}L \cos k_{\mathrm{l}z}L) = 0 \end{aligned} \tag{15}$$

where

$$k_{(\mathrm{l,h}),z} = \left(\frac{2E}{\gamma_1 \pm 2\bar{\gamma}}\frac{m_0}{\hbar^2} - k_y^2\right)^{1/2} \tag{16}$$

It is then possible to derive effective masses in the layer by

$$\frac{1}{m_{(\mathrm{l,h})z}} = 2\left(\frac{\partial^2 E_{(\mathrm{l,h})z}}{\partial k_\perp^2}\right)_{k_\perp = 0} \tag{17}$$

One finds then that for GaAs some of the heavy-hole subbands have positive (i.e., electronlike) masses, independently of the width of the well.

Only numerical results were obtained in the finite-well calculations,[89] but the features obtained in the infinitely deep well approximation (nonparabolicity, positive hole masses) are retained or even emphasized. Such effects have been considered to explain magnetic field measurements of absorption spectra,[90] luminescence,[90–92] and cyclotron resonance of holes in modulation-doped heterojunctions.[93–95] Tight-binding calculations have also led to nonparabolicities of hole dispersion curves,[96,97] in close agreement with the envelope wave-function approximation calculations (Fig. 9a). The effect of the symmetry of the confining potential on hole levels has been shown by Eisenstein *et al.*[98] by comparing modulation-doped single or double (quantum-well) heterostructures. The asymmetric single heterostructure reveals in magnetotransport a lifting of the spin degeneracy of hole bands.

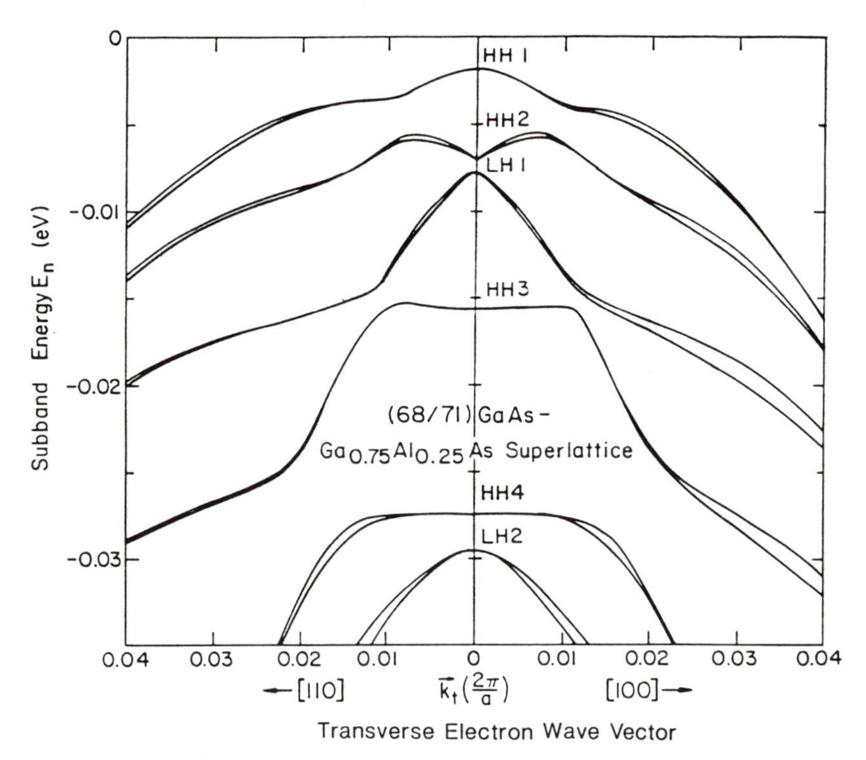

FIG. 9a. Calculated transverse dispersion curves in a GaAs/GaAlAs MQW in an LCAO model. The QW and barrier thicknesses are, respectively, 68 and 71 atomic layers. The double curves correspond to spin-orbit-split bands as the Kramers degeneracy is lifted at $k \neq 0$. Note the negative masses of some heavy-hole bands and the strong nonparabolicity (from Chang and Schulman[97]).

A number of properties of quantum wells show different behavior than in 3D structures, thanks to their bidimensionality[99,100]; we shall discuss them in the following paragraphs.

4. Triangular Quantum Well Energy Levels

Another often encountered potential is the triangular quantum well for which the potential $V(z)$ is linear for $z > 0$ and has an infinite barrier at $z = 0$. The Schrödinger equation for the envelope wavefunction is then

$$-\frac{\hbar^2}{2m}\frac{d^2\chi_n(z)}{dz^2} + eFz\chi_n(z) = E_n\chi_n(z),$$

where F is the electric field for $z > 0$ with the boundary condition $\chi_n(0) = 0$.

The differential equation $w'' - zw = 0$ has two independent solutions, and the one that is nonsingular for $z \to \infty$ is called the Airy function Ai. Figure 9b shows that the function oscillates for $z < 0$ and approaches 0 exponentially for $z \to \infty$. Direct insertion of $\mathrm{Ai}(\alpha + \beta z)$ in the above Schrödinger equation shows that

$$\mathrm{Ai}\left(\left(\frac{2m}{\hbar^2 e^2 F^2}\right)^{1/3}(eFz - E_n)\right)$$

is a solution that has the correct behavior for $z \to \infty$. The boundary condition at $z = 0$ then gives the eigenvalues as

$$E_n = -\left(\frac{e^2F^2\hbar^2}{2m}\right)^{1/3} a_n,$$

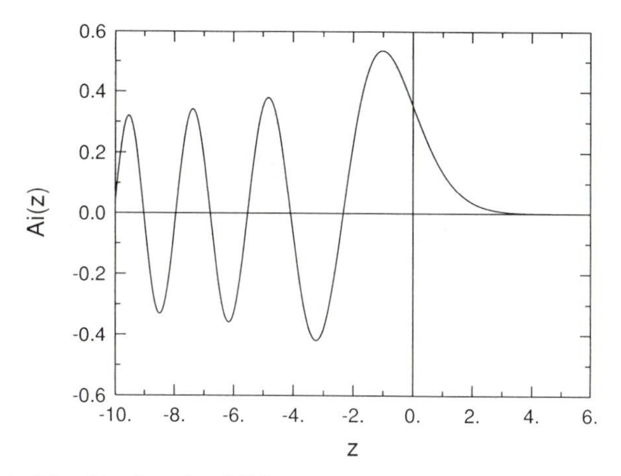

Fig. 9b. Plot of the Airy function Ai(z).

where a_n is the $n$the zero of Ai(z). Asymptotically, and to a very good approximation also for small n, one has

$$a_n \cong -\left[\frac{3\pi}{2}\left(n+\frac{3}{4}\right)\right]^{2/3}, \qquad n = 0, 1, \ldots$$

so that

$$E_n \cong \left(\frac{\hbar^2}{2m}\right)^{1/3}\left(\frac{3\pi eF}{2}\left(n+\frac{3}{4}\right)\right)^{2/3}.$$

Figure 9c shows the wavefunctions in a triangular quantum well.

5. Two-Dimensional Density of States

Besides the energy quantization along the z axis, the main property of thin quantizing films is the bidimensionality in the density of states (DOS).[100] As the motion along the z direction is quantized ($k_z = n\pi/L$ in the limit of an infinitely deep well), an electron possesses only two degrees of freedom along the x and y directions.

The spin-independent k-space density of states per unit area transforms into an E-space density of states through the usual calculation of k states allowed between the energies E and $E + dE$:

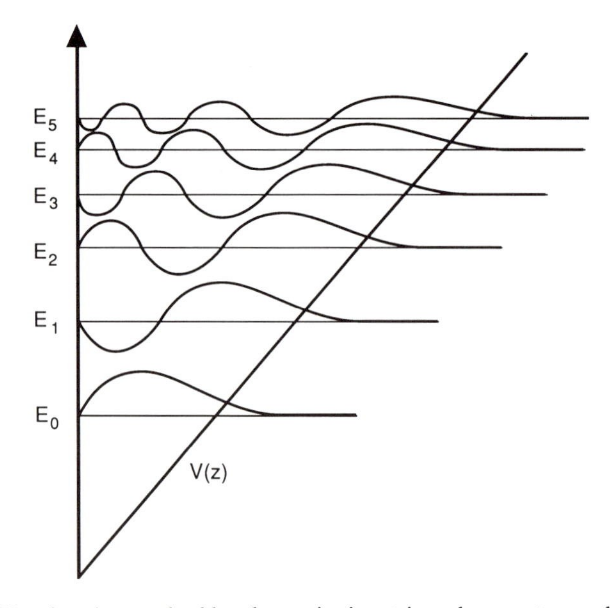

Fig. 9c. Wavefunctions and subband energies in a triangular quantum well.

$$\rho_{2D}(E)\,dE = \rho_{2D}(k_\perp)\,dk_\perp = 2 \times \frac{1}{(2\pi)^2} \times 2\pi k_\perp\,dk_\perp \tag{18}$$

In the parabolic approximation $E = \hbar^2 k_\perp^2/2m^*$, Eq. (18) yields

$$\rho_{2D} = m^*/\pi\hbar^2 \tag{19}$$

The density of states of a given quantum state E_n is therefore independent of E and of the layer thickness. The total density of states at a given energy is then equal to Eq. (19) times the number n of different k_z states at that energy (Fig. 10). The 2D DOS shows discontinuities for each E_n. It is interesting to compare the 2D density of states with the 3D areal DOS, calculated for a thickness equal to that of the layer. From

$$\rho_{3D} = 2^{1/2} m^{*3/2} \pi^{-2} \hbar^{-3} E^{1/2} \tag{20}$$

one finds that, in the infinitely deep well approximation, using the expression of E_n and (19), $\rho_{3D}(E_n)L = \rho_{2D}$; as shown in Fig. 10, the 3D and 2D densities of states are equal for energy E_n. Two remarks can be made:

(1) One should not conclude that there is little difference between 2D and 3D systems, even though one can always find an energy in a 3D system for which the DOS is equal to that of a 2D system. The important point here is that the DOS is *finite* even at the bottom of the 2D level, whereas it tends towards zero in the 3D system. This has fundamental consequences on the properties of 2D systems as it means that all dynamic phenomena

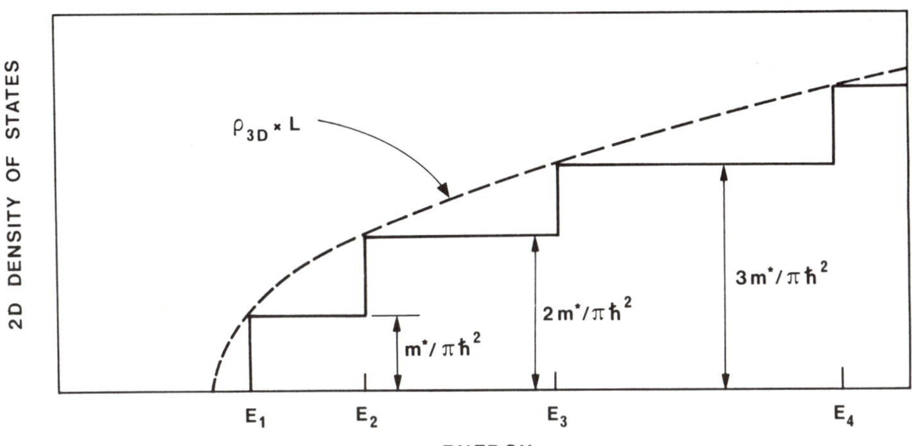

FIG. 10. 2D density of states (DOS) and comparison with the 3D DOS calculated for a layer with a thickness equal to that of the quantum well (after Dingle[1]).

remain finite at low kinetic energies and low temperatures, such as scattering, optical absorption, and gain.

(2) On the other hand, when numerous levels are populated or when one looks at transitions involving large values of n, such as in thick layers, nothing can distinguish between 2D and 3D behaviors, analogous to the correspondence principle between quantum and classical mechanics.

6. Excitons[101] and Shallow Impurities[102] in Quantum Wells[103]

From 3D semiconductor physics it is known that the absorption spectrum is not simply determined by the creation of a free electron and a free hole. The electron and the hole are correlated in their motion in a way that can be described as the simple Coulomb attraction of the electron and the hole. As in the case of the hydrogen atom, this attraction leads to bound levels, the lowest of which is one effective Rydberg (Ry*) below the continuum level and in which the electron and hole are bound to each other within an effective Bohr radius $a_B{}^*$. The effective Rydberg and Bohr radii are given by

$$\mathrm{Ry}^* = \frac{2m_{\mathrm{red}}e^4}{\hbar^2(8\pi\varepsilon)^2} = \frac{m_{\mathrm{red}}}{m_0}\frac{1}{\varepsilon_r^2}\mathrm{Ry}$$
$$a_B^* = \frac{4\pi\varepsilon\hbar^2}{m_{\mathrm{red}}e^2} = \varepsilon_r\frac{m_0}{m_{\mathrm{red}}}a_B \tag{20a}$$

where m_{red} is the reduced mass of electron and hole ($1/m_{\mathrm{red}} = 1/m_e + 1/m_h$); ε_r is the relative permittivity of the semiconductor; $\mathrm{Ry} = 13.6$ eV; and $a_B = 0.529$ Å. Note that in the limit of an infinitely heavy hole we find the shallow donor binding energy of an electron. From $a_B{}^* \approx 100$ Å, one infers that the wave function and energy levels of excitons and impurities are quite modified in a quantum well where the thickness is usually of the order of or smaller than the Bohr diameter $2a_B$.

In the limiting exact 2D case where $L \ll a_B$, one should obtain the usual 2D Rydberg value $R_{2D} = 4R_{3D}$ for the infinitely deep well.[100] The energy levels are then given by[104,105]

$$R_{n,2D} = R_{3D}[1/(n - \tfrac{1}{2})^2] \tag{21}$$

For finite thickness, exciton binding energies have been calculated through a variational method. The perturbative exciton Hamiltonian is,

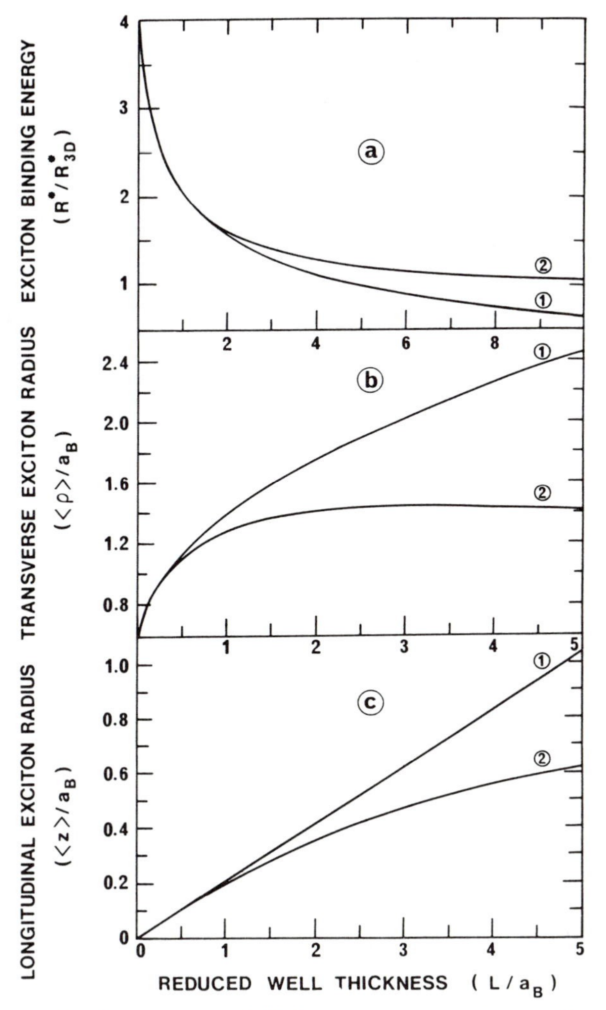

FIG. 11. Exciton binding energy and Bohr radius in the infinite-well approximation. Curves (1) and (2) display calculations for, respectively, separated or nonseparated [Eqs. (23) and (23a)] exponential factors in the wave function. The nonseparated wave functions of Eq. (23a) give back the usual 3D exciton quantities at large L. (a) Exciton binding energy; (b) $\langle \rho^2 \rangle^{1/2}$, (c) $\langle z_e - z_h \rangle$ (after Bastard[106]).

assuming nondegenerate, isotropic bands,

$$H = \frac{p_{ze}^2}{2m_e^*} + \frac{p_{zh}^2}{2m_h^*} + \frac{P_x^2 + P_y^2}{2(m_e^* + m_h^*)} + \frac{p_x^2 + p_y^2}{2\mu} - \frac{e^2}{4\pi\varepsilon_0\varepsilon_R[x^2 + y^2 + (z_e - z_h)^2]^{1/2}} \tag{22}$$

where m_e^*, m_h^*, z_e, z_h are the masses and z position of the electron and hole, respectively, P_x and P_y projections on the x and y axes of the center-of-mass exciton momentum, p_x and p_y the relative-momentum projections, and μ the reduced mass. Bastard *et al.*[106] used variational wave functions totally confined in the well such as

$$\psi_\lambda(\mathbf{r}) = N(L, \lambda) \cos(\pi z_e/L) \cos(\pi z_h/L) \exp[-(\rho/\lambda)] \tag{23}$$

or

$$\psi_{\lambda'}(\mathbf{r}) = N(L, \lambda') \cos(\pi z_e/L) \cos(\pi z_h/L) \exp\{-[\rho^2 + (z_e - z_h)^2]^{1/2}/\lambda'\} \tag{23a}$$

where λ and λ' are variational parameters, $\rho = (x^2 + y^2)^{1/2}$, and $N(L, \lambda)$, $N(L, \lambda')$ are the normalizing coefficients. The nonseparated exponential factor in the spatial coordinates[106] of Eq. (23a) ensures some amount of Coulombic binding even when the quantum well is wide as compared to a variable-separated factor. Binding energies and reduced Bohr radii are shown in Fig. 11 as a function of the reduced well thickness. At vanishing L, R_{2D} extrapolates to $4R_{3D}$ and ρ to $a_B\sqrt{3/8}$. More accurate exciton energies taking into account the well finiteness have recently been calculated, however using simple parabolic hole bands[107] (Fig. 12).

The increase in exciton binding energy (Fig. 13) has a profound influence on quantum-well properties. It allows GaAs-based quantum wells to have their optical properties dominated by exciton effects even at room temperature.[107a,b] This is a rather unique instance in standard semiconductors[85]: Usually large exciton binding energies are associated with large reduced masses,[101] i.e., large gaps (according to Kane's model), then to large ionicity (Phillips' theory of ionicity[108]) and therefore to strong LO-phonon coupling which ionizes excitons at room temperature. The room-temperature excitons in quantum wells allow very promising features such as optical bistability, four-wave mixing, and large electrooptic coefficients which are developed in the article by Chemla *et al.*[44]

Shallow impurity effects have been widely calculated using a number of approximations.[109–111] It should be noted that the problem is somewhat complicated by the degree of freedom brought about by the position of the impurity relative to the well interfaces. First, variational calculations[109]

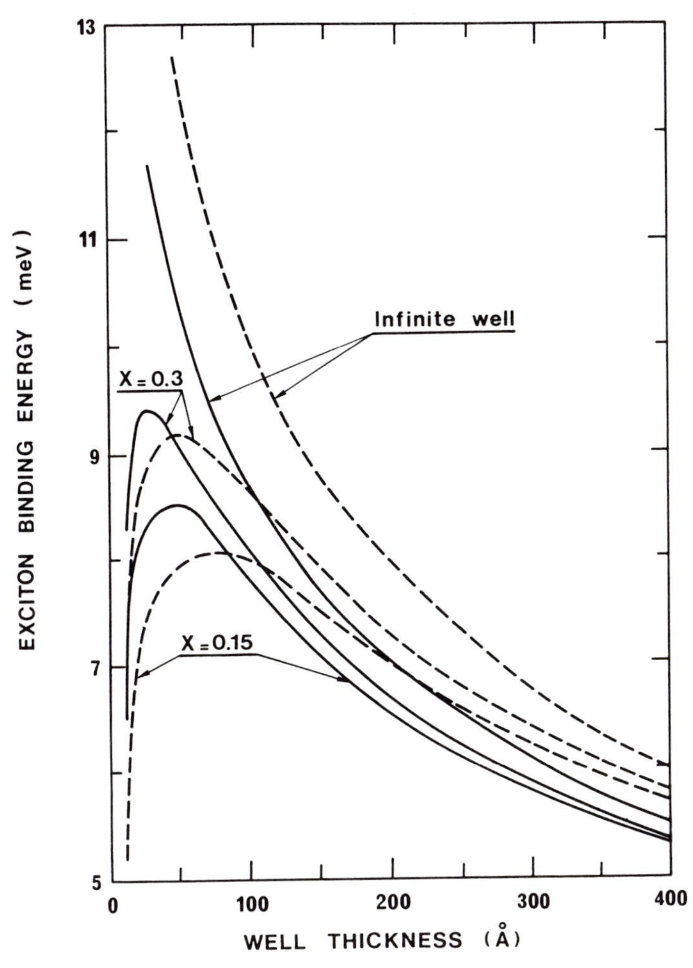

FIG. 12. Exciton binding energy for finite quantum wells as a function of well thickness. Three barrier potentials are shown corresponding to $x = 0.15$ and 0.30 and the infinite-barrier case. Heavy-hole excitons (——) and light-hole excitons (---) are displayed (from Greene *et al.*[107]).

considered an infinite potential at the interface. In this approximation, the wave function of bound particles must vanish in the barrier. The wave function for impurity atoms at the barrier must therefore be a truncated *p*-like state, whereas for impurity atoms located at the center of the well *s*-like wave functions are allowed. One expects that the ground-state binding energies for these two states in the *large-well limit* to be $R_{imp}/4$ and R_{imp} (R_{imp} being the 3D impurity binding energy).

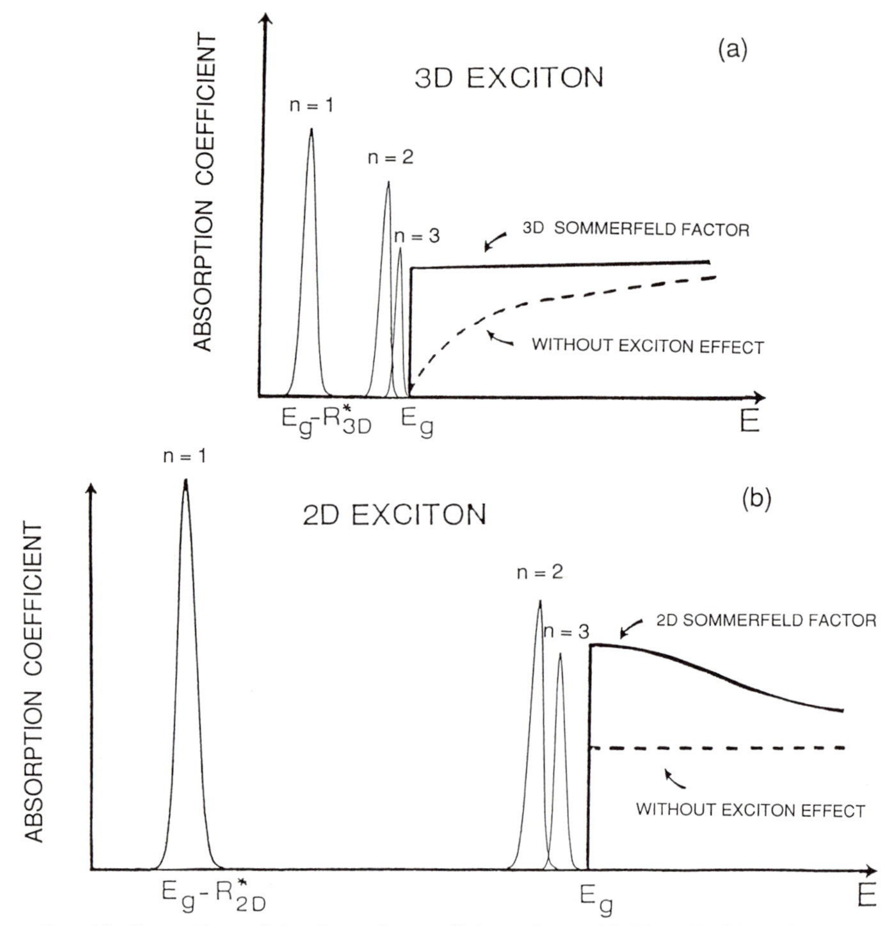

FIG. 13. Comparison of the absorption coefficients due to (a) 3D or (b) 2D excitons. The characteristic energy is $\simeq 4$ times larger for 2D excitons. Oscillator strengths are increased ($\sim a_B^{-3}$ in 3D, a_B^{-2} in 2D). For continuum states, the absorption coefficient is increased over the excitonless value (---) by the Sommerfeld factor, determined by the continuum wave functions of the hydrogen atom, which represents the effect of electron–hole correlation in unbound states.

The calculation starts from the perturbation Hamiltonian:

$$H = \frac{p^2}{2m^*} - \frac{e^2}{4\pi\varepsilon_0\varepsilon_R}[x^2 + y^2 + (z - z_i)^2]^{-1/2} + V_{\text{conf}}(z) \qquad (24)$$

z_i being the impurity atom position and $V_{\text{conf}}(z)$ the quantum-well confining potential (defined by energy-band discontinuities).

In the infinite-well approximation, trial wave functions are taken as

$$\psi_\lambda(\mathbf{r}) = N(L, \lambda, z_i)\cos(\pi z/L)\exp\{-[x^2 + y^2 + (z - z_i)^2]^{1/2}/\lambda) \quad (25)$$

if $|z| < L/2$, and $\psi_\lambda = 0$ otherwise.

The binding energy measured from the confined quantum state is given by

$$E(L, z_i) = \pi^2\hbar^2/2m^*L^2 - \min_\lambda\langle\psi_\lambda|H|\psi_\lambda\rangle$$

where $\min_\lambda$ means the minimum value of $\langle\psi_\lambda|H|\psi_\lambda\rangle$ with respect to the variational parameter λ.

Exact solutions are found when $L \to 0$ or ∞. For $L \to \infty$, one finds $E(L, 0) = R_{\mathrm{imp}}$ and $E(L, \pm L/2) = R_{\mathrm{imp}}/4$. When $L \to 0$, one finds the usual 2D result:

$$E(L, z_i) \to 4R_{\mathrm{imp}}; \qquad \psi_\lambda(\mathbf{r}) \to \frac{1}{a_{\mathrm{B}}}\left(\frac{8}{\pi}\right)^{1/2}\exp\left(\frac{-2[x^2 + y^2]^{1/2}}{a_{\mathrm{B}}}\right)$$

More detailed calculations[110,111] have taken the finite barriers into account. Mailhot *et al.*[111] have also considered ion image charges due to the different dielectric constants of GaAs and GaAlAs. In the limit of vanishing well thicknesses, one expects to recover the 3D GaAlAs donor energy, since in that case the confinment effect of the wave function due to the GaAs well becomes vanishingly small for finite barriers. Measurements of the donor energy levels by electronic Raman scattering[112] and infrared absorption[113] are in good agreement with the theoretical evaluations.

As in 3D, the calculation of the energy levels of the acceptor impurities is much more complicated than for donors due to the degeneracy of the valence band. Masselink *et al.* has recently provided a detailed calculation.[114]

Various other situations have been studied: Chaudhuri[115] considered the influence on the binding energy of the spreading of the impurity wave function in superlattices. The influence of high carrier densities on the impurity binding energies in modulation-doped QWs[116] and superlattices[117] was also calculated. Finally, the impurity bound states associated with excited quantum-well subbands were analyzed by Priester *et al.*[118] and observed by Perry *et al.*[119] in Raman scattering studies. Recent reviews on shallow impurities in quantum wells can be found in Refs. 119a and 119b.

7. Tunneling Structures, Coupled Quantum Wells, and Superlattices

Tunneling phenomena across barriers open the way to many fascinating effects, the most eagerly expected one being the Bloch oscillator (to be described in Section 19). The renewed interest in transmission across simple systems such as single barriers, double barriers, etc. also lies in the recent availability of the high-performance growth techniques developed for the multiple heterojunction superlattice. The advances in growth techniques are evidenced by the symmetric $I-V$ characteristics now observed. Whereas transport properties and related structures will be discussed with experimental results in Section 22 we develop here the energy-level schemes of these communicating multiheterointerface devices. Several calculation techniques can be used, such as the Kronig–Penney model,[3] successive multiple tunneling model,[120] perturbative tight-binding model,[121] or the LCAO model.[122] We will use here the simplest descriptions by a tight-binding model and a successive tunneling model.

a. The Double-Well Structure

Beyond the double-barrier single-well structure (Fig. 1f) which leads to zero-bias electronic properties very similar to the single QW previously described, the simplest structure is the double-well configuration (Fig. 14), which can be easily analyzed by the usual tight-binding perturbation model. As the barrier thickness is decreased, the exponentially decaying wave function in the barrier can have some finite value in the next well. Treating this wave-function overlap as a perturbation, one finds the perturbation matrix element to be, in a two-well configuration,

$$V_{12} = \langle \psi_1 | H | \psi_2 \rangle = \langle \psi_1 | V_2 | \psi_2 \rangle \tag{26}$$

where H is the electronic Hamiltonian, ψ_1 and ψ_2 the unperturbed wave-functions of single wells 1 and 2, and V_2 the confining potential of well 2.

Within the restricted basis of the functions ψ_1 and ψ_2 the Schrödinger equation is then

$$\begin{Bmatrix} E_1 + V_1 - \varepsilon & V_{12} \\ V_{12}^* & E_1 + V_1 - \varepsilon \end{Bmatrix} \begin{Bmatrix} a_1 \\ a_2 \end{Bmatrix} = \begin{Bmatrix} 0 \\ 0 \end{Bmatrix} \tag{27}$$

where $V_1 = \langle \psi_1 | V_2(z) | \psi_1 \rangle = \langle \psi_2 | V_1(z) | \psi_2 \rangle$, and $V_{12} = \langle \psi_1 | V_2(z) | \psi_2 \rangle = \langle \psi_2 | V_1(z) | \psi_1 \rangle$, so that $\varepsilon = E_1 + V_1 \pm |V_{12}|$, and the levels are split by the amount $2|V_{12}|$.

b. The Communicating Multiple-Quantum-Well Structure or Superlattice—Tight-Binding Calculations

Introducing more wells leads to the creation of a continuous band of states. The transition from single wells to multiple connected wells, as revealed by optical absorption, has been studied by Dingle *et al.*[123] For N wells, the N-degenerate levels give rise to bands with $2N$ states. The simplest way to analyze this is to consider a tight-binding model of the N-well

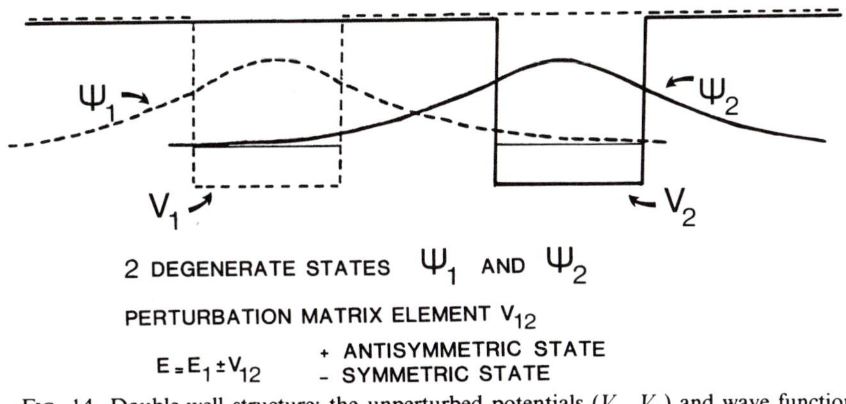

FIG. 14. Double-well structure: the unperturbed potentials (V_1, V_2) and wave functions (ψ_1, ψ_2) of the separate wells are represented by (---) and (——) respectively.

chain[125] (Fig. 15). The Bloch-like envelope wave function can be written as

$$\psi_q^{(i)}(z) = \frac{1}{\sqrt{N}} \sum_n e^{iqnd} \chi_{\text{loc}}^i(z - nd) \tag{28}$$

where $\chi_{\text{loc}}^i(z - nd)$ is the ith wave function of the quantum well centered at $z = nd$ and q is the Bloch wave vector. Assuming a nearest-well interaction, the energy is

$$\varepsilon_i(q) = E_i + s_i + 2t_i \cos qd \tag{29}$$

with

$$s_i = \int_{-\infty}^{+\infty} \chi_{\text{loc}}^i(z - d) V(z) \chi_{\text{loc}}^i(z - d)\, dz$$

$$t_i = \int_{-\infty}^{+\infty} \chi_{\text{loc}}^i(z) V(z) \chi_{\text{loc}}^i(z - d)\, dz \tag{30}$$

The factor of $2t_i$ in Eq. (29) that yields a bandwidth of $4t_i$ as compared with $2V_{12}$ in the double-well case comes from the interaction of one well with its two neighbors in the chain. The variation of the electronic bandwidth in GaAs MQW is shown in Fig. 16.

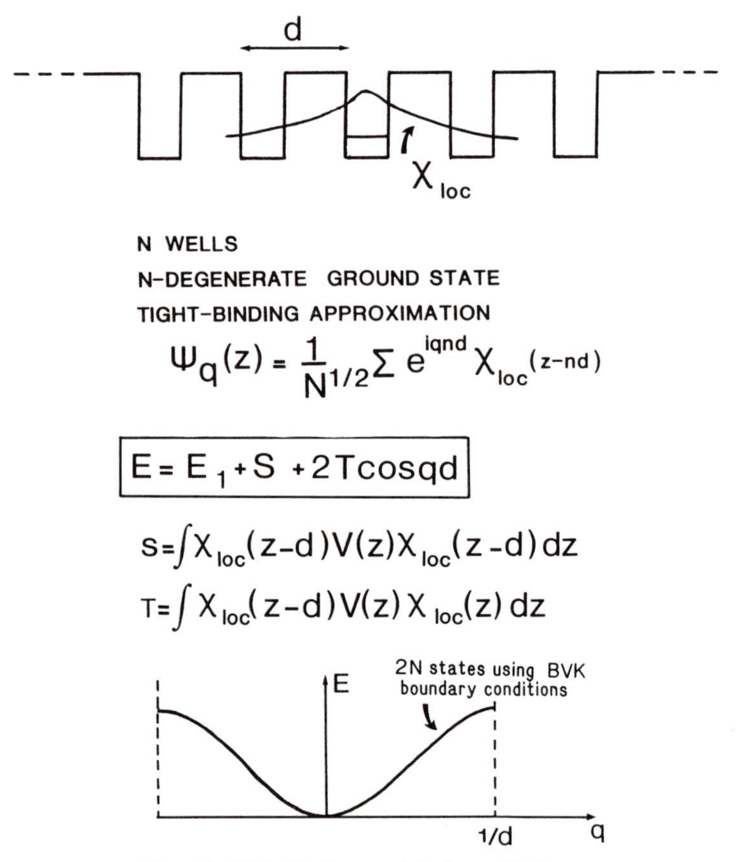

FIG. 15. Tight-binding model of superlattices.

Assuming usual Born – von Karman periodic conditions, one finds that q can only take discrete values which are integer numbers of $1/Nd$. Therefore, the superlattice band can accommodate $2N$ electrons with different quantum states [different q's in Eq. (28)].

The superlattice effect introduces a profound change in the 2D DOS. The dispersion of the N states in a band destroys the steepness of the square density of states. From the energy

$$\varepsilon_{n*}(q, k_{\perp}) = \hbar^2 k_{\perp}^2 / 2m + \varepsilon_n(q)$$

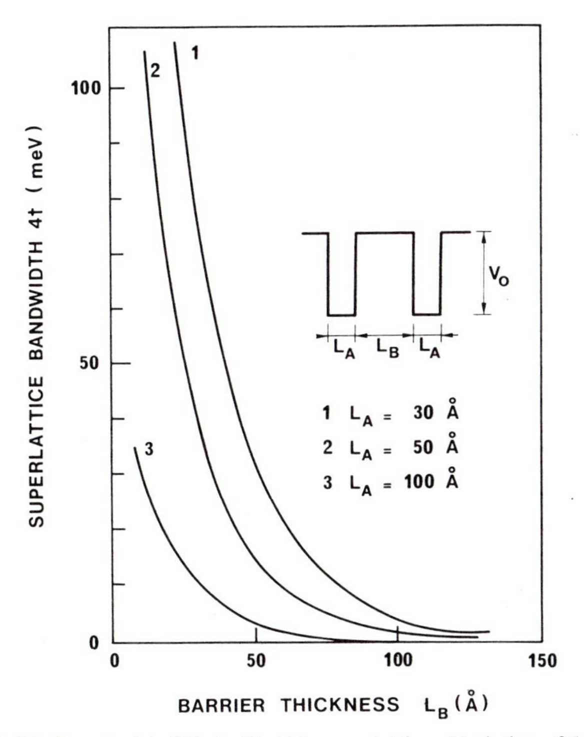

FIG. 16. Tight-binding model of GaAs/GaAlAs superlattices: Variation of the fundamental state bandwidth [$4t_i$ of Eq. (30)] in the tight-binding model as a function of barrier thickness for three different well thicknesses. $X = 0.2$; $V_0 = 212$ meV (after Bastard[125]).

one finds a density of states as represented in Fig. 17:

$$\rho_n(\varepsilon) = N \frac{m^*}{\pi\hbar^2} \arccos\left(\frac{\varepsilon_i - E_i - S_i}{2t_i}\right) \tag{31}$$

This is obtained by summing over q values and over the various bands the 2D DOS corresponding to the transverse free motion of a single q state. There are still two singularities of the density of states at both extrema of each band. One sees that the 2D limit for N independent wells is retained when the bandwidth goes to zero, i.e., when the overlap matrix element vanishes due to wide barriers.

More precise calculations of the band structure of superlattices have been carried out in the envelope wave function approximation, using the Kane model to describe the band structure within each well and barrier. Bastard[82] has shown that for $k_\perp = 0$, in the parabolic band approximation,

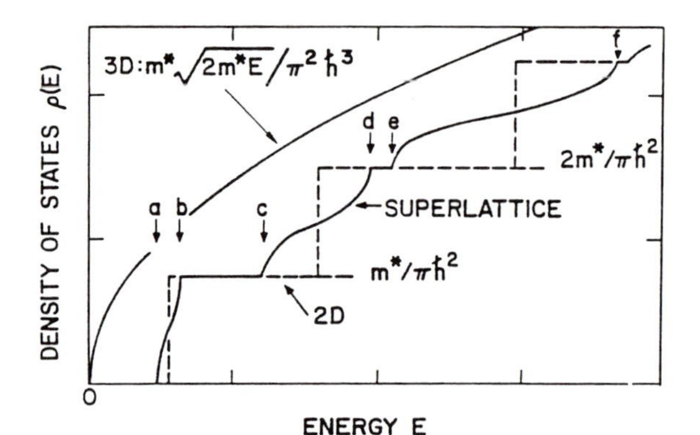

FIG. 17. Comparison of the DOS of a superlattice with that of a 2D system (---) and a 3D isotropic system. Note the broadening of the superlattice band with band index as the overlap of wave functions increases with energy E in the tight-binding description, increasing the transfer matrix element t_i (Reprinted with permission from World Scientific Pub. Co., L. Esaki, "Recent Topics in Semiconductor Physics" (H. Kamimura and Y. Toyozawa, eds.), 1983.)

the equation yielding the values of q takes the simple Kronig–Penney form

$$\cos qd = \cos k_A L_A \cosh \kappa_B L_B - \tfrac{1}{2}(1/\xi - \xi) \sin k_A L_A \sinh \kappa_B L_B \quad (32)$$

with $\xi = m_A^* k_A / m_B^* k_B$.

The allowed energy bands are given as usual by $-1 \leq \cos qd \leq 1$ (Fig. 18c).

The solutions of unbound states ($\varepsilon > 0$) are similarly given by

$$\cos qd = \cos k_A L_A \cos k_B L_B - \tfrac{1}{2}(1/\xi + \xi) \sin k_A L_A \sin k_B L_B \quad (33)$$

Equations (32) and (33) can be solved graphically.

The limit of noncommunicating wells is found in Eq. (32) for $\kappa_B L_B \rightarrow \infty$, which leads to $\cos k_A L_A - \tfrac{1}{2}(1/\xi - \xi) \sin k_A L_A = 0$, the usual single-well equation [Eq. (5)] after simple transformation.

For the hole bands, Bastard[82] considered *uncoupled* hole bands and showed that they obey Eq. (32) with a corresponding change of parameters. The situation is much more complicated if $k_\perp \neq 0$, as in the single quantum well. The heavy- and light-hole states are mixed by the boundary conditions. Only numerical calculations have been carried out.[82a,82d,89,94,95]

The limiting cases of very thin layers, where the envelope wavefunction approximation tends to break down, can be calculated using LCAO methods. Such calculations have been performed by Schulman *et al.*,[121,122] yielding results similar to the envelope approximation when the layer thickness is $\geq 6-8$ monolayers. Superlattice effects involving band extrema other than at the Γ point (X point of GaAs, for instance) have been shown by Mendez *et al.*[126]

c. Tunneling

Consider the potential $V(z)$ describing a barrier between two regions of constant potential (taken as 0) from $-\infty$ to z_l and from z_r to ∞. For a given energy $E > 0$, the solution to the Schrödinger equation outsider the barrier can be expressed as

$$\psi_l(z) = a_l e^{ikz} + b_l e^{-ikz} \tag{33a}$$

and

$$\psi_r(z) = a_r e^{ikz} + b_r e^{-ikz} \tag{33b}$$

respectively, where $E = \hbar^2 k^2/2m$.

Since the Schrödinger equation is linear and of second order, its solution across the barrier will establish two linear relations between the four coefficients a_l, b_l, a_r, b_r. These relations can be expressed as a transfer matrix **S**:

$$\begin{pmatrix} a_r \\ b_r \end{pmatrix} = \begin{pmatrix} s_{11} & s_{12} \\ s_{21} & s_{22} \end{pmatrix} \begin{pmatrix} a_l \\ b_l \end{pmatrix} \tag{33c}$$

or as a transmission matrix **T**:

$$\begin{pmatrix} a_r \\ b_l \end{pmatrix} = \begin{pmatrix} t^{\rightarrow} & r_r \\ r_l & t^{\leftarrow} \end{pmatrix} \begin{pmatrix} a_l \\ b_r \end{pmatrix} \tag{33d}$$

in which $T^{\rightarrow} = |t^{\rightarrow}|^2$ indicates the probability for transmission of a particle from the left through the barrier, and $R = |r_l|^2$ indicates the probability of reflecting the same particle off the barrier.

From the fact that the complex conjugate of a solution to the Schrödinger equation is also a solution, it is easily shown that $s_{11} = s_{22}{}^*$, and $s_{12} = s_{21}{}^*$. The same fact also implies that

$$\begin{aligned} t^{\rightarrow} t^{\leftarrow *} &= 1 - |r_l|^2 \\ |r_l|^2 &= |r_r|^2 \\ t^{\rightarrow} r_r^* &= -r_l t^{\rightarrow *} \\ t^{\leftarrow} r_l^* &= -r_r t^{\rightarrow *} \end{aligned} \tag{33e}$$

so that the transfer matrix can be written as

$$\mathbf{S} = \begin{pmatrix} \dfrac{1}{t^{\leftarrow *}} & \dfrac{r_{\mathrm{r}}}{t^{\leftarrow}} \\ \dfrac{r_{\mathrm{r}}^{*}}{t^{\leftarrow *}} & \dfrac{1}{t^{\leftarrow}} \end{pmatrix} \tag{33f}$$

with det $\mathbf{S} = t^{\rightarrow}/t^{\leftarrow} = t^{\leftarrow *}/t^{\rightarrow *}$. Finally, the conservation of probability current implies that

$$|t^{\rightarrow}|^2 + |r|^2 = |t^{\leftarrow}|^2 + |r|^2 = 1 \tag{33g}$$

so that $t^{\leftarrow} = t^{\rightarrow}$. It is worth noticing that the last relation is only valid when the potential energy is the same on each side of the barrier, whereas the former relations hold also when a potential difference exists.

For a square barrier of height V_B between $z_l = -a/2$ and $z_{\mathrm{r}} = a/2$ one finds by standard quantum mechanics that $t = |t| \exp(i\phi_t)$ is given by (for $E < V_B$):

$$T = |t|^2 = \frac{1}{1 + \dfrac{1}{4}\left(\dfrac{k}{\kappa} + \dfrac{\kappa}{k}\right)^2 \sinh^2 \kappa a} \tag{33h}$$

$$\phi_t = \psi - ka$$

$$\tan \psi = \frac{1}{2}\left(\frac{k}{\kappa} - \frac{\kappa}{k}\right) \tanh \kappa a \tag{33i}$$

where $E = \hbar^2 k^2/2m$ and $V_B - E = \hbar^2 \kappa^2/2m$.

By replacing z by $z - L$ in (33a) and (33b) one can easily convince onself that it is a property of the $\mathbf{S}$ matrix that a barrier displaced by the amount L, has an $\mathbf{S}$ matrix given by

$$\mathbf{S}_{\mathrm{L}} = \begin{pmatrix} e^{-ikL} & 0 \\ 0 & e^{ikL} \end{pmatrix} \mathbf{S} \begin{pmatrix} e^{ikL} & 0 \\ 0 & e^{-ikL} \end{pmatrix} \tag{33j}$$

Furthermore, the total matrix of two barriers in sequence is the product of their matrices:

$$\mathbf{S}_{\mathrm{tot}} = \mathbf{S}_2 \mathbf{S}_1 \tag{33k}$$

where barrier 2 is assumed to be to the right of barrier 1.

Two applications of these properties will now be made: We first calculate the transmission of a double barrier, and second show how the existence and width of minibands in superlattices follow from this description.

In a double barrier structure two identical barriers are separated by the distance L. Then, if the first barrier is described by the $\mathbf{S}$ matrix (33f) the $\mathbf{S}$

matrix of the double barrier is, from (33j) and (33k)

$$\mathbf{S}_{\text{tot}} = \mathbf{S}_{\text{L}}\mathbf{S} = \begin{pmatrix} \dfrac{1}{t^{*2}} + e^{-2ikL}\left|\dfrac{r}{t}\right|^2 & \dfrac{r}{t}\left(\dfrac{e^{-2ikL}}{t} + \dfrac{1}{t^*}\right) \\ \dfrac{r^*}{t^*}\left(\dfrac{e^{2ikL}}{t^*} + \dfrac{1}{t}\right) & \dfrac{1}{t^2} + e^{2ikL}\left|\dfrac{r}{t}\right|^2 \end{pmatrix} \tag{33l}$$

from which we can find the transmission by analogy with (33f) as

$$t_{\text{tot}} = 1/s_{22}^{\text{tot}} = \frac{t^2}{1 + \dfrac{t^2}{|t^2|}|r|^2 e^{2ikL}} = \frac{t^2}{1 + |r|^2 e^{2i(kL + \phi_t)}} \tag{33m}$$

$$T_{\text{tot}} = |t_{\text{tot}}|^2 = \frac{T^2}{|1 + |r|^2 e^{2i(kL+\phi_t)}|^2} = \frac{(1 - |r|^2)^2}{|1 + |r|^2 e^{2i(kL+\phi_t)}|^2} \tag{33n}$$

The interesting result is that for most energies the total transmission is of the order of the square of the transmission of a single barrier, but that for very special values T_{tot} can become 1, i.e., total transparency of the double barrier. This occurs when $2(kL + \phi_t) = (2n + 1)\ \pi$. For the important case of thick square barriers we find this condition from (33i) to be

$$\operatorname{cotan}(k(L - a)) = \frac{1}{2}\left(\frac{k}{\kappa} - \frac{\kappa}{k}\right) \tag{33o}$$

i.e., the condition (see Section 3a) for having a bound state in the well of width $L - a$ between the two barriers. The double barrier therefore serves as a filter, which only transmits electrons of energy close to the resonance values given by (33o). This effect is utilized in the double barrier diodes described in Section 19.

In a perfect superlattice the same barrier is repeated periodically with a period L. Suppose the **S** matrix of one barrier situated around $z = 0$ is given by (33f). The condition for an allowed state in the superlattice can now be expressed as the condition that the **S** matrix of the first barrier must be such that the wavefunction described by a_{r} and b_{r} (see Fig. 18a) seen from the second barrier has the same amplitude and relative phase as the wavefunction to the left of the first barrier:

$$\begin{aligned} a_{\text{r}} e^{ikL} &= a_{\text{l}} e^{i\phi} \\ b_{\text{r}} e^{-ikL} &= b_{\text{l}} e^{i\phi} \end{aligned} \tag{33p}$$

where ϕ is a common phase. If this condition were not fulfilled, the amplitude of the wavefunction would grow indefinitely for either increas-

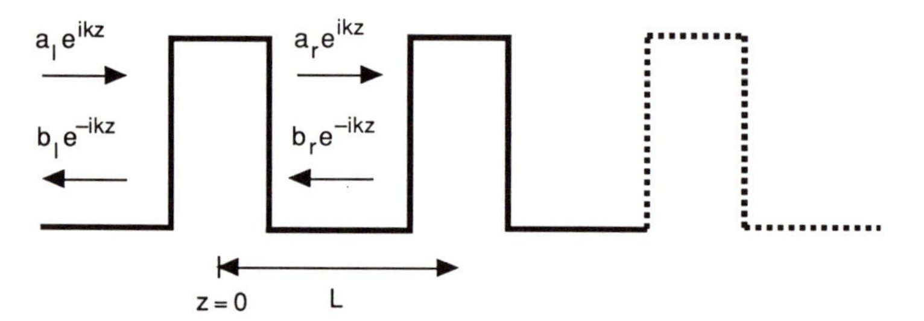

FIG. 18a. Transmitted and reflected waves in a superlattice.

ing or decreasing z. Combining (33f) with (33p) one finds the homogeneous set of equations

$$\begin{pmatrix} \frac{1}{t^*} - e^{i(\phi - kL)} & \frac{r}{t} \\ \frac{r^*}{t^*} & \frac{1}{t} - e^{i(\phi + kL)} \end{pmatrix} \begin{pmatrix} a_l \\ b_l \end{pmatrix} = \begin{pmatrix} 0 \\ 0 \end{pmatrix} \tag{33q}$$

which has a solution if and only if

$$e^{2i\phi} - e^{i\phi} \frac{2\,\mathrm{Re}(te^{ikL})}{|t|^2} + 1 = 0 \tag{33r}$$

or

$$e^{i\phi} = \frac{\mathrm{Re}(te^{ikL})}{|t|^2} \pm \sqrt{\left(\frac{\mathrm{Re}(te^{ikL})}{|t|^2}\right)^2 - 1} \tag{33s}$$

so that solutions are possible if and only if

$$\mathrm{Re}(te^{ikL}) \le |t|^2 \tag{33t}$$

or

$$\cos(\phi_t + kL) \le |t| \tag{33u}$$

Since $0 < |t| < 1$, there will be bands of allowed k values around energies for which $\phi_t + kL = \pi/2 + n\pi$. In general the bandwidths will increase when $|t|$ approaches 1; above the top of the barriers when $|t|$ is very close to 1 the bands are large and separated by small gaps in energy around $\phi_t + kL = n\pi$. Fig. 18b illustrates the condition (33u): For a single square barrier are drawn $\pm|t|$, ϕ_t, and $\cos(\phi_t + kL)$ as functions of energy; the allowed energy bands are indicated by the energy intervals in which (33u) is satisfied. These are the minibands of the superlattice already described in tight-binding approximation in Section 7b. Fig. 18c shows the evolution of the minibands as the barrier and well thickness are varied. For narrow barriers the minibands are large; above the barriers gaps still exist but they shrink rapidly when the barriers decrease.

d. *Continuum States*

It is a classic textbook[14a] example that continuum states ($\varepsilon > 0$) in a quantum well can play an important role in the dynamics of incident particles. The transmission and reflection coefficients of a quantum well display resonances every time the condition $kL = n\pi$ is fulfilled. This is the quantum analog for the electronic de Broglie waves of Fabry–Perot resonances in classical wave optics. The particle spends a longer time in the quantum-well region, which should have important consequences regarding the particle capture by the well. One should note in Fig. 7 that, with decreasing well thickness, a new resonant continuum state pops out of the well whenever a bound state reaches the well top for $kL = n\pi$. These states have been calculated by Bastard[127] in the envelope wave function framework, and by Jaros and Wong[128] using pseudo-potential calculations. The

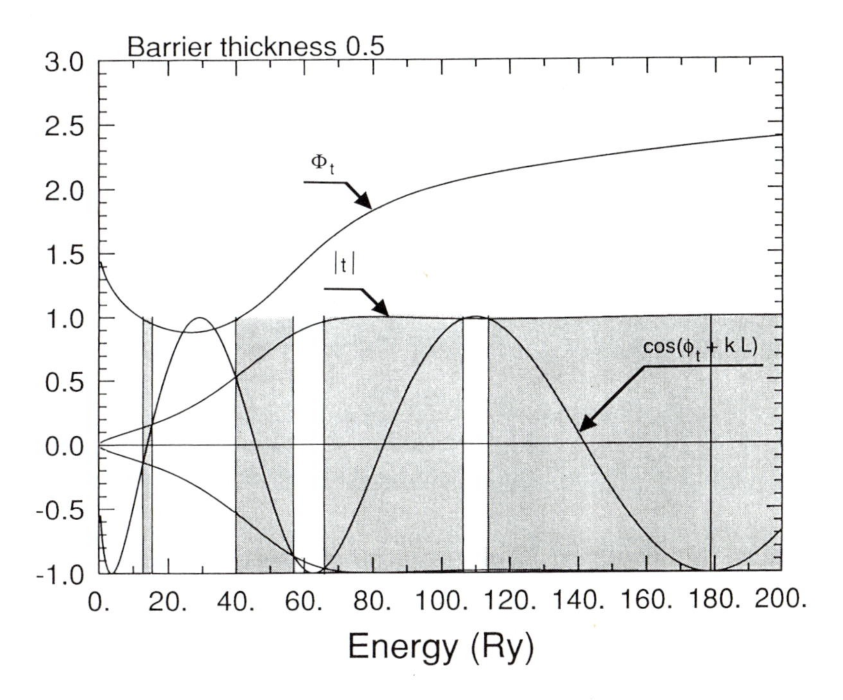

FIG. 18b. Minibands in a rectangular barrier superlattice. The transmission of one barrier has amplitude $|t|$ and phase Φ_t. The barriers and wells have the same thickness $L = 0.5a_B^* = 5$ nm, the barrier height $V_B = 40\text{Ry}^* = 240$ meV. The minibands determined by the condition (33u) are shown shaded.

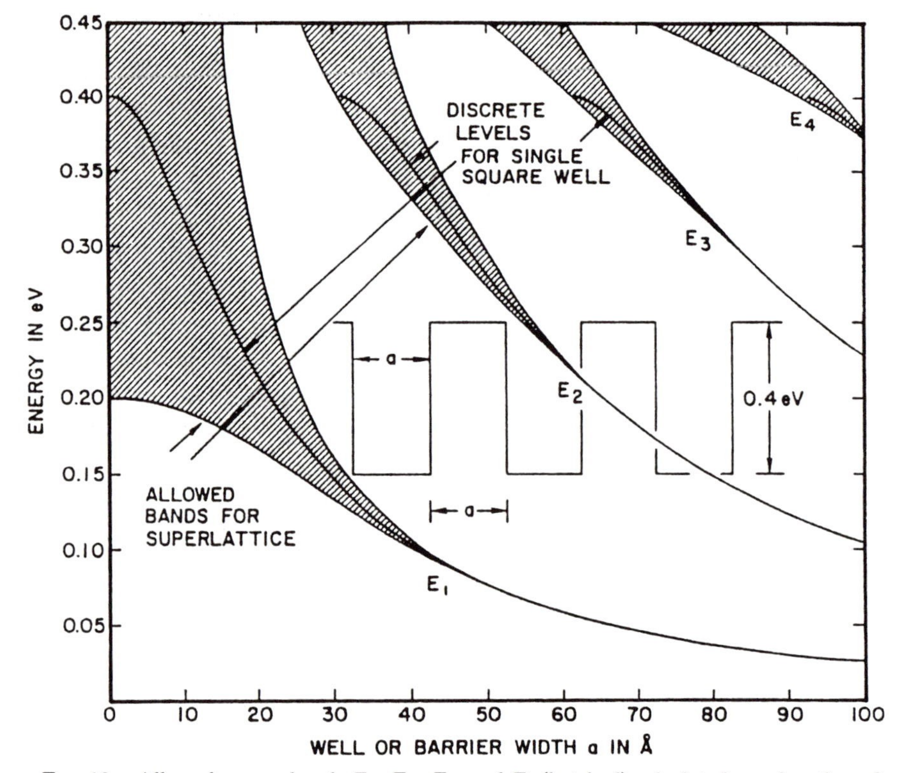

FIG. 18c. Allowed energy bands E_1, E_2, E_3, and E_4 (hatched) calculated as a function of well or barrier width ($L_z = L_B = a$) in a superlattice with a barrier potential $V = 0.4$ V. Note the existence of forbidden gaps even above the barrier potential (Reprinted with permission from World Scientific Pub. Co., L. Esaki, "Recent Topics in Semiconductor Physics" (H. Kamimura and Y. Toyozawce, eds.), 1983.)

resonant continuum states should be of greatest importance in the carrier capture of QW, as the reflection coefficient is near unity for such states.[129] The continuum states have recently been observed in optical studies of coupled wells[124] and by resonant Raman scattering.[130]

8. Modulation Doping of Heterostructures[130a]

A major advance for potential high-performance devices was made when Störmer and Dingle *et al.*[131,132] introduced *n*-type modulation-doped samples (Fig. 19). The underlying idea is that, at equilibrium, charge transfer occurs across a heterojunction to equalize the chemical potential (i.e., the Fermi level[133]) on both sides. Doping the wide-bandgap side of a GaAlAs/GaAs heterojunction, electrons are transferred to the GaAs layer until an equilibrium is reached; this occurs because electron transfer raises the

Fermi energy on the GaAs side due to level filling and also raises the electrostatic potential of the interface region because of the more numerous ionized donors in the GaAlAs side. The charge-transfer effect makes possible an old dream of semiconductor technologists, i.e., getting conducing electrons in a high-purity, high-mobility semiconductor without having to introduce mobility-limiting donor impurities. Since then, modulation doping has been applied to a number of situations involving various semiconductor pairs and also to hole modulation doping. The impressive development of the subject is due to the applicability both to basic science (2D physics, quantum Hall effect, . . .) and to very high-performance devices called equivalently HEMT (high-electron-mobility transistor), TEGFET (two-dimensional electron-gas FET), SDHT (selectively doped heterostructure transistor), and MODFET (modulation-doped FET).

a. Charge Transfer in Modulation-Doped Heterojunctions

The understanding of the mechanism of charge transfer in heterojunctions is of the utmost importance, as it determines the GaAs channel doping and sets the design rules for the growth sequence of the doped and undoped layers. There are three main phenomena to be determined in a

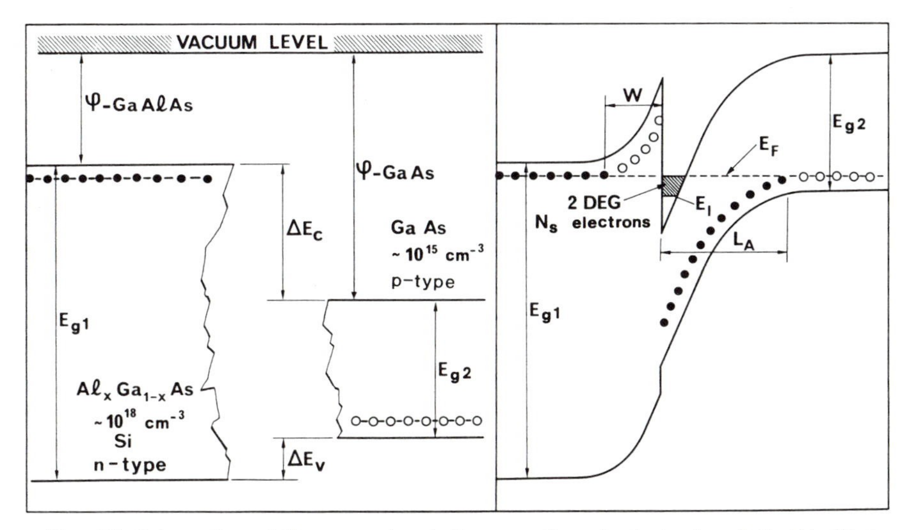

FIG. 19. Schematics of the energy-band diagram of a selectively doped GaAlAs/GaAs heterostructure before (left) and after (right) charge transfer has taken place. The relative energy bands are, as usual, measured relative to the vacuum level situated at an energy φ (the electron affinity) above the conduction band. The Fermi level in the $Ga_{1-x}Al_xAs$ bulk material is supposed to be pinned on the donor level, which implies a large donor binding energy ($x > 0.25$).

self-consistent manner to calculate the Fermi energy throughout the structure and therefore the charge transfer.

(1) The electric charges and field near the interface determine the energy-band bendings in the barrier and in the conducting channel.
(2) The quantum calculation of the electron energy levels in the channel determines the confined conduction-band levels.
(3) The thermodynamic equilibrium conditions (constant Fermi energy across the junction) determine the density of transferred electrons.

A very crude calculation can show the interplay of the various factors in a simple situation (Fig. 20). Assume that before the charge transfer occurs the potential is flat-band; after charge transfer of N_S electrons, the electric field in the potential well created can be taken as constant to first order, given by $F = N_S e/\varepsilon_0 \varepsilon_R$ (Gauss's law). The electrostatic potential is then $\phi(z) = -Fz$ for $z > 0$. The Schrödinger equation for the electron envelope wave function is then (see Section 4)

$$[p^2/2m^* - e\phi(z)]\chi(z) = \varepsilon\chi(z) \tag{34}$$

A quantum calculation of the energy levels in the infinite triangular quantum well gives the ground state[134,135]

$$E_1 = (\hbar^2/2m^*)^{1/3}(\tfrac{9}{8}\pi e^2 N_S/\varepsilon_0\varepsilon_R)^{2/3} \tag{35}$$

One usually writes phenomenologically $E_1 = \gamma N_S^{2/3}$, with γ to be determined experimentally. As charge transfer increases, the electrostatic con-

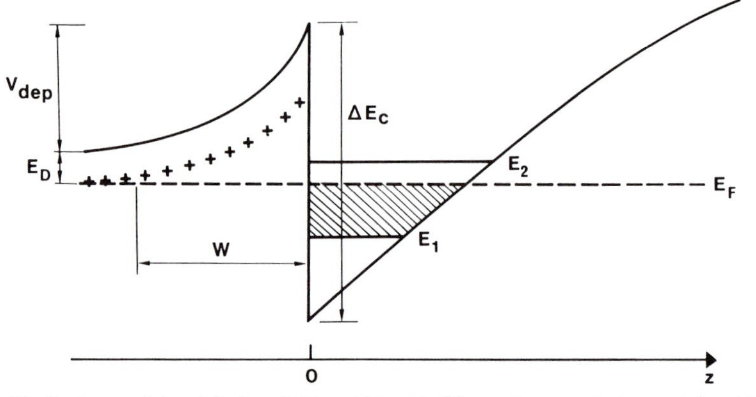

FIG. 20. Scale-up of the right-hand side of Fig. 19. The various symbols are defined in the text.

fining potential created by the transferred electrons also increases, leading to the raising of the bottom of the conduction band E of channel electrons. Equilibrium at $T = 0$ occurs when the top of the filled states, given by (if only one confined subband is occupied)

$$E = E_1 + \frac{N_S}{\rho_{2D}} = E_1 + \frac{\pi\hbar^2 N_S}{m^*} \tag{36}$$

is equal to the Fermi level on the GaAlAs side of the heterojunction. That level is equal to the bulk GaAlAs Fermi energy level, pushed downwards by the electrostatic potential V_{dep} built up at the interface due to the depleted donor atoms. Assuming a constant doping in the GaAlAs, V_{dep} is

$$V_{\text{dep}} = -\int_{z=0}^{-W} F\,dz = \int_0^{-W} \frac{eN_D z}{\varepsilon_0 \varepsilon_R}\,dz = \frac{eN_D W^2}{2\varepsilon_0 \varepsilon_R} \tag{37}$$

where W is the depleted thickness. Counting the energies from the bottom of the conduction band at $z = 0$, one finds that

$$\Delta E_c = E_1 + \frac{\pi\hbar^2 N_S}{m^*} + \varepsilon_d + eV_{\text{dep}} \tag{38}$$

where ε_d is the donor binding energy in GaAlAs. This assumes that the donor level in GaAlAs is sufficiently deep so that the Fermi level is pinned there (In other cases one has to calculate through standard procedures the Fermi level position in GaAlAs). Remembering that $N_S = N_d W$, one obtains the implicit equation in N_S:

$$\Delta E_c - \frac{e^2 N_S^2}{2\varepsilon_0 \varepsilon_R N_D} - \varepsilon_d = E_1(N_S) + \frac{N_S \pi\hbar^2}{m^*} \tag{39}$$

Various more exact calculations have been provided, relying on more or less rigorous bases and providing analytical or numerical results. They, however, rely on the very simple, although well-justified, assumption that the GaAs electron wave function has a negligible penetration in the barrier. Thus the electron energy levels are unaffected by the possible changes of the barrier electrostatic potential induced by charge transfer. As will be seen below, the wave function penetrates at most 20 Å in the barrier, which is much less than the depleted thickness in the barrier. Therefore, the wave function is only determined by the barrier height, and only at second order by the electric field at the interface due to the depletion charges. It is then possible to uncouple the equilibrium conditions from the electrostatic and energy-level calculations. The additional ingredients of the more exact calculation follow.

b. Electrostatic Potential

The different parameters entering the calculations of the electrostatic potentials are as follows:

(1) The various layers have some degree of compensation that must be taken into account for evaluations of the charge transfer. As will be discussed in Section 17, these uncontrolled ionized impurities play a crucial role in the ultimate performance of devices.

(2) Residual doping in the GaAs layer creates an electrical field in the resulting depleted region of GaAs, but also contributes to the potential in the barrier.[135a] For *p*-type residually doped GaAs, this doping is described by the depletion charge due to the interface band bending in the depletion length L_A of the material:

$$N_{\text{dep}} = N_A L_A \approx N_A \left(\frac{2\varepsilon_0 \varepsilon_R E_G}{N_A e^2} \right)^{1/2} = \left(\frac{2 N_A \varepsilon_0 \varepsilon_R E_G}{e^2} \right)^{1/2} \tag{40}$$

As the doping is usually quite small, the depletion width in GaAs is much larger than all other dimensions in the system and the potential due to these charges can be considered triangular (constant E field) in the region of interest. In the limit of large charge transfer, this potential can be almost neglected when compared to the field F_S of transferred electrons: For $N_A = 10^{14}\ \text{cm}^{-3}$, $N_{\text{dep}} = 4.6 \times 10^{10}\ \text{cm}^{-2}$, $L_A = 4 \times 10^{-4}$ cm, $F_{\text{dep}} = 7.5 \times 10^3$ V/cm, whereas for $N_S = 5 \times 10^{11}\ \text{cm}^{-2}$, $E_S = 7.5 \times 10^4$ V/cm. On the other hand, the detailed knowledge of the residual impurities potential is extremely important in the case of small charge transfer. It also dominates in all cases for the determination of the electron excited states, as their wave function extends far away from the interface.

(3) The case of *n*-type residual doping in GaAs is not so easy to solve, as the Fermi level away from the interface cannot be evaluated independently of the charge transfer occurring at the interface. One usually treats this case as a quasi-accumulation case, considering that the situation is the limit of either a very small *p*-type residual doping ($N_{\text{dep}} = 10^9\ \text{cm}^{-2}$ in the case of Ando[136,137]) or that the Fermi energy far away from the interface lies 1 eV below the conduction band. Detailed calculations actually show that one or the other of these choices does not influence the calculated energy levels.

(4) An undoped GaAlAs "spacer" layer of thickness W_{sp} is usually used to separate the ionized donor atoms further from the channel electrons: increasing this spacer layer diminishes the Coulomb interaction between the ionized donors and the electrons, resulting in an increased mobility.[138,139] There is, however, a limit: since the electric field is constant in the spacer layer (no space charge in the absence of ionized impurities),

the electrostatic potential builds up there, although it does not correspond to transferred charges. The consequence is that increasing the spacer layer width W_{sp} tends to decrease the channel electron density N_S. At low temperatures, the equilibrium Eq. (38) can be rewritten as

$$\Delta E_C = eV_{dep} + eV_{sp} + \varepsilon_d + E_1 + \pi\hbar^2 N_S/m^* \tag{41}$$

with $V_{sp} = eW_{sp}N_S/\varepsilon_0\varepsilon_R$.

c. Energy-Level Calculation

The electron energy levels can be self-consistently calculated using approximations of various degrees of sophistication. The most widely used scheme is the Hartree approximation calculated using variational Fang–Howard-type wave functions.

At lowest order, the electron–electron interaction $V_{ee}(z)$ is described by the Hartree approximation; i.e., $V_{ee}(z)$ is given by

$$\frac{d^2 V_{ee}(z)}{dz^2} = \frac{e^2 N_S}{\varepsilon_0 \varepsilon_R} |\chi(z)|^2 \tag{42}$$

which expresses that an electron feels the average electrostatic field created by all others electrons.

The Schrödinger equation is then

$$\left[-\frac{\hbar^2}{2m^*}\frac{d^2}{dz^2} + V_0(z) + V_{imp}(z) + V_{ee}(z)\right]\chi(z) = \varepsilon\chi(z) \tag{43}$$

where V_0 and V_{imp} are, respectively, the heterojunction and channel ionized impurity electrostatic potentials.

The usual Fang–Howard[140] functions used for the Si–SiO_2 case (no penetration in the SiO_2 barrier) are modified to account for the penetration in the GaAlAs barrier[137,141]

$$\chi(z) = Bb^{1/2}(bz + \beta)\exp(-bz/2), \qquad \text{if } z > 0 \tag{44}$$

$$= B'b'^{1/2}\exp(b'z/2), \qquad \text{if } z < 0 \tag{45}$$

where B, B', b, b', and β are variational parameters. The usual boundary and normalization conditions leave only two variational parameters, b and b'. A very good approximation for b' is actually $b' = 2\hbar^{-1}(2m^*\,\Delta E)^{1/2}$, the standard wave-function penetration in the barrier.

More refined values of the electron–electron interactions have been considered, such as the local exchange correlation potential. Ando gave numerical solutions for the Schrödinger equation in that case[137] (Fig. 21). Very complete calculations, including finite-barrier effects, effective mass

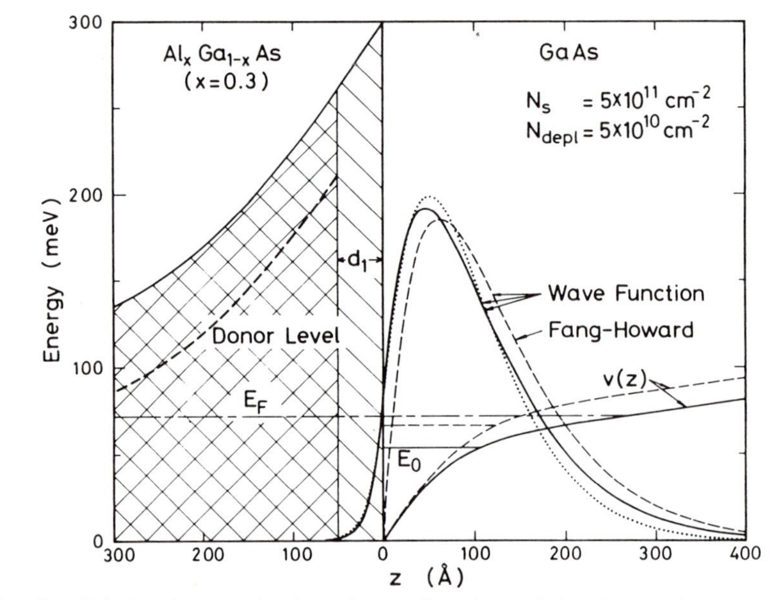

FIG. 21. Calculated energy levels and wave functions of the GaAlAs/GaAs selectively doped interface. The Fang–Howard variational wave function (no penetration in the barrier) is shown (---). The variational wave functions [Eqs. (44) and (45)] are also shown (——). The line (· · ·) represents the numerical calculation, which includes correlation effects. The different confining potentials $V(z)$ are shown. The spacer thickness is 50 Å and the donor binding energy of GaAlAs has been chosen as 50 meV (Reprinted with permission from the *J. Phys. Soc. Jpn.* **51**, 3893, T. Ando (1982).)

and dielectric constant discontinuities, interface grading, and nonzero temperature, were performed by Stern and Das Sarma.[142]

d. Thermodynamic Equilibrium

In a real device situation one has to consider the finite temperature through Fermi–Dirac distribution functions for level occupancy. The determination of the relation between the Fermi energy and the channel density N_S is very simple because of the constant DOS in 2D. From the usual expression

$$N_S = \int_{E_1}^{\infty} \rho_{2D}(E) f(E)\, dE$$

where $f(E)$ is the usual Fermi–Dirac probability function. One deduces

$$N_S = \frac{kTm^*}{\pi\hbar^2} \ln\left[1 + \exp\left(\frac{E_F - E_1}{kT}\right)\right]$$

which can be used in Eq. (41).

The finite extension of the GaAlAs layer also plays an important role in devices (Fig. 22): The full MESFET-type structure must therefore be analyzed, taking into account the Schottky built-in voltage at the metal–GaAlAs interface. Vinter has made quantum-mechanical calculations of the gated heterojunction[143,144]: As can be seen in Fig. 22b, quantum levels exist both in the GaAs channel and in the barrier material. The potential barrier at the GaAs/GaAlAs interface is actually so thin that coupling of the quantum states occurs between the two regions. Two technologically interesting solutions of the whole structure can arise, such as the two situations of normally-on or normally-off GaAs channel (Fig. 22a); the latter situation occurs because the rather large Schottky voltage ($\approx$1 eV) can completely deplete the GaAlAs layer and the GaAs channel for thin enough GaAlAs layers. One can therefore have on the same chip normally-on and-off devices by controlled etching of some GaAlAs thickness from a normally-on layer. The voltage control of devices is also tailored by the layer thickness and doping. One of the challenges raised by LSI and VLSI components is actually the required layer uniformity for constant switching characteristics across the whole circuit. Only MBE at its best seems now acceptable for the needed level of control.

The device characteristics of TEGFET/HEMT/SDHT/MODFET structures due to the gate control of the charge densities, which in turn induce such quantities as transconductance, etc., have been calculated using various models and approximations for various structures[19–21,143,145–150] (Fig. 23).

Several authors have addressed the problem of the design rules needed to obtain optimum performance of single heterostructures: Increasing the doping level N_D of the barrier material increases charge transfer as W varies as $N_D^{-1/2}$ [Eq. (37)], leading to a charge transfer $WN_d \propto N_D^{1/2}$. Such a high transfer leads to efficient channel impurity screening, but it also leads to more scattering by the impurities located in the barrier. Inclusion of an undoped spacer layer leads to a decrease of this last scattering mechanism, but also diminishes charge transfer. Considerations on the optimal doping have been produced by Stern.[151] The experimental determination of the density of transferred electrons as a function of the doping level N_d and the spacer thickness W_{sp} is shown in Fig. 24a. As can be seen, there is not much room for obtaining densities larger than 10^{12} cm^{-2} in single GaAs/GaAlAs interfaces. In order to obtain higher carrier densities in the channel, several authors[152,153] have studied the double-heterojunction field-effect transistor where the active layer consists in a wide GaAs undoped layer imbedded between two selectively doped GaAlAs barriers. This configuration produces a double hetero-interface situation, thus allowing one to double the channel carrier density (Fig. 24; see also Section 17). More recent studies have been made on modulation-doped multiquantum wells

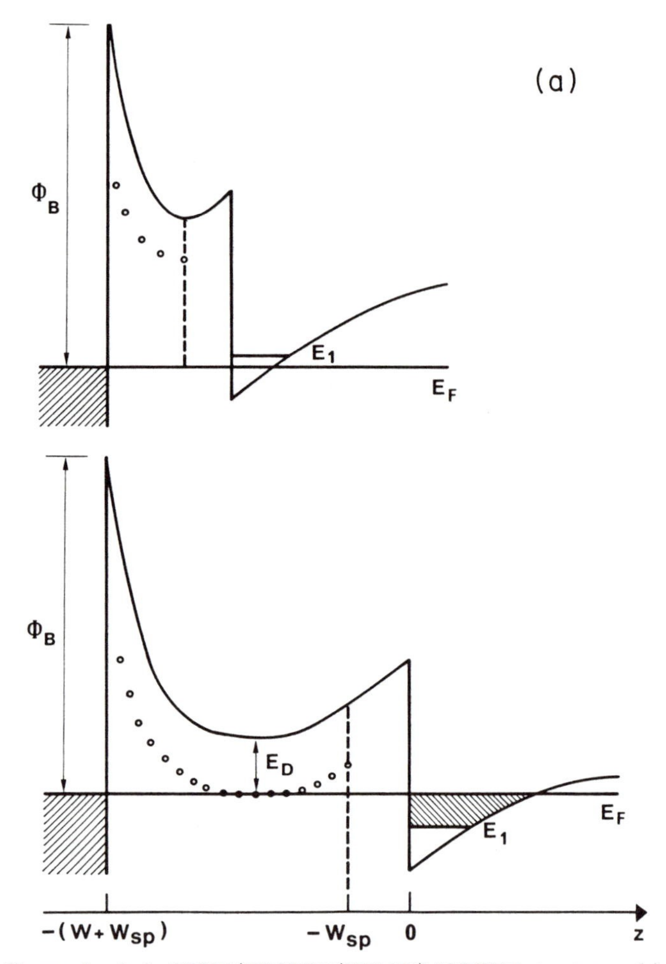

FIG. 22 Energy levels in SDHT/TEGFET/HEMT/MODFET structures. (a) Schematic band diagram for normally off (top) and normally on (bottom) transistors. (b) Calculated self-consistent potentials for conduction electrons in two TEGFETs for two gate voltages V_G. Left part: normally on device; right part: quasi-normally off device at room temperature. The Fermi level is at $E = 0$. Horizontal lines: bottom energy of the lowest four subbands. The energies of the donor levels are shown dashed for the higher gate voltage (from Vinter).[144]

which have still higher densities and allow high-current operation.[153a]

It might well seem that we have at hand enough theoretical mastery to be able to design heterostructures with great precision. Unfortunately, a number of precautions must be taken, in order to precisely design a desired structure:

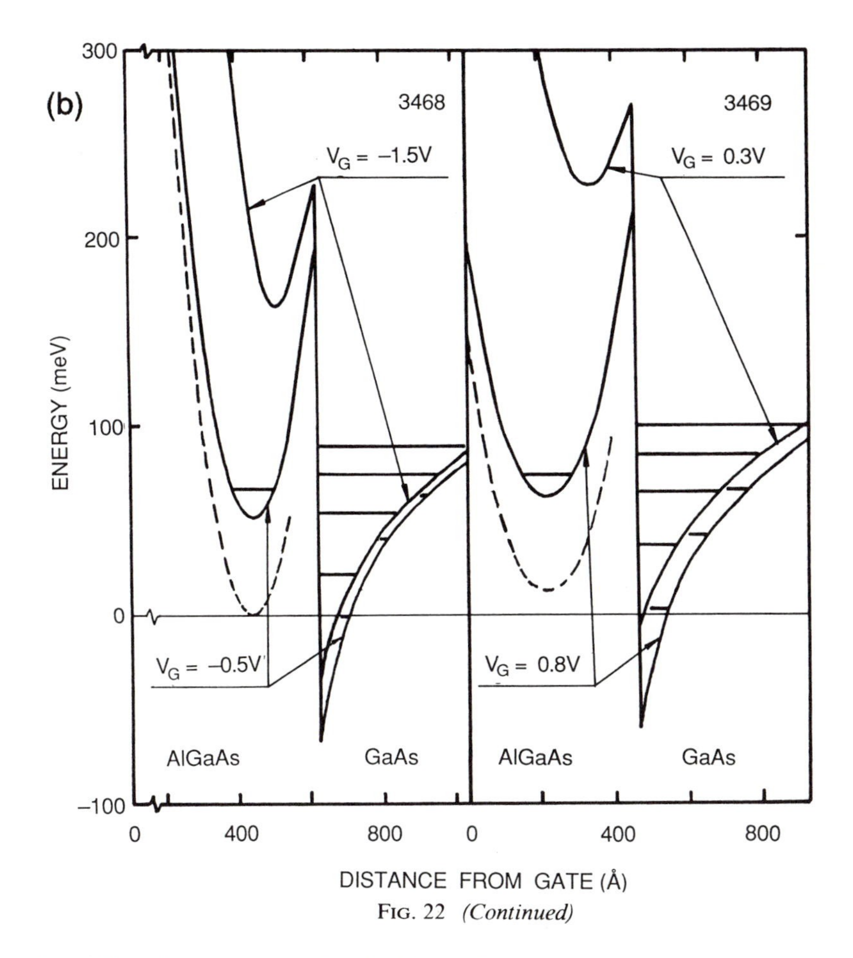

FIG. 22 *(Continued)*

(1) The donor energy levels E_d in $Ga_{1-x}Al_xAs$ are quite unreliable, changing from only 6 meV at $x \leq 0.1$ to more than 160 meV for indirect-gap material[154–156] (Fig. 25). The transition from shallow to deep donor level occurs at $x \approx 0.235$. This increase in E_d is very detrimental to the transfer of large charge densities [see Eq. (39) and Fig. 20] and to good operation at low temperatures (77 K), because of carrier freeze-out. On the other hand, one requires large values of x in order to increase ΔE_c, thus increasing the charge transfer. A very elegant way to solve this problem has recently emerged[157]: the charge-transferring side of the heterojunction is made up from a GaAs–GaAlAs superlattice, where only the GaAs layers are strongly n-type doped and the GaAlAs barriers are thin enough to allow charge tunneling. Due to the large carrier confinement effects in the thin

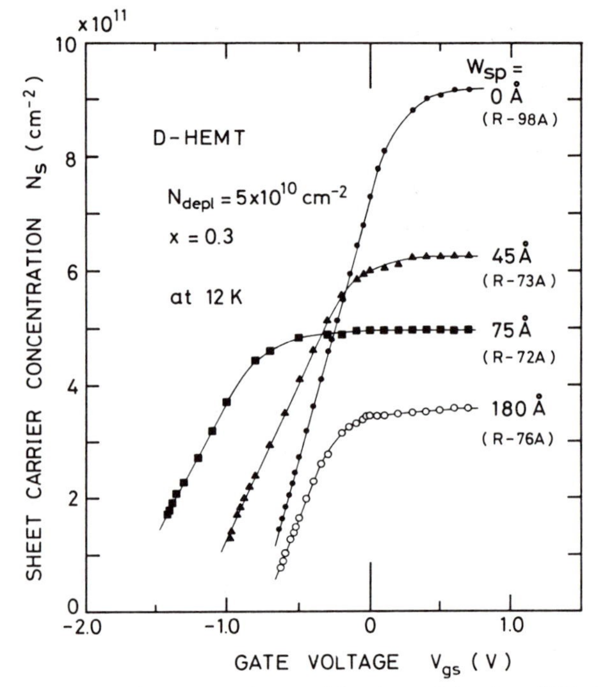

FIG. 23. Measured gate-voltage dependence of the channel density of 2D electrons in the GaAlAs/GaAs system. W_{sp} represents the spacer thickness. All samples have $N_D = 4.6 \times 10^{17}$ cm^{-3}, except sample #R-76 A, which has $N_D = 9.2 \times 10^{17}$ cm^{-3}. The channel carrier density saturates with V_{gs}, as electrons tend to accumulate in the barrier subbands once they can be populated (see also Fig. 22) (from Hirakawa *et al.*[149]).

(<20 Å) GaAs layers, the donor ground level associated with the lowest confined level is raised well above the bulk GaAs conduction-band level, almost to the GaAlAs barrier level. Charge transfer can therefore occur between these GaAs confined donor levels and the GaAs channel (Fig. 26). A number of bothersome low-temperature effects have thus been eliminated, such as source-drain polarization effects, persistent conductivity, and electron freeze-out at 77 K.[158,159]

(2) The number of electrically active donors is not unambiguously related to the number of metallurgical donors. Some Si atoms seem to be associated with deep defects, giving rise to persistent photoconductivity centers. These centers are also responsible for the collapse of device characteristics at low temperatures.

(3) Recent detailed experiments[160-161] point out the segregation of Si towards the GaAlAs growing front, the more so for larger values of x. This observation tends to provide a satisfactory explanation for a number of

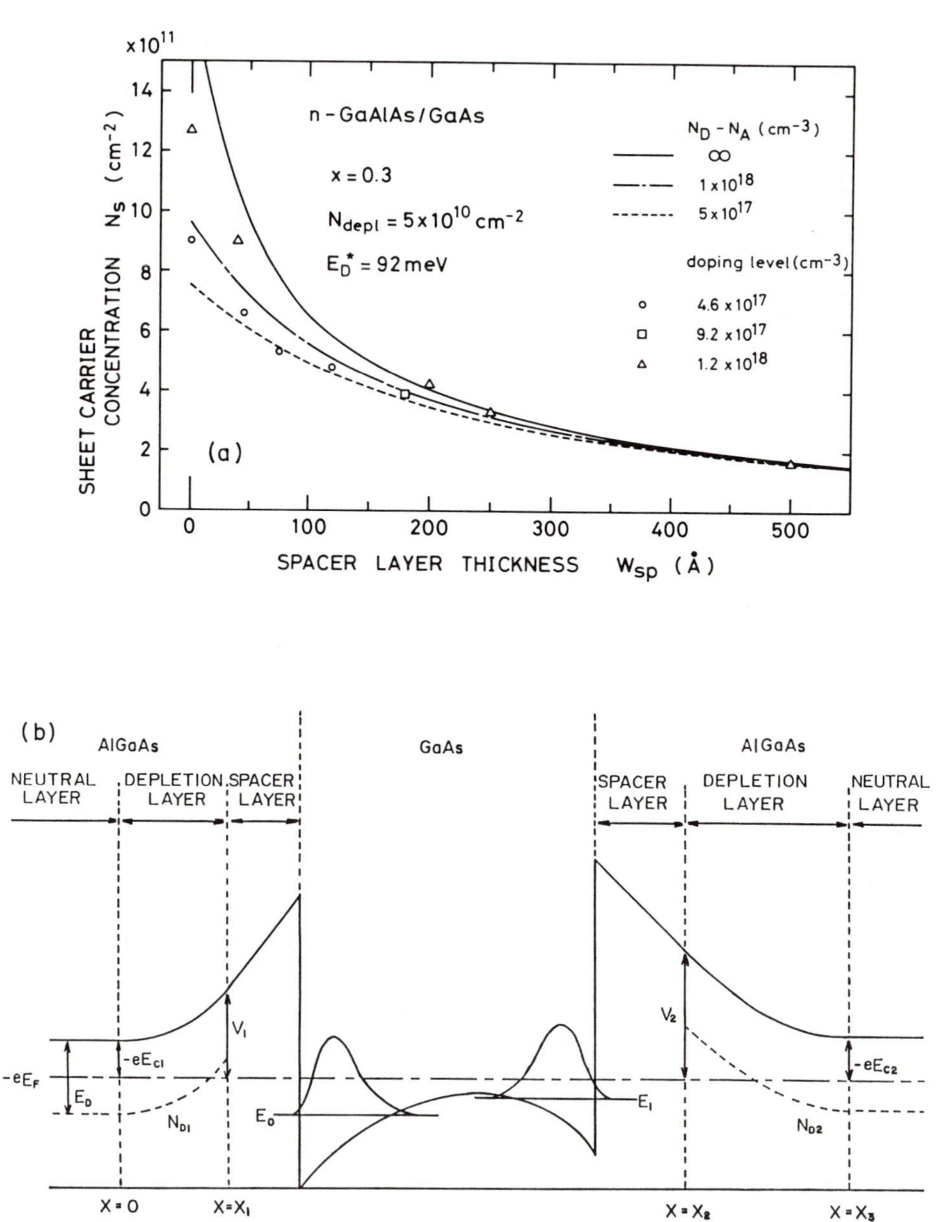

FIG. 24. (a) Sheet electron concentration N_S at an n-type GaAs/GaAlAs single interface as a function of the spacer layer thickness W_{sp} and of the doping level in the doped portion of the GaAlAs layer. The lines are theoretical calculations for the three doping levels indicated (after Hirakawa *et al.*[149]). (b) Schematics of the double-heterojunction transistor (Reprinted with permission from *Superlattices and Microstructures* **1**, 43, K. Miyatsuji, H. Hihara, and C. Hamaguchi (1985).)

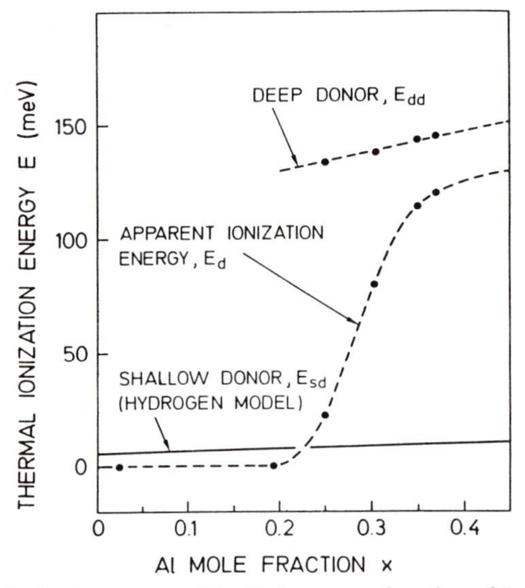

FIG. 25. Thermal ionization energy of the Si donor as a function of the Al mole fraction in $Ga_{1-x}Al_xAs$ (from Schubert and Ploog[155]).

hitherto puzzling results: (a) The smaller-than-expected transfer of charges in normal heterostructures (GaAlAs over GaAs) is due to the actual wider spacer than metallurgically grown. This in turn explains the sometimes observed higher mobility than theoretically expected. (b) On the other hand, the past poor performance of inverted structures (GaAs over GaAlAs) can also be explained: the segregation of Si impurities to the GaAlAs surface brings impurities near or in the GaAs channel. The excellent characteristics of the inverted MODFET by Cirillo *et al.*[161a] evidence the high quality of inverted interfaces now attainable. Heiblum[161] observed that for $x \approx 0.1$ and low growth temperatures the segregation effect was much reduced.

(4) The compensation and residual doping of the various layers are not too well controlled even in a given growth chamber, and might not be uniform in a given sample depth or reproducible from sample to sample. This situation is best exemplified in the sequential growth of high-quality samples by Hwang *et al.*[162]: charge transfer as a function of spacer thickness follows a reasonable behavior and allows a determination of the donor energy level in the GaAlAs barrier taken as an adjustable parameter. On the other hand, the mobility does not show a maximum as predicted from the decreasing barrier and increasing channel scattering probabilities. This is probably due to the erratic residual doping in the channel, evidencing

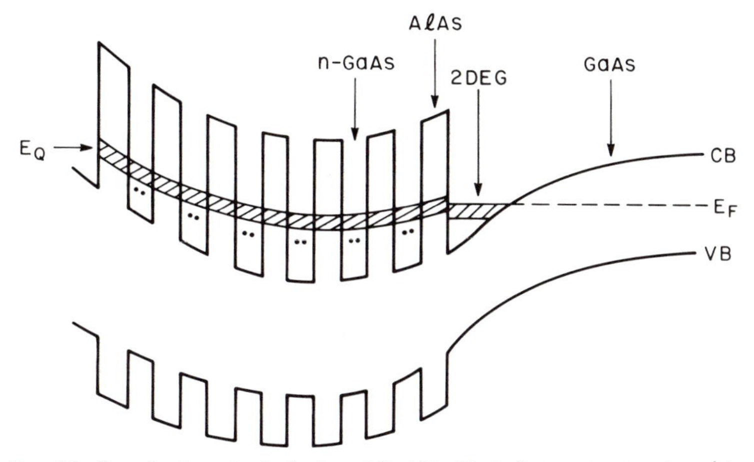

FIG. 26. Superlattice selectively doped GaAlAs/GaAs heterostructure transistor.

that this doping is the limiting factor in these high-quality samples.[151] Earlier results on lower-quality samples did exhibit a well-behaved mobility variation due to the more consistently reproducible background channel doping.[163] The progress in growth control is, however, such that one can expect to achieve more systematically the well-behaved charge transfer and mobility obtained recently (see Fig. 56).[149,161,163a]

9. $n-i-p-i$ STRUCTURES[164]

As was proposed in the original paper by Esaki and Tsu,[3] a spatial modulation of the doping in an otherwise homogeneous lattice can produce a superlattice effect, i.e., a spatial modulation of the band structure which induces a reduction in the Brillouin zone of electrons and new energy bands in the superlattice direction. The realization of such structures was achieved using periodic *n*-doped, undoped, *p*-doped, undoped, *n*-doped, . . . , multilayer structures, hence the acronym $n-i-p-i$ for such doping superlattices (as opposed to compositional superlattices).

By comparison with the modulation-doped heterostructure, the appearance of the doping superlattice effect is easy to understand (Fig. 27): charged particles are subject to a self-consistent potential:

$$V(z) = V_{\text{imp}}(z) + V_{\text{H}}(z) + V_{\text{xc}}(z) \tag{46}$$

where $V_{\text{imp}}(z)$ is the electrostatic potential of the ionized impurities, $V_{\text{H}}(z)$ the Hartree potential of electrons and holes, and $V_{\text{xc}}(z)$ the exchange and correlation potentials. The first term, $V_{\text{imp}}(z)$, can be calculated from the Poisson equation

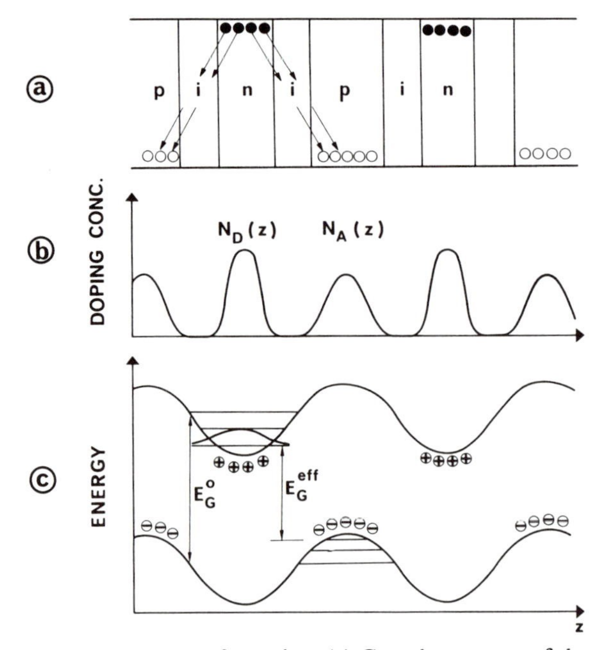

FIG. 27. n–i–p–i band-structure formation. (a) Growth sequence of the structure; electrons from neutral donors recombine with holes located on the neutral acceptors, leaving a net space charge associated with ionized impurities shown in (b); the resulting band-gap variation and carrier confinement are shown in (c).

$$\frac{d^2 V_{\text{imp}}(z)}{dz^2} = \frac{e^2}{\varepsilon_0 \varepsilon_R} [N_{\text{D}}(z) - N_{\text{A}}(z)] \tag{47}$$

Similarly, the Hartree potential is

$$\frac{d^2 V_{\text{H}}(z)}{dz^2} = \frac{-e^2}{\varepsilon_0 \varepsilon_R} [|\chi_e(z)|^2 - |\chi_h(z)|^2] \tag{48}$$

The exchange and correlation terms for electrons have been calculated by Ruden and Döhler[165] in a density-functional formalism:

$$V_{\text{xc}}(z) \simeq 0.611 \frac{e^2}{4\pi\varepsilon_0\varepsilon_R} \left(\frac{4\pi}{3N(z)} \right)^{1/3} \tag{49}$$

where $N(z)$ is the electron density.

Energy levels for the z-quantized motion have to be calculated self-consistently through the one-dimensional Schrödinger equation

$$\left(-\frac{\hbar^2}{2m^*} \frac{\partial^2}{\partial z^2} + V(z) \right) \chi_{e,h}(z) = \varepsilon \chi_{e,h}(z) \tag{50}$$

where χ_e and χ_h are the envelope functions of the electron or hole wave functions.

A number of straightforward features can be extracted from Eqs. (46) and (50)[164]:

(1) In the case of exact compensation (equal numbers of donors and acceptors),

$$\int_{-d/2}^{+d/2} N_D(z)\, dz = \int_{-d/2}^{+d/2} N_A(z)\, dz$$

where d is the superlattice period, no free carrier exists in the unexcited sample at low temperatures.

(2) For equal uniform doping levels $N_A = N_D$ and zero-thickness undoped layers, the periodic potential consists of parabolic arcs and has an amplitude

$$V_0 = \frac{e^2}{8\varepsilon_0\varepsilon_R} N_D d_m^2 \tag{51}$$

(for GaAs with $N_A = N_D = 10^{18}\ \text{cm}^{-3}$ and $d_m = 500$ Å, one has $V_0 =$ 450 meV). The quantized energy levels in the potential wells are approximately the harmonic oscillator levels:

$$\varepsilon_{e,h} = \hbar\left(\frac{e^2 N_D}{\varepsilon_0\varepsilon_R m^*_{e,h}}\right)^{1/2}\left(n + \frac{1}{2}\right) \tag{52}$$

For electrons, for instance, the subband separation is 40.2 meV for the above parameters. Since the effective bandgap E^{eff} shown in Fig. 27 is given by $E^{\text{eff}} = E_G - 2V_0 + \varepsilon_e + \varepsilon_h$, it is reduced below the bulk material value.

(3) When there is unequal doping, free carriers will accumulate in the corresponding potential well (Fig. 28). Equations (46) and (50) must then be solved self-consistently. The Fermi level can be located at will (Fig. 28b).

(4) For large enough spacings and dopings, the effective bandgap can become negative (i.e., $d \geq 700$ Å for $N_A = N_D = 10^{18}\ \text{cm}^{-3}$). There then exists charge transfer from hole wells to electron wells until a zero gap is attained due to three factors: band filling, diminishing of the periodic superlattice potential thanks to the charge neutralization by the transferred charges, and quantized energy-level modification (Fig. 28c).

(5) Under nonequilibrium conditions such as photoexcitation or carrier injection, electron and hole populations can build up in the wells, leading to charge neutralization and an effective bandgap increase.

(6) Under such nonequilibrium conditions, electrons and holes are

spatially separated and the radiative recombination rate is strongly diminished as compared to the bulk case as for an indirect-bandgap semiconductor. At the same time, nonradiative recombination rates are also strongly decreased, leading to reasonable quantum efficiencies. This justifies the hopes for tunable light sources expected from doping superlattices, even though they are real-space indirect semiconductors. They should also lead to excellent photodetectors, as the photoconductive gain should be very large.

Many of the features that are expected from the $n-i-p-i$ structure have indeed been observed: variation of the bandgap with increased excitation, change of absorption features with light intensity, tunable luminescence, etc. The reader is referred to the review articles by Ploog and Döhler[165] and Abstreiter[166] for a very exciting description of doping superlattices. An

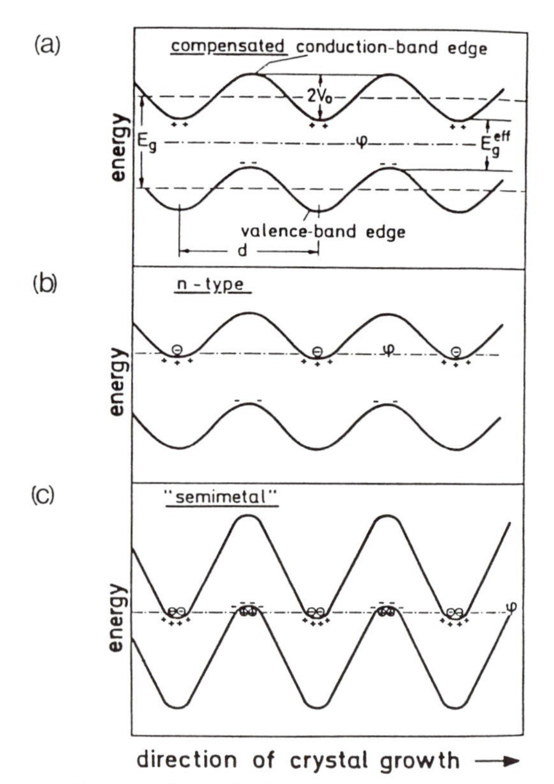

FIG. 28. Three types of $n-i-p-i$ superlattices: (a) compensated intrinsic superlattice with $2V_0 < E_g$ and $N_D d_n = N_A d_p$; (b) n-type superlattice with $2V_o < E_g$ and $N_D d_n > N_A d_A$; (c) semimetal superlattice with $2\ V_0 > V_g$ (Reprinted with permission from Taylor and Francis Ltd., K. Ploog and G. H. Dohler, *Adv. in Phys.* **32**, 285 (1983).)

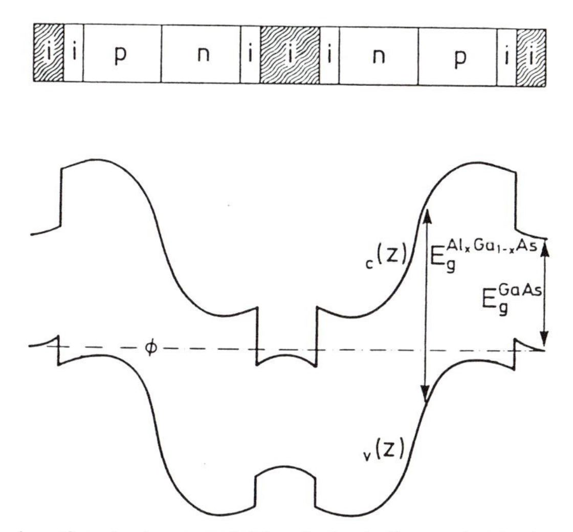

FIG. 29. Heterojunction doping superlattice: Instead of occurring in doped (low-mobility) regions as in the standard structure of Fig. 27, carrier transport occurs here in the high-purity undoped potential wells (Reprinted with permission from Taylor and Francis Ltd., K. Ploog and G. H. Dohler, *Adv. in Phys.* **32**, 285 (1983).)

interesting recent development is the heterojunction doping superlattice (Fig. 29); in the standard doping superlattice the transport of electrons and holes occurs in doped regions and therefore mobility is rather poor. By introducing undoped small-gap semiconductor layers in the middle of the *n*- and *p*-doped layers of the superlattice, the carriers are transferred in the undoped small-gap material where they experience high mobilities as in usual modulation-doped structures.

CHAPTER III

Optical Properties of Thin Heterostructures

The most general and surprising feature of the optical properties of quantum wells is the strength of the intrinsic optical effects as compared to bulk optical properties: in many circumstances one measures features of comparable size for a single quantum well of ≈ 100 Å as for bulk samples of thickness of the order of an absorption length, a few 1000 Å (Fig. 30a).[167] In particular, the quantum efficiency of luminescence has been observed to be larger in QW structures for all the systems reported up to now: GaAs/GaAlAs,[168] GaInAs/AlInAs,[169] GaSb/GaAlSb,[170] GaInAs/InP,[61a] CdTe/CdMnTe,[79] GaAsSb/GaAlSb,[170a] ZnSe/ZnMnSe,[170b]. . . .

10. Optical Matrix Element

a. *Interband Transitions*

The interband transition probability for particles confined in quantum wells can be calculated by perturbation theory and is, as usual, the product of an optical matrix element times a density of states[15,16,170c,d]. The transition rate is given by Fermi's golden rule:

$$W = \frac{2\pi}{\hbar} \sum_{f,i} |\langle f | H_{\mathrm{I}} | i \rangle|^2 \delta(E_f - E_i + \hbar\omega) \tag{53a}$$

where i,f are transition initial and final states with energies E_i and E_f, respectively, H_{I} is the interaction Hamiltonian, which in the electric dipole approximation is $-e\mathbf{r}\cdot\mathbf{E}$. The summation over initial and final states

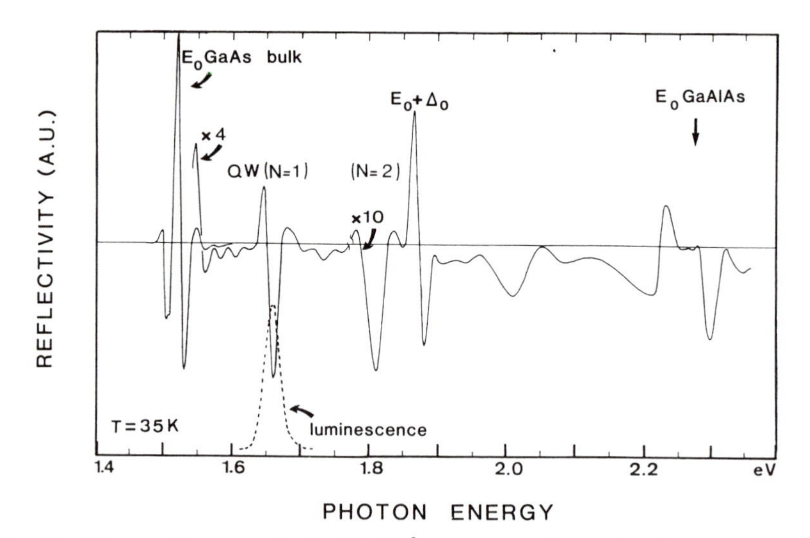

FIG. 30a Electroreflectance of a double 49 Å quantum-well sample. The remarkable feature is the size of the $n = 1$ QW exciton electroreflectance peak, quite similar (factor of 4) to that of the GaAs substrate. The luminescence spectrum is used for the peak assignment (Reprinted with permission from Editions de Physique, C. Alibert, F. Jiahua, M. Erman, P. Frijlink, P. Jarry, and J. B. Theeten, *Rev. Phys. Appl.* **18**, 709 (1983).)

introduces the reduced density of states. The modification to the usual 3D probability stems from the 2D density of states, because we can easily show that the optical matrix element is hardly changed as compared to 3D. The interband optical matrix element[15,16] has the form

$$M \propto |\langle f|\mathbf{r}\cdot\eta|i\rangle| = \int \chi_e(z)e^{i\mathbf{k}_{e\perp}\cdot\mathbf{r}\perp}u_{ck_e}(\mathbf{r})\boldsymbol{\eta}\cdot\mathbf{r}\,\chi_h(z)e^{i\mathbf{k}_{h\perp}\cdot\mathbf{r}_\perp}\,u_{vk_h}(\mathbf{r})\mathbf{dr} \quad (53b)$$

where $\chi_e(z)$ and $\chi_h(z)$ are the electron and hole envelope wave functions, $\mathbf{k_e}$, $\mathbf{k_h}$ are electron and hole wave vectors, $\boldsymbol{\eta}$ is the polarization vector of light, $u_{ck_e}(\mathbf{r})$ and $u_{vk_h}(\mathbf{r})$ are the usual Bloch functions. The integral contains fast-varying functions over unit cells (u_{ck} and u_{vk}) and slowly varying functions (Fig. 30b). Using the usual procedure, one transforms Eq. (53b) in a summation of localized integrals involving only Bloch functions over the N crystal unit cells labeled by their centers $\mathbf{R_i}$:

$$M \simeq \sum_{R_i} \chi_e(\mathbf{R_i})\,\chi_n(\mathbf{R_i})\,e^{i(\mathbf{k}_{h\perp}-\mathbf{k}_{e\perp})\cdot\mathbf{R_i}} \int_{\text{cell}} u_{\mathbf{ck_e}}(\mathbf{r})\boldsymbol{\eta}\cdot\mathbf{r}u_{vk_h}(\mathbf{r})\,d\mathbf{r}. \quad (54a)$$

The latter integral is independent of $\mathbf{R}_i$ and is ΩP where Ω is the unit cell volume and P the usual 3D matrix element that contains the selection rules due to the band symmetries and light polarization. When summing for the transverse directions, exponential factor gives a null contribution unless $\mathbf{k}_{e\perp} = \mathbf{k}_{h\perp}$, which is the vertical transition selection rule.

The only difference between Eq. (54a) and the usual 3D summation lies in the z-direction summation, which produces a factor $\Sigma\chi_e(\mathbf{R}_i)\chi_h(\mathbf{R}_i)a$, where the R_i's are the lattice cell centers in the z direction and a is the lattice constant.

Transforming back into an integral, $\int\chi_e(z)\chi_h(z)\,dz$, one finds a unity factor for the transitions between electron and hole states with the same quantum number n, as the χ's are identical $[\sim\sin(n\pi z/L)]$ and normalized to unity. The optical matrix element is therefore the same in 2D and 3D. In the absence of exciton effects, the absorption coefficient should reflect the

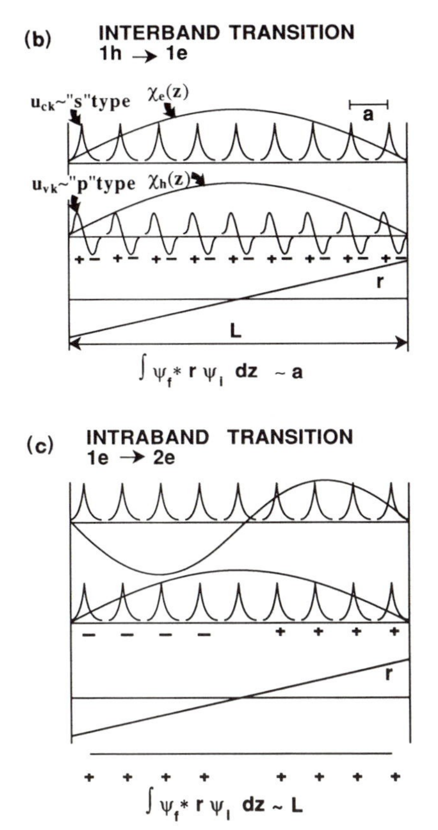

FIG. 30b,c. (b) Schematics of the integrand function of the interband matrix element as the product of the functions, i.e., $\chi_e(z)$, $\chi_h(z)$, u_{ck}, u_{vk}, and r. Note the inversion of u_{vk} at each lattice site, which determines equal contributions to the total dipole moment. (c) Same for intraband transitions in the conduction. Sites that are quite away from the well center and extremity contribute more to the dipole moment, which leads to the giant dipole effect.

reduced 2D DOS, i.e., should consist of square steps corresponding to the various confined states. It has been calculated by Voisin[172] to be $\simeq 10^{-3}$ per layer using the known parameters of GaAs, independent of layer thickness. This situation is usually obscured by exciton effects, and has only been observed in standard absorption measurements in GaSb–AlSb system[171].

b. Oscillator Strength of Interband Transitions

A useful quantity to characterize the strength of an optical transition is the oscillator strength f, defined by

$$f = 2\frac{|\langle f|\boldsymbol{\eta}\cdot\mathbf{p}|i\rangle|^2}{m\hbar\omega} = \frac{2m\omega}{\hbar}|\langle f|\boldsymbol{\eta}\cdot\mathbf{r}|i\rangle|^2 \tag{54b}$$

where m is the free electron mass and i and f are initial and final states of a given transition.

The oscillator strength can be used in many physical systems, such as atoms, molecules or solids, since it is simply related to many quantities such as the dielectric function[172b]

$$\varepsilon(\omega) = 1 + \sum_j \frac{4\pi e^2}{m}\frac{f_j}{(\omega j^2 - \omega^2) - i\gamma_j\omega} \tag{54c}$$

where the summation is performed over all transitions j and γ_j is a phenomenological damping constant.

For instance, for an isolated transition, the integral absorption line is[172a]

$$\int_{\text{line}} \alpha(E)\,dE = \frac{\pi e^2 h}{mnc} f$$

where m is the free electron mass, n the index of refraction.

When dealing with solids, Eq. (54c) evidences a major effect. Although the oscillator strength retains an atomic value f_{at} through the Bloch function matrix element in Eq. (54a), transition energies are widely spread over whole energy bands instead of occurring at a single energy when dealing with identical, noninteracting atoms. Therefore, optical effects described by $\varepsilon(\omega)$ are quite diminished when dealing with nonresonant excitations due to this spreading, and much more so for resonant transitions (i.e., near a band edge or within a band). An alternate description relies on the distribution of quantum states in k-space of a solid: due to k-conservation in optical transitions, only very few states are coupled within a given energy range, i.e., only a very small fraction of the total number of electrons of the solid participate to the transitions.

As is clear, the lower dimensionality of quantum wells does not increase the oscillator strength per electron-hole pair. However, when comparing 3D and 2D systems, as is clear from the density-of states curves, quantum confinement leads to a better matching of electron and hole wavefunctions: in the z-direction all electrons and holes in a same quantized subband have same k_z wavevector which "concentrates" the oscillator strength when compared to the 3D case. This point becomes more evident and important when going to still lower dimensions. It will be discussed in section 27.

c. Intraband (Intersubband) Transitions

The matrix element M given in Eq. (53) takes a much different shape when dealing with intersubband transitions, i.e., between confined electron states or hole states only (Fig. 30c). In that case, the fast-varying integrals involve the same periodic part of the wavefunction, which have then zero matrix element with $\mathbf{r}$. Equation (53) becomes

$$M \sim \int_{\text{crystal}} \chi_e(z)\boldsymbol{\eta}\cdot\mathbf{r}\chi_e'(z)\,dr \int_{\text{cell}} u_{cke}(\mathbf{r})u_{cke}^*(\mathbf{r})\,d^3r \tag{54e}$$

when using the usual procedure of transforming the M integral into a summation over fast varying contributions and transforming back into an integral. The second integral yields unity when using normalized Bloch functions. The matrix element has now large values, of the order of L, dimension of the quantum well, instead of an atomic dimension, as in interband transitions. This "giant" dipole effect was first observed by West and Eglash.[172c]

The value of the dipole moment between the ground state and first excited state is, in the infinite well approximation,

$$e\langle z\rangle = (16/9\pi^2)eL \tag{54f}$$

Applications of the giant intersubband dipole are already numerous. Levine and his team[172d] have developed a number of detectors on this basis.[172e] This detection scheme, however, seems to be intrinsically inferior to interband-based detectors such as HgCdTe alloys.[172f,g] This is due to the large generation-recombination noise originating in the very fast LO-phonon recombination processes of excited electrons within quantum wells or recaptured from the confining layers. Intersubband excitations are also extremely efficient for optical rectification[172h] and second harmonic generation.[172i,j] Efficiencies of optical rectification up to six orders of magnitude larger than for bulk GaAs have been obtained for optimized asymmetric quantum-well structures.[172k]

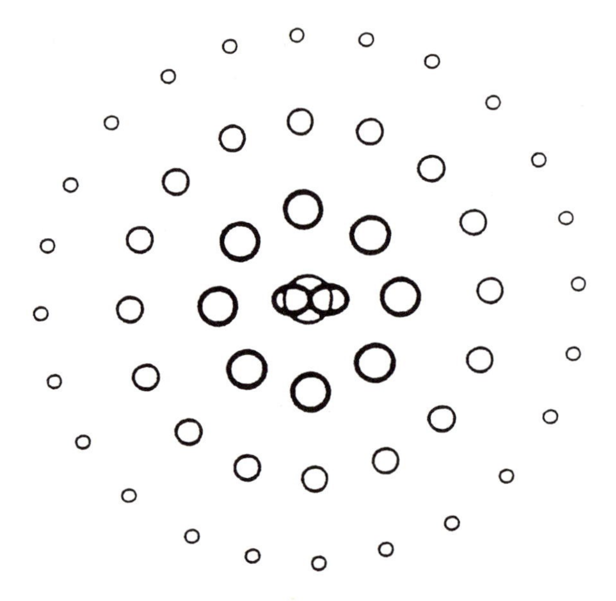

FIG. 30d. Schematics of the 2D extension[101] and 1D amplitude of the relative exciton wavefunction. The hole has been considered much heavier than the electron, hence its relative localization.

d. Excitonic Effects

Besides providing an energy shift of the absorption band to well-defined exciton absorption peaks, the electron-hole correlation concentrates the light-matter oscillator strength. This is most readily seen by examining the relative motion of the electron and hole in the exciton (Fig. 30d): Clearly, electron and hole wavefunction overlap much more strongly than in the case of delocalized free electron-hole pairs. The oscillator strength *per crystal unit volume* can be shown to be in 3D, after a lengthy but straightforward calculation:[101]

$$f = \frac{2m\omega|\langle u_c|\boldsymbol{\eta}\cdot\mathbf{r}|u_v\rangle|^2}{\hbar}\cdot\frac{1}{\pi a_{\mathrm{B}}^3} = f_{\mathrm{at}}\cdot\frac{1}{\pi a_{\mathrm{B}}^3} \tag{54g}$$

where $\langle u_c|\boldsymbol{\eta}\cdot\mathbf{r}|u_v\rangle$ is the standard valence band to conduction band matrix element and a_{B} the Bohr radius.

This value can be qualitatively justified from examination of the exciton wavefunction (Fig. 30d): Compared with delocalized free electron and hole wavefunctions, the dipole matrix element of the exciton state is increased in one direction by a factor of N_x (concentration increase of the hole wavefunction on a single site), times a factor N_x/n_x (concentration

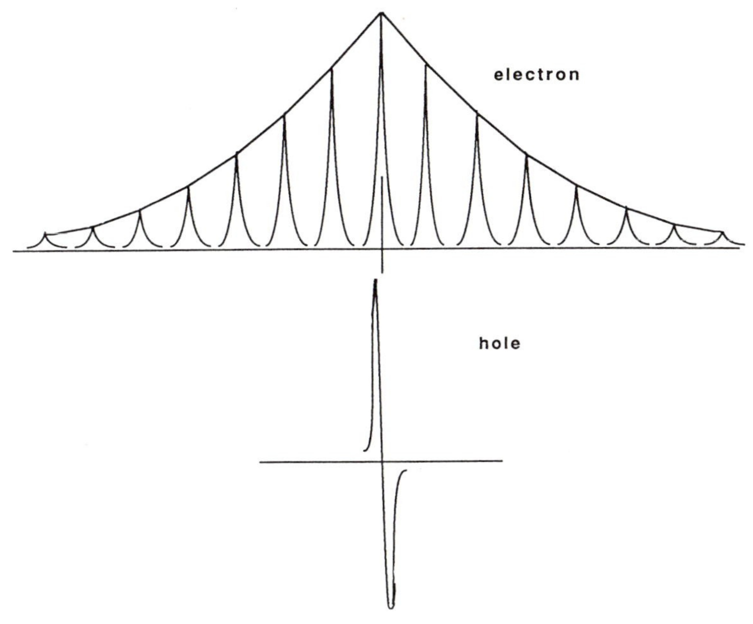

FIG. 30d

increase of the electron wavefunction on the site where it overlaps with the hole), diminished by a factor N_x (only one site integral is considered instead of N_x for delocalized pairs). Here N_x represents the number of crystal sites in the x-direction and n_x the number of crystal sites contained within the exciton volume. One therefore finds an overall increase in oscillator strength per transition given by N/n, where N is the number of unit cells in the crystal and n the number of unit cells in the exciton volume. This represents the number of excitons that can be closely packed in the crystal, as in Eq. (54g) for a crystal with unit volume. In other words, the oscillator strength due to the exciton effect in a solid is the same as that of closely packed "atoms" of size that of the exciton. It can also be shown that the exciton oscillator strength (54g) is that of all the free electron-hole pairs contained within the exciton binding energy (the Rydberg) from the band edge. This is not surprising, since the exciton state is a linear combination of electron and hole states with k vectors up to those yielding the Rydberg energy. Therefore, if the exciton line is sharper than the Rydberg energy, oscillator strength has been concentrated at the exciton energy by the electron-hole interaction.

In 2D, a similar concentration of oscillator strength from free states into the exciton state also exists. The infinite well approximation yields, in the

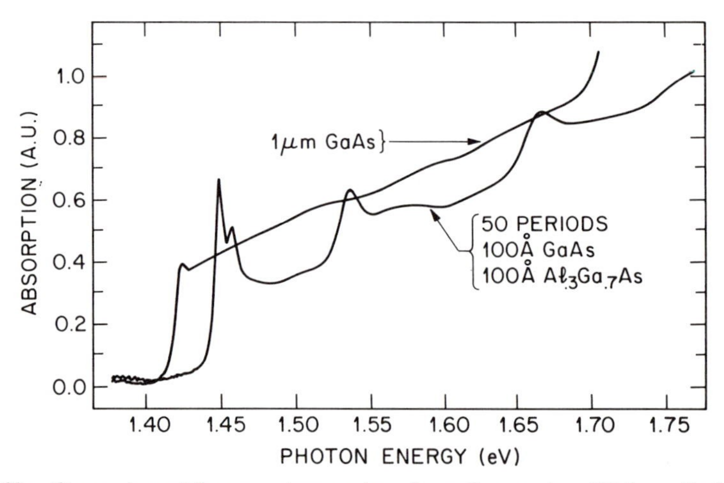

FIG. 30e. Comparison of the room-temperature absorption spectra of high-quality bulk and $L_z = 100$ Å MQW GaAs samples. The bump at the onset of bulk absorption is the remnant of the thermally dissociated exciton. The sharp exciton peak in MQWs denotes a stronger resonant light-matter coupling. (Ref. 254.)

exactly 2D limit, an oscillator strength *per unit surface*[104,1721]

$$f = \frac{2m\omega|\langle u_c|\boldsymbol{\eta}\cdot\mathbf{r}|u_v\rangle|^2}{\hbar}\cdot\frac{8}{\pi a_{\mathrm{B}}^2} = f_{\mathrm{at}}\cdot\frac{8}{\pi a_{\mathrm{B}}^2} \tag{54h}$$

where use of the four-times-smaller 2D Bohr radius when compared with 3D has been made, a_{B} retaining here its 3D value.

One therefore expects two main effects from the diminution of dimensionality when comparing 3D and 2D excitons:

(1) As already mentioned in Section 6, the increase in binding energy leads to resolved exciton peaks at room temperature, as the broadening due to exciton ionization through LO-phonon collision is smaller than the measured Rydberg (1.5 meV versus 9 meV).

(2) The concentrated oscillator strength is much larger than in 3D, due to the reduced Bohr radius. The integrated absorption peak is therefore ~16 times larger in the exact 2D limit, never reached as evidenced by the more precise calculations and measurements of the exciton binding energy (9 meV instead of 16 meV). However, as seen on Fig. 30e, the observed absorption peak at room temperature depicts a large resonant enhancement of the light-matter interaction, which is used in many applications (see below, Section 16).

11. Selection Rules

a. Interband Transitions

One should first notice that the quantum well and superlattice potentials are symmetric under space reflection changing z into $-z$. Therefore, parity is a good quantum number; i.e., the envelope wave functions are characterized by their even or odd character under space reflection.

Considering the electric-dipole matrix element in Eq. (53), the factorization procedure leads to the following results.

(1) The usual change of parity of electric-dipole transitions appears in the Bloch integral matrix element.

(2) Transitions are then allowed for confined states with the same envelope wave function symmetry under space reflection (even or odd).

(3) In his original paper, Dingle[14] remarked that in the *infinite-well* approximation, due to the orthogonality of the envelope wave function, only transitions between confined valence and conduction states with the same quantum number n were allowed ($\Delta n = 0$ rule). It is actually true that these transitions are the strongest observed features in the absorption and excitation spectra.

However, a number of additional transitions have been observed, originating in the breakdown of the simplifying assumptions made.

(1) For *finite* quantum wells the envelope wave functions are not exactly orthogonal, which leads to the observation of transitions with different n (such as the $n = 3$ heavy hole to $n = 1$ electron line; see Fig. 36).

(2) Even the parity selection rule has been broken in the case of many-particle spectra either optically created[173] or due to modulation doping.[174] In that case, particle–particle interaction breaks the single-particle picture used to derive the selection rules.

Light polarization matrix elements have also been calculated at $k = 0$, where the quantum-well potential acts as a simple perturbation to the Kane description of the bands. The split-valence states retain their symmetry characterized by the angular momentum of the Bloch wave functions: the heavy-hole level at $k = 0$ has $J_z = \pm\frac{3}{2}$; the light-hole level has $J_z = \pm\frac{1}{2}$. The various allowed transitions can be calculated as in the atomic physics case of transitions between ground levels with $J = \frac{3}{2}$, $J_z = \pm\frac{3}{2}$ or $J_z = \pm\frac{1}{2}$ and excited levels $J = \frac{1}{2}$, $J_z = \pm\frac{1}{2}$. The various *absorption* transitions are shown in Fig. 31 with the corresponding light polarizations, respective to the momentum quantization axis. From the correspondence

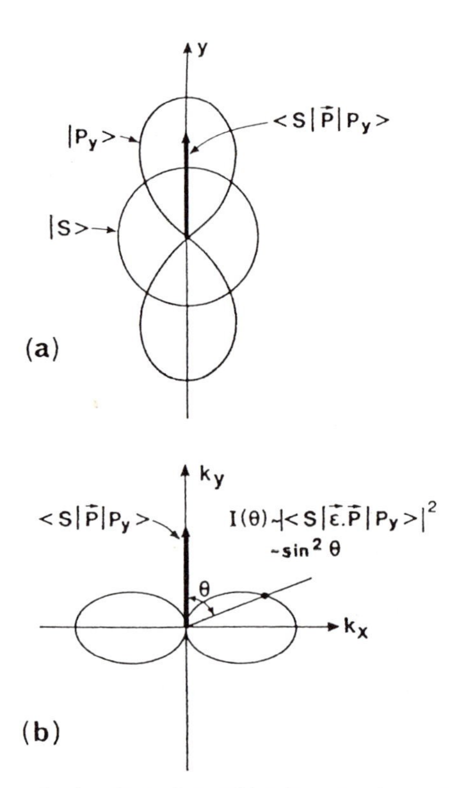

FIG. 31. Optical selection rules for absorption and luminescence between atomiclike states (Bloch states) of valence and conduction bands: (a) Dipole matrix element between an *s* and a *p* state; (b) Emission diagram of that dipole according to the correspondance principle; (c) Possible dipole moments between conduction and valence band states; $p+(-)$ indi-

principle, the polarization vector also describes the electron dipole motion.[175] Using the classical description of radiation emission, which states that an electric dipole radiates mainly perpendicular to its own motion and does not radiate in the parallel direction, the following selection rules can be deduced for light absorption or emission[175a].

Light Propagating Perpendicular to the Layers

Only those dipole moments in the plane can absorb or radiate. Free electron–hole absorption (no exciton effect) must be three times larger for the HH band than for the LH band transitions. Under circularly polarized light excitation, 100% spin polarization occurs when electrons are only excited from one of the HH or LH band. This is to be compared with the 50% polarization obtained in the bulk case, where one excites at once both transitions and creates electron spins with opposite directions, the net

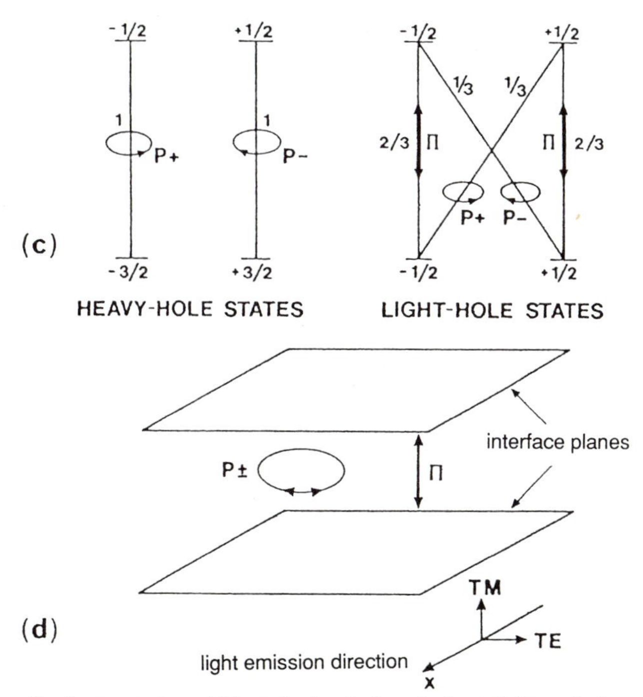

cates rotating dipole moments, which emit circularly polarized light; π indicates a linearly polarized dipole. Relative dipole strengths are indicated; (d) Geometry of the dipoles in the quantum-well situation. One can see that heavy holes only emit TE polarized light in the x-direction.

polarization occurring only because of the unequal transition probabilities.[176] The possibility to obtain 100% electron spin polarization should be of great interest for the production of photoemitted spin-polarized electrons.[177–179] Preliminary experiments have not yet succeeded in yielding higher free-electron spin polarizations than for bulk GaAs, although photoluminescence measurements of electron spin have evidenced high polarizations (~ 70%) within the crystal.[180] No specific linear polarization effects are expected. The selection rules have helped to ascertain the quantum states participating in QW luminescence (Fig. 32).

Light Propagating along the Layers

The HH transition can only occur with light polarization parallel to the layer (TE mode). The LH transitions occur both for TE and TM light polarizations (see Fig. 31). A remarkable feature is the relative intensities of

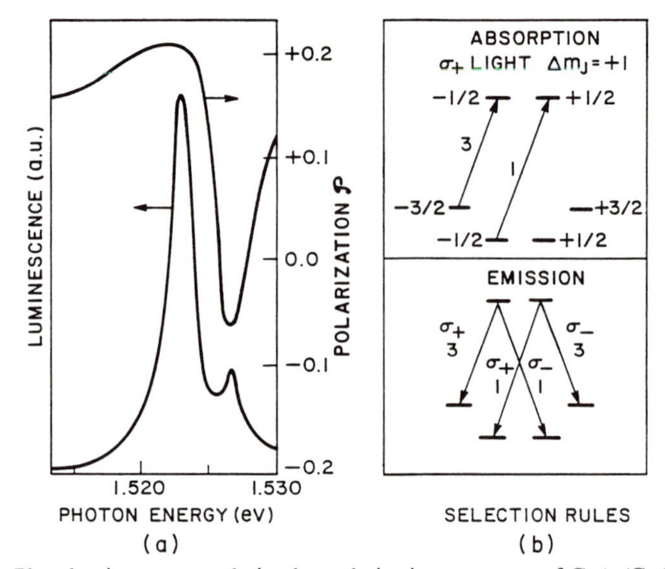

FIG. 32. Photoluminescence and circular polarization spectrum of GaAs/GaAlAs quantum wells under circularly polarized excitation (a). For σ_+ excitation well above the bandgap, where exciton effects become negligible, one creates three times more spin electrons than $+\frac{1}{2}$. Assuming some spin memory at the moment of recombination, these dominant electrons will emit σ_+ light when recombining with heavy holes and σ_- light when recombining with light holes (b). The observation of opposite signs for the polarizations of the two lines in part (a) ascertain the 1.522 eV peak as being related to heavy holes and the 1.527 eV peak as being due to light holes. Resonant light excitation experiments were also done to confirm these assignments (Reprinted with permission from Weisbuch *et al.*[196], © 1981, Pergamon Press plc.)

the TE and TM modes (which from Fig. 31 should be in the ratio $1:\frac{2}{3}$). It was already reported by Dingle that the TE mode luminescence due to the HH transition was much larger than the TM emission.[181] Although thermalization effects at low temperatures could reasonably explain the effect, it is not so at room temperature, where both the TE–HH luminescence and gain have been shown to be larger (≈ 4 times) than for the TM mode.[182–184] This is the more surprising as one could expect from the simplest 2D-DOS analysis the HH transition probability (corresponding to the *light*-transverse hole mass) to be significantly smaller than the LH transition probability. The analysis clearly requires a more profound analysis of exciton effects, valence-band symmetry, selection rules, and density of states, as the valence-band anticrossing discussed in Section 3 should play an important role.

Luminescence of modulation-doped QW samples shows a similar breakdown of selection rules. Pinczuk *et al.*[174] were able to show in z

propagation a rather strong $n = 2$ electron to $n = 1$ HH luminescence, where the ratio between the parity forbidden and allowed transitions amounts to 0.5. More recently, [91,92] observations of the luminescence along the QW plane in such samples show a strong HH-forbidden TM luminescence, which extends over a wide spectral range. This observation allows the k spectroscopy of the HH valence band (symmetry and DOS) to be performed, since the electron plasma acts as a supply of well-defined excitations (i.e., well-known dispersion curve, DOS, and band filling.

b. Exciton Effects

The exciton matrix element yielding Eq. (54g) is classically calculated[101] as

$$M \sim \langle u_c | r | u_v \rangle | \psi(0) |^2 \, \delta_{0,\mathbf{K}} \tag{54i}$$

where $|\psi(0)|^2$ is the square of the wavefunction for the relative motion of the correlated electron and hole at zero relative motion and $\mathbf{K}$ is the exciton total wavevector. This equation indicates that only transitions for $\mathbf{K} = 0$ and excitons states with s-symmetry ($\psi(0) \neq 0$) are permitted. Therefore, for such allowed transitions, all the polarization dependence of transitions is contained in the interband matrix element discussed at length in Section 11a for electron-hole transitions, and excitonic effects do not introduce further selection rules.

c. Intraband Transitions[172c–k]

The selection rules are determined by the optical dipole moment

$$M \sim \int \chi_e'(z) \boldsymbol{\eta} \cdot \mathbf{r} \chi_e(z) \, dr \tag{54j}$$

The envelope functions $\chi_e(z)$ being orthogonal, M is nonzero due to the component of $\boldsymbol{\eta} \cdot \mathbf{r}$ along z. Therefore, for exact perpendicular incidence, the transition probability is zero. If q is the internal angle of incident light with the normal to the quantum well, $(\boldsymbol{\eta} \cdot \mathbf{r})_z$ is proportional to $\sin q$, and the absorption will vary as $\sin^2 q$ for a given allowed transition determined by the dipole matrix element, which can furthermore only relate states with different or undefined parities under space inversion (such as for an asymmetric quantum well).

12. Energy Levels, Band Discontinuities, and Layer Fluctuations

As mentioned earlier, the 2D exciton has a stronger exciton–photon coupling due to the increased overlap of the electron and hole wave functions. Reflectivity,[167] ellipsometric,[167,185,185a] and photoreflectance[185b,c]

measurements evidence this enhancement, as the quantum-well peaks appear strong as compared to the much thicker confining and buffer layers. The well thickness and spectral dependences of the index of refraction have also been observed near resonance.[186] Strong room-temperature excitonic effects have now been reported for a number of semiconductor pairs. One of the surprising features of absorption and excitation spectra is the observation of exciton states related to the higher-lying confined states, as they are degenerate with the continuum electron states of lower confined states. This has been theoretically explained by the weak coupling between these states, which leads to long disintegration times of the excitons of the higher confined states and therefore a weak broadening.[118]

Exciton absorption corresponding to the various confined levels has been the first optical evidence of quantum size effects in semiconductor thin layers. The transition energies are given by

$$E = h\nu^{h,l,m,n} = E_{G} + E_{\mathrm{conf}}^{e,n} + E_{\mathrm{conf}}^{h,l,m} - E_{\mathrm{exc}} \tag{55}$$

where E_G is the bandgap, $E_{\mathrm{conf}}^{e,n}$, $E_{\mathrm{conf}}^{h,l,m}$ are the electron, heavy-, or light-hole nth confinement energies, respectively, and E_{exc} is the exciton binding energy. Prominent peaks are those with $n = m$. The influence of layer quality and its progress can be traced through time. Until the middle of 1975, when liquid nitrogen shrouding of the MBE chamber began, no heavy- and light-hole $n = 1$ transitions could be observed for samples with well thicknesses > 150 Å. A major improvement occurred in 1978, when introduction of samples in the growth chamber through UHV interlocks was used. Growth interruption for interface smoothing has led to the present state of the art of atomically flat interfaces and ultrasharp peaks in optical spectra, which, however, can now also be reproduced in high-quality MBE systems without interruption.[186a]

The fit of the early absorption measurements led to the determination of the bandgap discontinuity ΔE between the conduction bands ΔE_c and valence bands ΔE_v.[81] Calling $Q = \Delta E_c/\Delta E$, Dingle[14] found that $Q = 0.85 \pm 0.03$, assuming standard values for the [100] electron and hole masses, i.e., $m_e = 0.067\ m_0$, $m_{\mathrm{HH}} = 0.45\ m_0$, $= m_{\mathrm{LH}} = 0.08 m_0$. In recent similar experiments on square[187] and parabolic[188] quantum wells, Miller was led to a reexamination of this partitioning and evaluated $Q \simeq 0.60$, using a heavy-hole mass $m_{\mathrm{HH}} = 0.34 m_0$. It is remarkable that the two sets of parameters can explain all the standard features of the $\Delta n = 0$ transitions for the square wells.[187] It is only for the case of parabolic wells (Fig. 33), and for the $n = 1$ e to $n = 3$ HH forbidden transitions in square wells (Fig. 34), that the need arises to consider the newer set of parameters. In an elegant method, using separate-confinement heterostructure QWs, Meynadier *et al.*[188a] were able to measure combination absorption lines

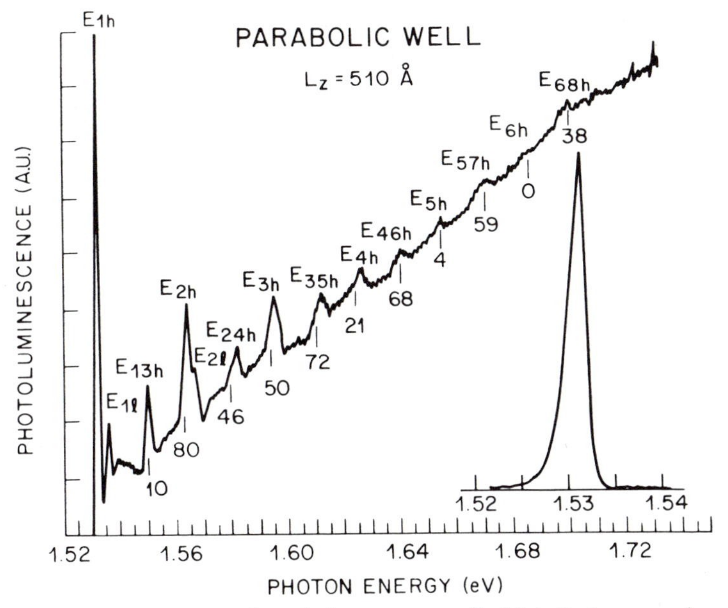

FIG. 33. Excitation spectrum of parabolic quantum wells. Note the large number of peaks observed when compared to square wells, due to the relaxation of the $\Delta n = 0$ selection rule. $E_{nl,h}$ refers to a transition from an electron state with quantum number n to a heavy- or light-hole state with same n: $E_{nml,h}$ refers to a transition with a change in quantum number (from Miller *et al.*[188]).

between a narrow well embedded in a wider confinement well. Their transition energy is strongly dependent on bandgap discontinuities and allows a determination of $Q \simeq 0.6$. It should be remarked that a small value of Q, i.e., a small conduction-band discontinuity, tends to account for the smaller-than-expected charge transfer in electron MD heterostructures[168,190] and conversely the good properties of hole MD heterostructures.[191]

Early experiments were also carried out on double and multiple interacting quantum wells.[93] Interwell coupling leads to the lifting of the degeneracy of the degenerate ground state, evidencing the formation of a superlattice band according to Eq. (29). The data of Fig. 35 show the transitions from the coupled double well with two states, one symmetric and one antisymmetric, to a quasi-continuum of states due to superlattice band formation in a 10 coupled-well sample. More recently, superlattice formation has been shown from electroreflectance measurements for Brillouin zone points away from the center of the zone.[126]

The data fitting must also include the exciton binding energy [Eq. (55)] as an adjustable parameter, a function of well thickness and exciton (heavy or light hole). Dingle[14] extracted a value of 9 meV for the exciton Rydberg

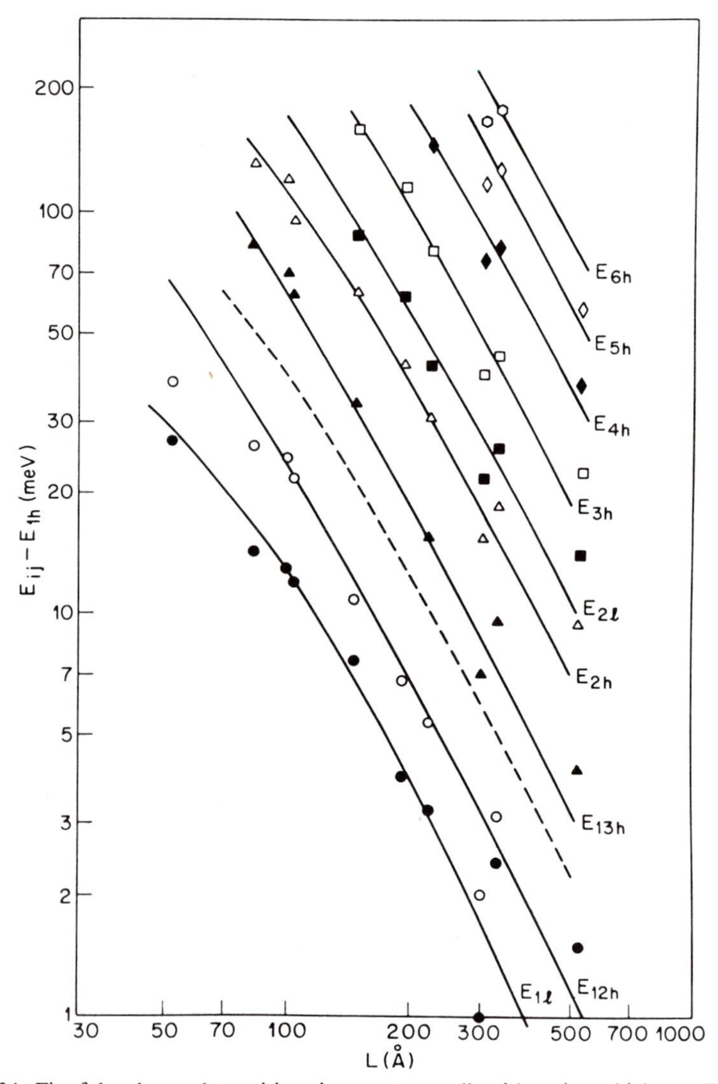

FIG. 34. Fit of the observed transitions in quantum wells with various thickness. Full lines are obtained with $Q = \Delta E_c/\Delta E = 0.57$, $m_e = 0.0665m_o$, $m_{hh} = 0.34m_o$, $m_{lh} = 0.094$. The dashed line is obtained using previous parameters $Q = 0.85$ and $m_{hh} = 0.45m_o$, and is only shown for the E_{13h} transition ($n = 1$ electron to $n = 3$ heavy hole), as all other transitions would be satisfactorily fitted by this set of parameters (from Miller *et al.*[187]).

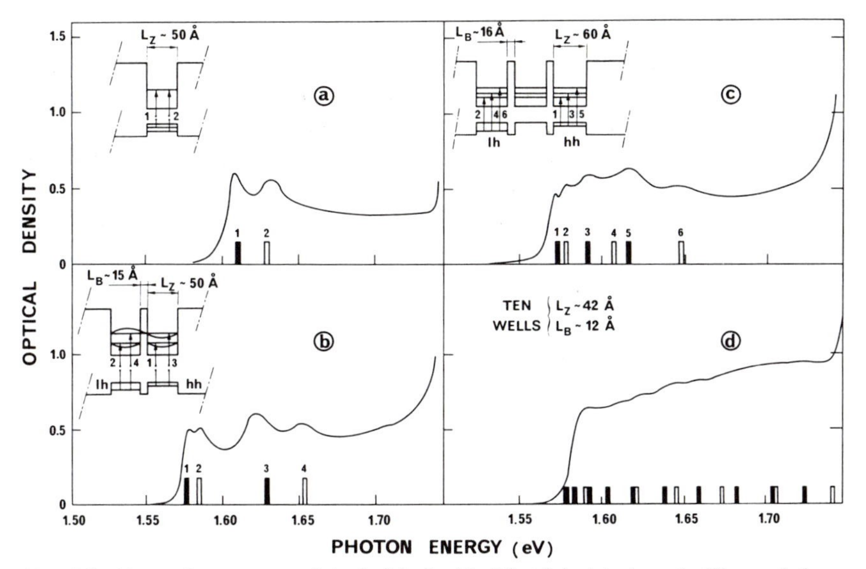

FIG. 35. Absorption spectra of single (a), double (b), triple (c), decuple (d) coupled quantum wells. The positions of the expected transitions in the perturbative approach [Eq. (26)] are indicated. The appearance of bonding and antibonding states is well evidenced in (b). The inserts show the structures under measurement (after Dingle *et al.*[123]).

R^* in thin wells (< 100 Å), which is to be compared to the 4.2 meV value in the bulk. This is, however, a rather imprecise measurement due to its dependence upon the overall fitting procedure. More recently, Miller was able to extract similar values from the onset of the $n = 2$ exciton absorption edge in excitation spectra.[192] Another measurement has recently been carried out by Maan *et al.*,[193] in which the unbound electron and hole state levels are determined by extrapolation from their high magnetic field value. The heavy-hole exciton binding energy can be as large as 17 meV for 50 Å wells, and 10 meV for the light-hole exciton in 100 Å wells. The heavier hole mass of the heavy exciton (as determined from the apparent μ), contradictory to the light transverse mass of the heavy-hole band, is a proof of the strong perturbation of the valence band from the simplest pictures. From the exciton radius a_B it is clear that one needs to know the dispersion of the valence band up to $k \approx a_B \approx 10^6$ cm^{-1} to construct the exciton wave functions and deduce the exciton Rydberg.

It might be thought that interband transitions should provide a convenient way to measure interface grading. It has, however, been shown theoretically[194,195] that grading does not modify the energy-level structure for grading extending up to a few atomic layers, unless the wells are extremely thin.

A very convenient way to deduce absorption spectra without any sample preparation (in particular thinning) is the photoluminescence excitation spectra (ES) method[196]: observing the photoluminescence at a given wavelength, one scans the exciting light wavelength (with a tunable dye laser, for instance). Peaks will appear in the spectrum as a result of increased absorption coefficient (Fig. 36). Actually, there is another contribution to luminescence ES peaks which is due to more efficient relaxation/coupling to the luminescent level under observation, such as resonant LO-phonon relaxation.[197] One should carefully watch whether this effect occurs as it could lead to erroneous assignments in ES peaks. Although very efficient in II–VI compounds, this mechanism has proven negligible in QW structures, thanks to the smallness of the LO phonon coupling and to the very efficient nonresonant relaxation mechanisms to the luminescent channels.

A first use of the ES method was to assess the layer–to–layer thickness reproducibility using a sample with noncommunicating wells (barriers ≈ 150 Å).[198,199] For such an MQW structure, luminescence peaks and their ES are characteristic of a given well and its thickness. In the case of varying QW thicknesses, the various portions of the overall luminescence spectrum

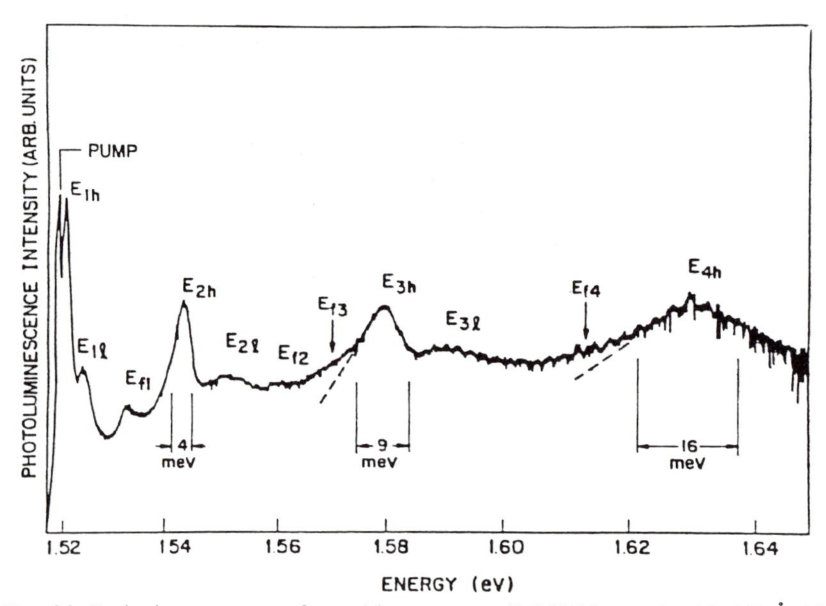

FIG. 36. Excitation spectrum of a multi-quantum-well (MQW) sample with 260 Å thick barriers and wells. The various observed peaks are labeled according to their origin. Several forbidden peaks (E_{fn}) are also observed. The peak labeled E_{f1} has since then been assigned to E_{13h} and yields crucial data for the determination of bandgap discontinuities (see Fig. 34 and discussion in the text). The detection monochromator is set at the energy marked "pump" in the figure, where there is a signal peak due to elastically scattered light (after Miller *et al.*[189]).

should be inhomogeneous, i.e., should have different origins in space and have different ES. On the contrary, for equal QW thicknesses, one expects the same ES whatever the luminescence observation energy (Fig. 37). The detection of a single ES in good MQW samples, to a precision better than a tenth of the ES linewidth, allows one to assess the reproducibility of the average layer thickness as better than one-tenth of a monolayer.

For such an optimally grown sample, it was observed that the ES linewidth would increase for decreasing layer thickness (Fig. 38). This was interpreted as being due to variations of the confined energies due to intralayer thickness fluctuations.[199] In a layer-to-layer growth mode, one expects to find islands where the thickness varies by $\simeq 0.5$ of a monolayer from the average monolayer thickness. Therefore, the various zones correspond to various confinement energies, which leads to broadening of the absorption and excitation spectra due to the spatially disordered exciton absorption band. The simple fit of Fig. 39 represents quite well the results of a series of samples grown sequentially at the optimum temperature. The model assumes that the lateral size of the exciton is smaller than the island size, so that the confinement energy change is that calculated in the usual infinitely wide layer model. Bastard *et al.*[200] calculated the confinement energy variation with the lateral size of islands or holes in otherwise atomically perfect layers. The next step in a detailed analysis would be the determination of the interface topology, either theoretically or experimen-

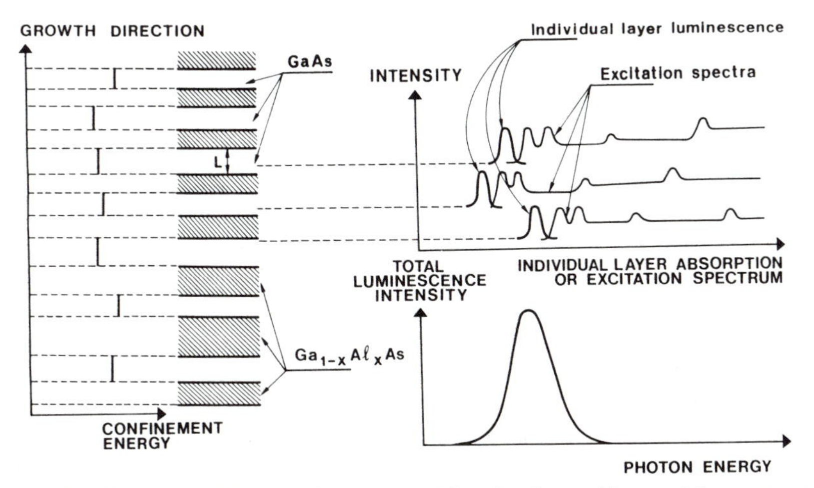

FIG. 37. Schematics of the excitation spectra (ES) to be observed in a multi-quantum-well structure where the wells are unequal, leading to an inhomogeneous luminescence line due to different wells with varying thicknesses. Each recombination wavelength, corresponding to a different well with its own confined energy spectrum, gives rise to different ES.

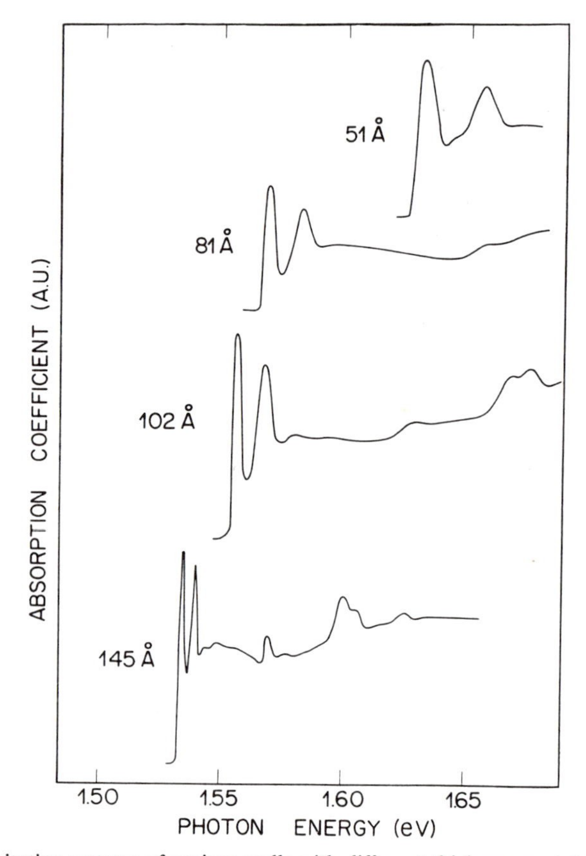

FIG. 38. Excitation spectra of various wells with different thicknesses at 1.8 K.

tally. In the Bell Laboratories series of samples,[199] the linewidth data could be interpreted assuming a majority of island sizes larger than the exciton diameter (e.g., ≈300 Å), which is in agreement with X-ray diffuse scattering observations[66] and TEM imaging techniques.[65] More knowledge of the topology of the interfaces as revealed by X rays and TEM and of the spatially disordered DOS is required to be able to describe the detailed correlation between the interfaces and the DOS, as revealed by absorption and ES in such samples. In some more perfect crystals, one could expect almost atomically flat layers. A few experiments tend to show such perfection, as deduced from luminescence experiments[201,202] in samples grown by standard procedures. More recently, *interrupted MBE growth*[202a–202c] has been used in order to allow for atomic migration and island coalescence at

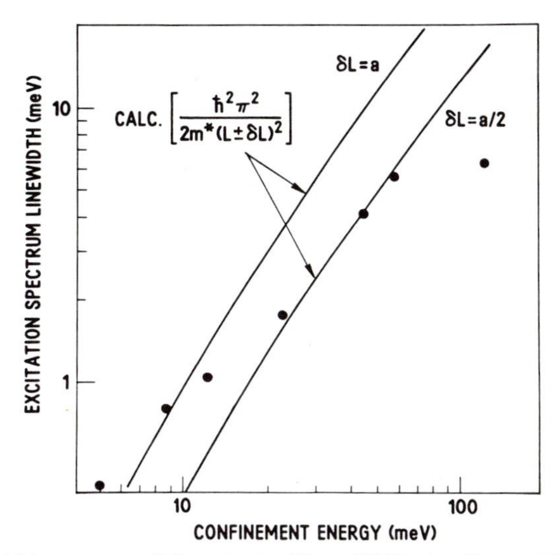

FIG. 39. Linewidth versus confining energy: The solid line is a fit assuming fluctuations of each interface equal to $\pm a/2$, where a is a monolayer. The fit is not good for large confining energies where the energy fluctuations are larger than the exciton Rydberg. At low energies, other broadening mechanisms come into play (Reprinted with permission from Weisbuch *et al.*[199], © 1981, Pergamon Press plc.)

interfaces. In that case, the island size can become much larger, and discrete exciton lines display the exact quantized energy corresponding to the various layer thicknesses equal to an integer number of atomic layers.

The linewidth assessment has been used to optimize growth conditions. Varying growth temperatures, Weisbuch *et al.*[168] were able to identify three different growth regimes in an MBE system (Fig. 40); at the optimum temperature ($\approx 690°C$), the growth occurs in a layer-to-layer mode, a layer being completed through island extension from nucleation (impurity?) centers and coalescence of the islands into a complete layer. At lower temperatures, surface atom mobilities are not large enough to ensure lateral size growth of islands and instead islands with a height higher than a monatomic height can occur. At the same time, periodic macroscopic fluctuations of the surface can be observed with an optical microscope. At high temperatures, the atom's kinetic energy is large enough to overcome the binding energy at the island coast which leads to a roughening transition, which here appears temperature broadened by impurities instead of abrupt, as in the case of the helium phase transition.[203] This interpretation is supported by the earlier TEM imaging measurements of ultrathin structures (down to alternate monolayers)[65] and by Monte Carlo calculations of

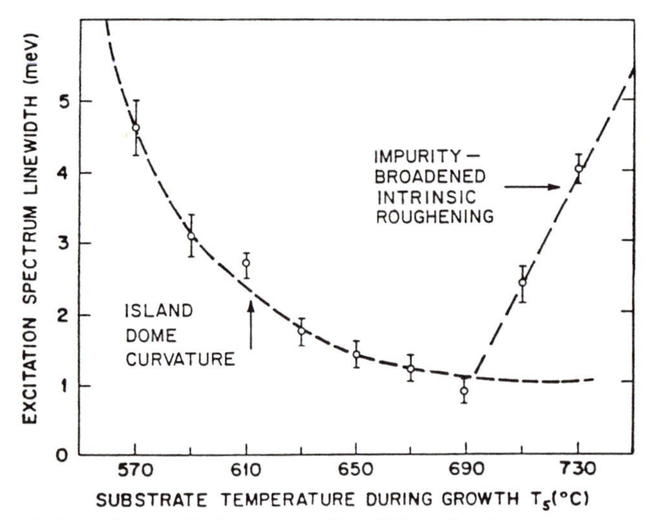

FIG. 40. Variation of the excitation spectra linewidth with substrate temperature for samples grown under equivalent conditions (after Weisbuch *et al.*[168] and Phillips[203]).

the growth mechanism.[204–206] MOCVD growth exhibits an opposite effect of line narrowing with increased T_S,[206] as an island size decrease leads to fluctuation averaging.

Absorption and ES have been used to study various processing methods of GaAs–GaAlAs QWs. Thermal interdiffusion was studied optically,[181,207] by x rays,[208] and by TEM.[209] In the optical method, one observes the upwards energy shift of the quantum well levels as Al interdiffuses in the well material. Dingle was able to determine a diffusion constant for Al, $D = 5 \times 10^{-18}$ cm^2 s^{-1} at 900°C, in good agreement with X-ray and TEM measurements, which justifies the neglect of interdiffusion during MBE growth under standard conditions ($T_s < 750$°C). A very promising technique for controlled interdiffusion is provided by Zn-assisted Al diffusion[210]; during their diffusion, Zn atoms induce Al atom diffusion. This opens the way to low-temperature spatially controlled smear-out of quantum wells. This has already been used in the selective protection of laser facets, which enables higher optical powers without facet damage;[211] the higher-gap regions obtained near the facets confine the recombining electron–hole pairs, thus suppressing radiation-induced defect formation, the main mechanism for facet degradation. Although requiring higher temperatures and longer times, Si-assisted Al diffusion can also be observed.[212]

The 2D DOS and the excitonic nature of the absorption coefficient lead to an absorption edge much sharper in QWs than in the usual 3D double

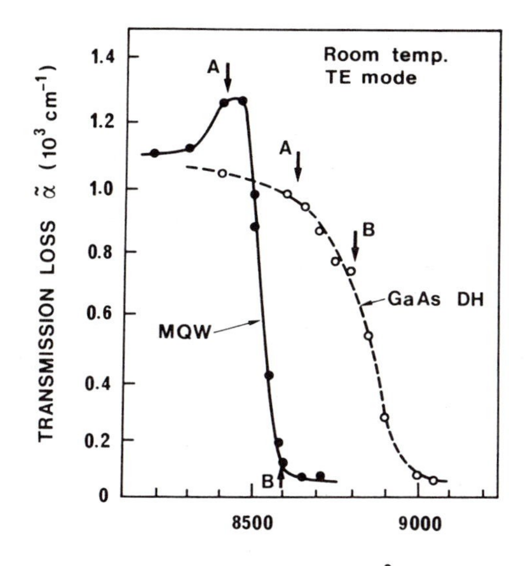

FIG. 41. Transmission curves for passive waveguides with MQW and conventional DH structures. Arrows A and B denote the spontaneous (higher-energy) and lasing wavelengths, respectively, when each waveguide is current injected (Reprinted with permission from *Jpn. J. Appl. Phys.* **22**, L482, S. Tarucha, Y. Horikoshi, and H. Okamoto (1983).)

heterostructures. This results in a lower transmission loss in MQW waveguides as compared to DH waveguides at the lasing wavelength of the structures[213] (Fig. 41). One therefore expects better performance of monolithic integrated optoelectronic circuits made on QW material.

13. LOW-TEMPERATURE LUMINESCENCE[184a]

The luminescence of undoped GaAs/GaAlAs quantum wells at low temperatures consists of a single narrow line,[196,198] which is at first quite different from the observed multiple-impurity-related lines observed in bulk material of similar quality.[214,215] It is also brighter than in thick 3D layers such as typical double heterostructures.[168] A number of factors, occurring simultaneously or not, tend to this single recombination process and large quantum efficiency.

(1) Carrier collection in QW at low temperatures is extremely efficient. Carriers created in the overlayer barrier material are largely captured by quantum wells, as shown by the usually small luminescence of the barrier material as compared to quantum-well luminescence.

(2) The 2D exciton enhancement leads to efficient exciton formation.

The accumulation of photocreated carriers in the small phase space of QWs should also increase the bimolecular formation rate of excitons.

(3) Exciton luminescence is to first order a forbidden process, as the k selection rule of the optical matrix element only allows excitons with exactly the photon k vector to radiate.[215-217] In the pure bulk material, polariton phenomena (i.e., the coupled exciton–photon excitation) relaxes this rule but transforms the exciton fluorescence mechanism into the transport of the coupled excitation to the surface. In this picture, exciton fluorescence is no more an intrinsic phenomenon described by the exciton–photon coupling, but is a transport problem described in terms of excited depth, energy and momentum relaxation, group velocity, etc.[216,217] In another picture of strong damping (impure material, high temprature), polaritons do not propagate but luminescence occurs due to the scattering of an exciton state to a photonlike state, followed by the transformation of the exciton into a photon thanks to the exciton–photon coupling. In this picture, the exciton luminescence is a second-order process involving exciton interactions with impurities or phonons *and* the exciton–photon interaction. In quantum wells, excitons cannot propagate along the z axis as they are localized in the well. However, luminescence should be very efficient as the k conservation rule should be lifted thanks to the scattering by confining energy fluctuations. Exciton-mediated luminescence should also play a role at high carrier densities as demonstrated by the sharp ES peaks observed by Pinczuk *et al.*[174] in modulation-doped samples.

(4) Impurity gettering can occur during multilayer formation, which diminishes the number of nonradiative centers. The effect was first evidenced in GaAs–GaAlAs MQWs, where the usual dark spots in photoluminescence of double heterostructures associated with dislocations could not be observed.[218] The effect was shown to be due to impurity gettering by the GaAlAs barrier material in the first layers.[57] A smoothing of the interface roughness can also be observed as growth proceeds (see Fig. 2). Similar material improvement was also observed in MOCVD material[61] (see Fig. 3). However, material grown in other optimally set systems seems to be exempt of impurities even in single quantum wells.[219,220]

The free-exciton nature of the pure-material luminescence line was established through a careful study.[196] Its energy position coincides almost exactly with the exciton peak observed in ES. The possibility of excitons bound to neutral shallow impurities (donors or acceptors) is ruled out by the spin memory measurements under circularly polarized excitation light, which ascertains the symmetry of the luminescent state as that of a correlated single electron and single hole. The dependence of the high-energy slope of the line on temperature and excitation intensity points out the free-moving nature of the excitation, ruling out excitons bound to isoelectronic impurities. For thick enough QWs (>150 Å) the light-hole exciton can also be observed at low temperatures. More higher-lying levels were observed either at higher temperatures or under high-intensity excitation. A careful study of the transition from 3D luminescence features to 2D behavior in a series of samples with varying thicknesses has been given by Jung *et al.*[219] Theoretical calculations tend to support the dominance of free exciton over bound exciton recombination in quantum wells: Herbert and Rorison[220a] have shown that, whereas confinement increases free-exciton oscillator strength as $1/L$, it slightly decreases that of donor bound excitons due to the decrease in carrier correlation.

It is obvious that the quality of interfaces as revealed by ES will influence the luminescence line shape and width (Fig. 42). Whereas the ES directly probes the DOS and marks the peak of the disordered exciton energy band, the luminescence line shape does not represent directly the DOS of the exciton band (see Fig. 44 below). It cannot even be simply positioned relative to the center of the DOS, as the luminescence line shape results from the competition between the energy relaxation time of excitons down the disordered exciton energy band and the recombination time. Therefore, the discussion of the exciton line shape and the shift between luminescence and ES requires a detailed understanding of the disordered exciton band which we lack at the present time. The linewidth of the luminescence peak can however, be used as an indication of the quality of the interface. Sakaki *et al.*[202a] have used it to indicate the atomic flatness of growth-interrupted quantum wells.

Whereas the early optical measurements were limited to MBE material, measurements in MOCVD material show a very similar quality of such material and interfaces as compared to the best MBE material.[58–60] A very useful scheme has proven to be the sequential growth on the same substrate of several QWs with different layer thicknesses.[59] This allows one to compare different QWs grown under exactly similar conditions. This is also very useful to ascertain the spatial homogeneity in the case of alloy quantum wells such as GaInAs/InP[61a]; when the alloy composition varies across a multiwell sample, it acts as a constant shift of the ground-state energy of the well, independent of the well thickness. If the growth rate is spatially varying, this is reflected in unequal shifts for the various confining energies due to the nonlinear ($E_{conf} \sim L^{-2}$ in the infinite-well approximation) relation between well thickness and confining energy.

In some high-purity MQW samples a double peak is observed around the $n = 1$ heavy-exciton position (Fig. 42c). First interpreted as a reabsorp-

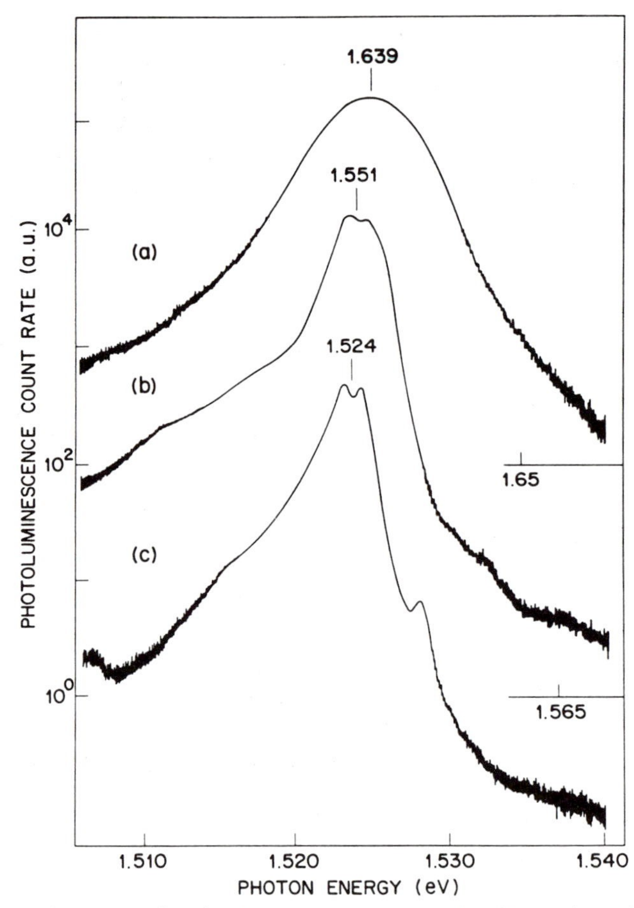

FIG. 42. Luminescence of optimally grown quantum wells with varying thicknesses [(a), 51 Å; (b), 102 Å; (c), 222 Å]. For the sake of clarity, the energy and count-rate scales have been shifted with respect to one another. Note the log scale for the count rate. The luminescence linewidth is to be compared with the excitation spectrum linewidth of Fig. 39.

tion feature, this structure was later[221] attributed to the heavy exciton band at high energy associated with a biexciton recombination line at low energy.

Although vanishingly small in undoped structures due to the efficiency of the free-exciton recombination, impurity-related luminescence can easily be observed for deliberately doped samples. Miller *et al.*[222] were able to observe the dependence on QW thickness and impurity position of acceptor binding energies through the detection of the electron-to-neutral acceptor transitions. Excitons bound to neutral acceptors were also shown.[223] Shanabrook and Comas used spike doping at the center of quantum wells to measure donor-related levels.[224]

14. Carrier and Exciton Dynamics

There has recently been a surge of optical transient measurements of carrier[224a] and exciton dynamics in QWs. Some care must be used in analyzing the results as compared to the 3D case, as a number of parameters are strongly altered. In particular, the rather small 2D DOS induces band-filling effects even at moderate exciting powers.

The pump- and -probe experiments were first performed on MQW structures.[225] At low densities (Fig. 43) the absorption spectra show a washout of the exciton peaks due to exciton screening. Studies of this bleaching allow one to measure the exciton band filling and ionization time of excitons by hot electron or phonon collisions.[226] An ionization time of $\simeq 300$ fs due to phonons is deduced at room temperature. The remaining absorption displays the exciton-less absorption curve, i.e., the 2D-DOS step structures. Higher pumping rates show the large amount of band filling and gain at the higher densities. Carrier relaxation rates have been estimated and shown to be very similar to those observed in 3D. At still higher excitation rates, energy relaxation was shown to be slower, as evidenced by the hot luminescence correlation-peak method.[227] It is not yet clear whether this slowing is due to phonon accumulation in the well or band-filling effects. Femtosecond experiments[228] have recently found similar energy relaxation rates in 2D and 3D, even in the higher-density regime, which, however, does not rule out the phonon-accumulation model[227] for longer times as the phonon population should be low at the early stages of relaxation.

Photoluminescence carrier dynamics has been studied through time-resolved luminescence.[229,230] The relaxation times of excitons in the disordered band have been measured by Masumoto *et al.*[231] showing the importance of the spatial disorder that controls the energy migration. Takagahara[232] made an analysis of the experimental results using a model of energy transfer in a disordered band. The shortening of the exciton lifetime with decreasing well thickness has been traced to the increase of electron–hole wave function overlap in the 2D exciton optical matrix element.[229,229a]

The straightforward application of light-matter coupling leads to some problems when calculating the exciton lifetime. Its value is[232a]

$$\tau = \frac{2\pi\varepsilon_o m_o c^3}{ne^2\omega^2 f}$$

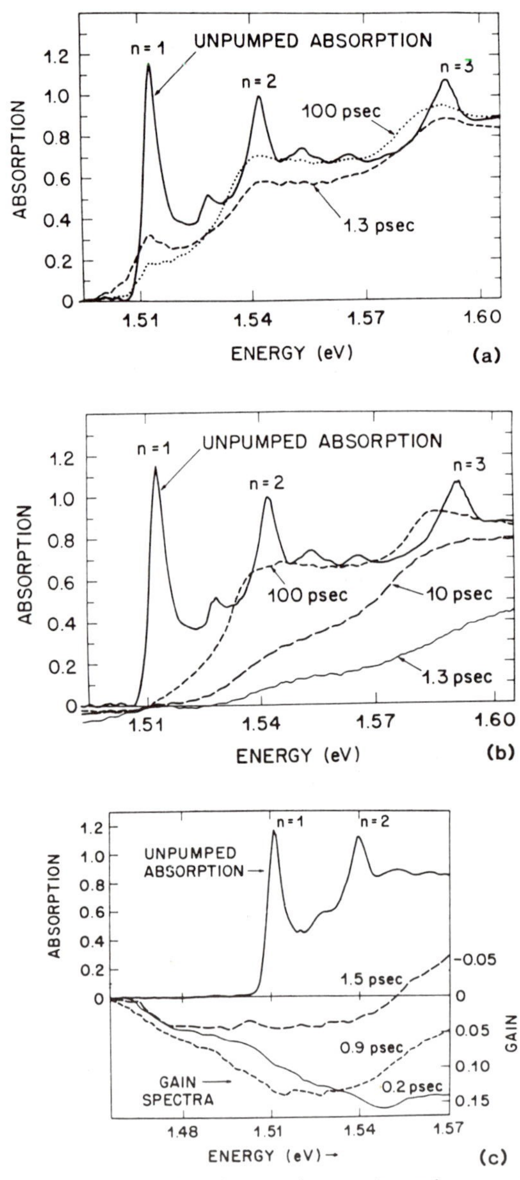

FIG. 43. Subpicosecond pump-and-probe experiment: absorption spectra of a 250 Å MQW sample are shown before (——) and at later times (broken curves) following excitation. Carrier density is (a) 5.10^{11} cm^{-2}, (b) 10^{12} cm^{-2}, (c) 10^{13} cm^{-2}. Curve (a) shows quite well the almost square-shaped absorption edges of 2D systems when the exciton effects are washed out at the densities used here. Curve (b) shows the large band-filling effects at short times. Curve (c) shows the large structureless gain curve (Reprinted with permission from Shank *et al.*,[225] © 1983, Pergamon Press plc.)

which should become vanishingly short with increasing crystal volume as the exciton oscillator strength *f* increases with the crystal volume. This has been long recognized in 3D where the concept of coupled exciton–photon excitations, the excitonic polariton, has been introduced.[232b] In that picture, the lifetime is just the time of flight of the excitonic polariton to the crystal surface.[232c] In quantum wells, the bounded motion of excitons prevents their free propagation as polaritons in all directions as soon as the layer thickness is submicronic,[232d] except in the layer-plane where they have indeed been observed through the induced slowing of light-wave propagation.[232e] Due to the weak coupling of light with a single quantum well, coherent propagation over a distance up to 1 mm has been observed. The overall exciton lifetime is due, however, to nonpropagating modes. In the limit where the mean free path of excitons is very short (i.e., due to collisions on interface roughness) the coherent volume entering the oscillator strength is considerably reduced compared with the crystal volume. Feldmann *et al.*[232f] were able to evidence the temperature variation of this coherence volume through the change in exciton lifetime.

A detailed and profound analysis of exciton dynamics has been carried out by Hegarty *et al.*[233–237] Resonant Rayleigh scattering[235] has been used as a probe of the homogeneous exciton linewidth within the inhomogeneous exciton absorption band due to interface disorder. As shown in Fig. 44, the exciton DOS as revealed by ES, the luminescence line, and the intensity of elastically (Rayleigh) scattered light are shifted relative to one another. The downward shift of the Rayleigh intensity shows the transition from localized exciton states (undamped, i.e., efficient for light scattering) to delocalized exciton states (less efficient). This is therefore an optical measurement of a mobility edge for excitons on the interface disordered band in the sense of Mott. Hole-burning experiments[233] in absorption at higher intensities yielded very similar behavior for excitons in the band; finally, transient grating experiments,[234] at still higher densities, directly indicate spatial exciton transport, with again the observation of a mobility edge near the center of the exciton energy band. As expected, the mobility edge appears to be strongly dependent on thermally activated exciton hopping, as revealed by increasing the temperature.[236] The whole picture of a disordered 2D exciton band appears well justified from this set of experiments. Quantum wells seem to be a good prototype for 2D disordered systems,[237] with the strong exciton–photon coupling allowing for the use of very convenient optical probes.

15. Inelastic Light Scattering

Inelastic light scattering by electronic excitations is a very powerful tool for the investigation of 2D systems, although perhaps not widely used. We cannot describe here all the important results which have been obtained

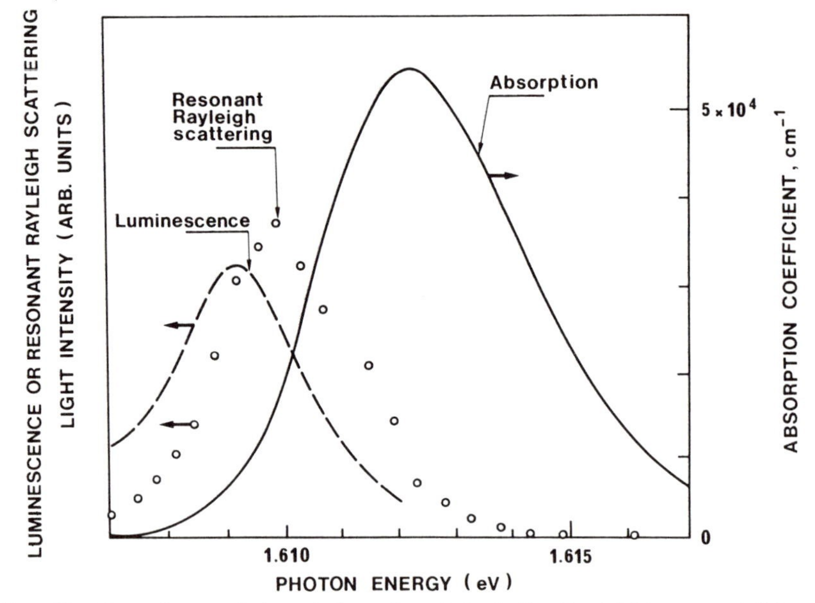

FIG. 44. Absorption peak (——), intensity of elastically (Rayleigh) scattered light (O), and luminescence intensity (---) of a 51 Å MQW sample. Whereas the absorption peak represents the DOS of the 2D disordered exciton band, the Rayleigh efficiency curve is shifted downward because of the inefficient light scattering by delocalized excitons above the center of the band. The shift of the luminescence peak represents the relaxed state of recombining excitons (from Hegarty *et al.*[235]).

with this type of experiment and strongly recommend two recent reviews of the subject,[238,239] still in active progress.

At first glance, it might appear that the electron *absolute number* at an interface might be too small to allow for any sizable inelastic scattering of light by electrons. However, Burstein *et al.*[240] pointed out that, thanks to resonant enhancement of the efficiency, signals from the standard electron density $N_S \simeq 10^{11}$ cm^{-2} should be observed. The resonance at the spin-orbit split-off gap at $E_0 + \Delta_0$ is usually used in order to prevent hot luminescence signals to obscure the light scattering spectrum. As in the bulk,[238,241] two sorts of signals are to be observed: the *single-particle* spectrum, corresponding to uncorrelated particles, which involves a spin-flip and is observed in the orthogonal-polarization configuration. The efficiency is then due to spin-density excitations. Excitations of the *collective modes* of the electron gas (plasmons) are due to the charge-density fluctuations of the gas and are observed in the parallel-polarization configuration. The theoretical considerations leading to these selection rules have been described in detail by Burstein *et al.*[242]

Single-particle intersubband scattering has been studied widely as it provides an excellent tool to directly measure energy levels in modulation-doped heterostructures, either single interfaces,[243] quantum wells,[244] multiple quantum wells or $n-i-p-i$'s.[245] The measured energy shifts provide good tests for the evaluation of energy-level calculations. Light scattering measurements of the 2D hole gas[246] confirm the ΔE_c determination of Miller *et al.*[187] They also indicate the nonparabolicity of the various valence bands (see Section 3). Using modulation-doped MQWs with varying spacer thicknesses, Pinczuk *et al.*[247] were able to show a striking correlation between light scattering linewidth and electron mobility. This is interpreted by assuming that the same collision mechanisms determining electron wave-vector changes are responsible for the mobility value and light scattering k-conservation rules (Fig. 45).

Collective excitations observed in the parallel polarization configuration allowed the determination of the LO phonon–plasma coupled modes.[248,249] Optically created plasmas have been detected by their induced light scattering in undoped MQW structures.[250,251] Carrier densities have been determined from the measured shifts. In $n-i-p-i$ structures, light scattering experiments on photocreated carriers have revealed their 2D character, and the transition to 3D at high intensities when light-induced photoneutralization of impurities destroys the superlattice potential.[252] Finally, inelastic light scattering in MQW under strong magnetic fields yields inter-Landau level transitions as observed by Worlock *et al.*[253]

16. Nonlinear and Electro-optic Effects[253a]

Nonlinear and electro-optic properties of materials are linked to the modification of their optical constants by high-intensity optical fields or by DC electric fields respectively. They usually rely on the dielectric response of ions and electrons. In semiconductors, a dominant role is usually played by the dielectric response of uppermost valence band and conduction band, i.e., excitations across the forbidden gap.

The properties of quantum wells make them unique materials in the field of nonlinear optics and electro-optics. This has given rise to an extremely rich literature[254] and we only outline the most salient features of the field, leaving detailed calculations and descriptions of most applications outside this short text, with the exception of a few clear-cut applications.

The outstanding properties of quantum wells stem from several effects, which can appear separately or reinforce each other:

(1) the increased light–matter interaction, due to exciton effects,[172d,255] which was discussed in Section 10c;

(2) the quantum-confined wavefunction, which yields large electric-field induced Stark shifts[256];

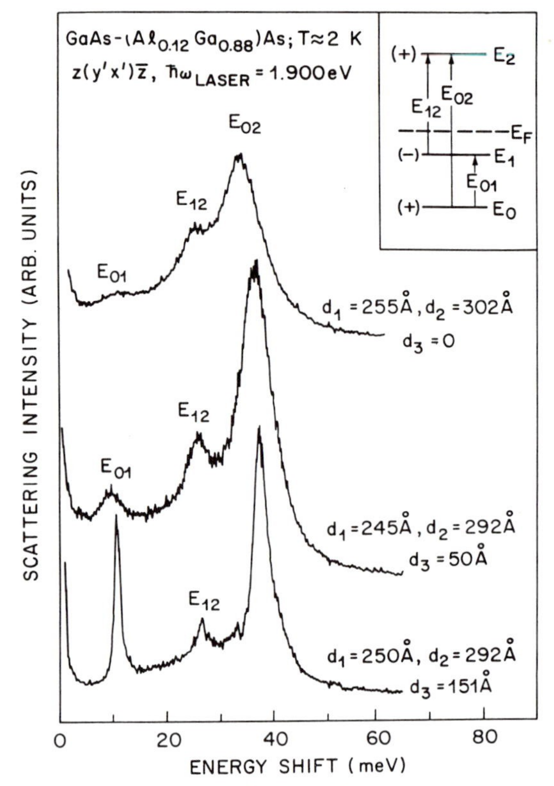

FIG. 45. Single-particle light scattering spectrum [depolarized backscattering $Z(y'x')\bar{Z}$] of three MQW MD samples. The varying spacer thicknesses of 0, 50, and 151 Å are correlated with increasing spectra sharpness and 4.2 K mobilities of 12,500, 28,000, and 93,000 $V^2\ cm^{-1}\ s^{-1}$, respectively. The observed transitions are shown in the insert (after Pinczuk *et al.*[247]).

(3) the diminished accessible phase space, which diminishes any threshold, as in nonlinear effects due to band-filling.

We first treat separately the two latter effects, and then their application to nonlinear and electro-optics.

a. Quantum-Confined Wavefunctions and Electro-optic Effects

Electro-optic coefficients determine the change of material optical constants while applying an electric field. The most well-known effect in semiconductors is the Franz–Keldysh modification of the absorption edge due to the fact that under an applied electric field the band structure is tilted;[172b] absorption can then occur below the band edge due to the fact that conduction and valence band wavefunctions have evanescent tails in the forbidden gap with some overlap (Fig. 46, left). An oscillatory struc-

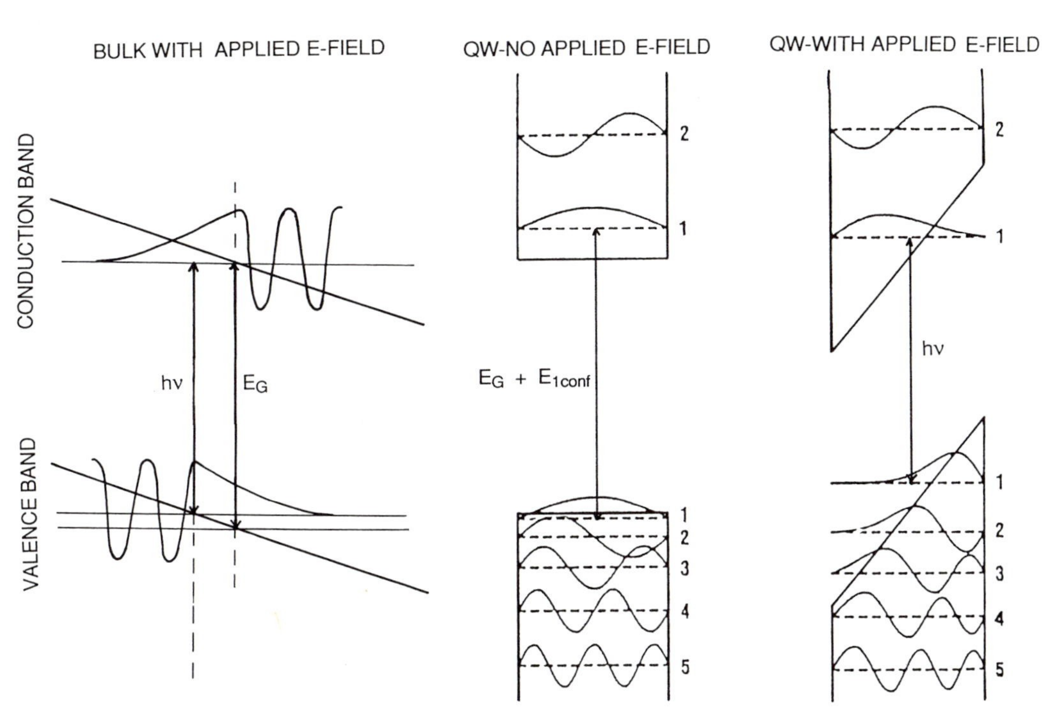

Fig. 46. Schematics of the action of an electric field on a bulk semiconductor (Franz–Keldysh effect, left) and on a quantum well (quantum-confined Start effect, right, after Ref. 254).

ture with energy is observed due to interference effects of the wavefunctions. The optical effects associated with the Franz–Keldysh mechanism in 3D are usually quite limited, however, because large electric fields lead to the smearing out of the absorption edge as electron and hole wavefunctions have diminished spatial overlap with increasing field. Exciton effects in bulk materials play a minor role as they are readily dissociated for rather weak fields[254,256–7], besides the usual thermal dissociation at room temperature.

The situation is quite different in quantum wells. As displayed in Fig. 46 (right), a large energy shift does occur due to the major change in potential well. In addition, large electric fields can be applied, while retaining finite electron-hole overlap: both wavefunctions can collapse only on opposite sides of the quantum-well. There is then enough electron-hole correlation to retain most of the large oscillator strength due to excitons. This is well depicted in Fig. 47, where the effects of electric fields parallel and perpendicular to the quantum wells are shown.[258] The former case represents the Quantum-Confined Stark Effect (QCSE) of an electric field on 2D ex-

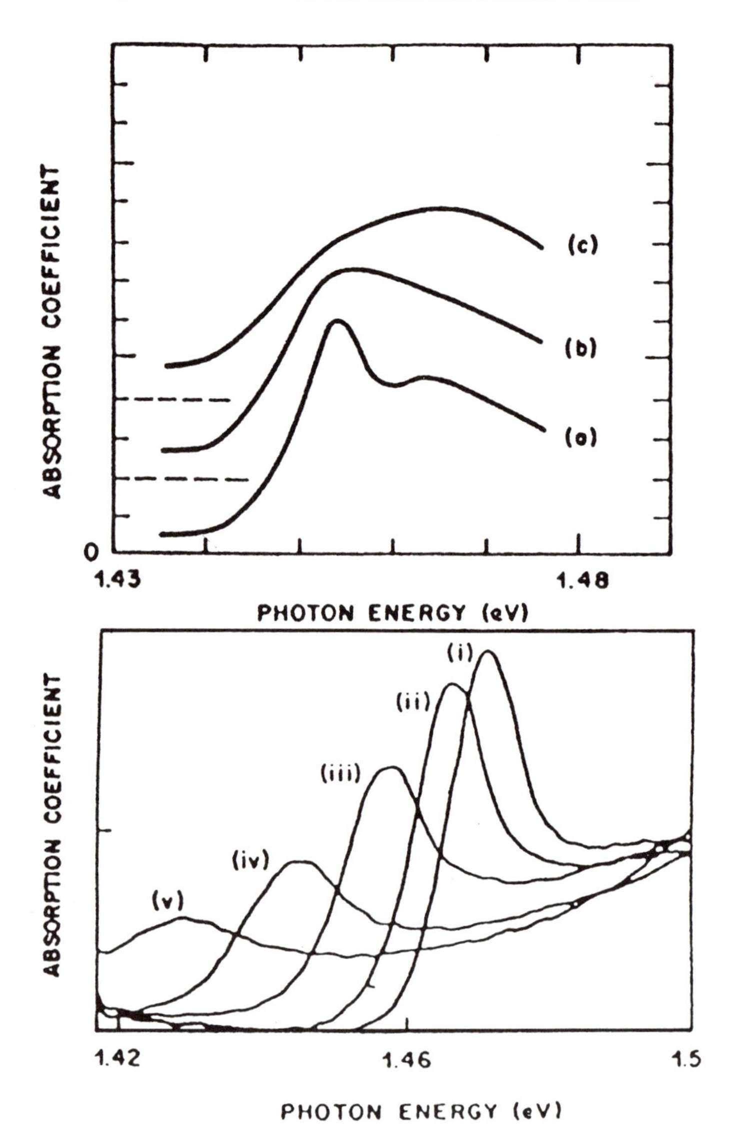

Fig. 47. Action of electric fields on 94 Å quantum wells when the field is parallel (top) or perpendicular (bottom) to the layers (a), (b), (c): Fields are 0, 1.6 10^4, 4.8 10^4 V cm^{-1}, respectively. (i)-(v): fields are 0, 6.10^4, 1.1 10^5, 1.5 10^5, 2. 10^5 V cm^{-1} respectively. (Reprinted with permission from Taylor and Francis Ltd, S. Schmitt-Rink, D. S. Chemla and D. A. B. Miller, *Adv. in Phys.* **38,** 89 (1989).)

citons, the latter the action of an electric field on "3D-like" unconfined excitons, leading essentially to broadening at moderate fields. Along with the decrease in oscillator strength associated with the smaller integrated exciton peak due to diminished electron-hole correlation, a large decrease in recombination rate is observed,[259] although with little impact on the radiative efficiency, as in *n–i–p–i*'s (Section 9), as both the radiative and nonradiative rates decrease due to change separation.

The energy shifts can be calculated in a perturbative or exact manner (in that case Airy functions are to be used in the quantum well). The calculation is usually simplified as one deals with the strong confinement case where the confinement energy is larger than the exciton binding energy. In that case, the z-motion is only determined by the confinement and the Schrödinger equation is separated in transverse motion (dominated by exciton effects) and longitudinal motion. The oscillator strength can also be calculated and is in good agreement with experiments.[258] It is remarkable that the applied fields can conserve strong exciton features up to 10^5V/cm whereas excitons disappear at $\sim 10^3$V/cm in bulk GaAs[256–7]. This is due to both the stronger exciton binding and conserved electron-hole correlation due to wavefunction confinement.

The exact calculation also provides a continuous transition to the 3D Franz–Keldysh effect when considering wide wells, neglecting exciton effects and incorporating all electric-field allowed transitions in addition to the zero-field allowed transitions (Fig. 48). There again the usual transition from quantized to classical physics at large quantum state numbers involved is demonstrated. The difference between QCSE and Quantum-Confined Franz-Keldysh Effect (QCFKE) is well-evidenced: [254] in QCSE, one observes mainly a shift in the ground state. In the QCFKE, large modifications also appear at high energies (yielding the usual oscillations in 3D).

The electro-optic effects described here give rise to two main modifications of materials parameters: (1) absorption changes, usually called electro-absorption and used through the modification of transmitted beams; (2) changes of index of refraction, usually called electro-optic effect, used in light-phase-sensitive situations such as in interferometers (such as Mach–Zehnders) or reflective single or multilayers. This latter effect can be used both in the absorption or transparent ranges of photon energies, whereas electro-absorption operates best near the absorption edge. Of course, both effects are related through Kramers–Kronig relations.

The above-described Quantum-Confined Stark Effect (QCSE) is not the only efficient mechanism for electro-optic effects. Band-filling of the electron or hole states is also very useful, since the 2D density of states is quite smaller.[260] To suppress the exciton effects, it is clearly sufficient to have a band-filling by carriers up to the Rydberg energy, i.e., ~ 10 meV in GaAs

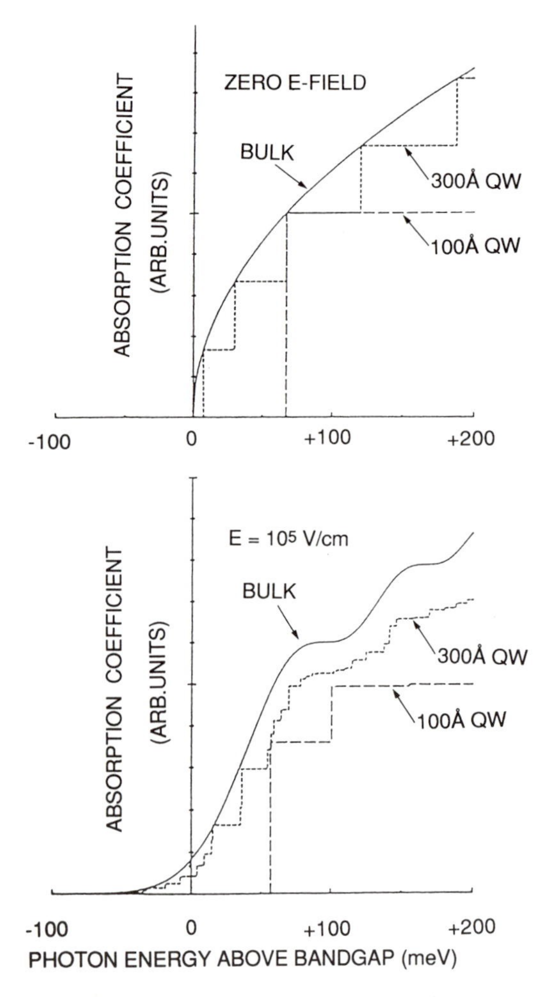

Fig. 48. Calculations of absorption changes for bulk GaAs and quantum wells (300 Å and 100 Å thick, respectively). Top: absorption without electric field; bottom electric field applied. Note that the numerous allowed transitions in an electric field for a 300 Å thick QW make its absorption curve smooth and almost 3D-like (Reprinted with permission from Taylor and Francis Ltd, S. Schmitt-Rink, D. S. Chemla and D. A. B. Miller, *Adv. in Phys.* **38**, 89(1989).)

quantum wells The size of the carrier-induced change in absorption coefficient is similar to that due to the QCSE. The refractive-index change can, however, be somewhat larger in that case due to the more efficient Kramers–Kronig transformation of the absorption-change, which retains a constant sign, opposite to that of the QCSE[261] (Fig. 49). At variance with

QCSE where no or few carriers are photocreated, this effect usually called Phase-Space Filling (PSF) is time-controlled by carrier lifetime, and can have quite long response times.

Many other electro-optic effects have been evidenced. More effective structures than the standard quantum well have been used: the usual symmetric quantum well yields a quadratic Stark effect. Using asymmetric quantum wells with a variable composition, a linear Stark effect can be obtained.[262] Coupled quantum wells have also been studied, because they yield a linear Stark effect $\sim eEd$, where d is the wells' center-to-center distance, while retaining some exciton oscillator strength provided the barrier thickness is small enough.[263] They also provide a system where numerous studies of electron and hole dynamics have been carried out, as luminescence line labeling allows one to monitor carrier transfer from one well to another.[264] The possibility to control the lifetime, by increasing it through charge separation of electrons and holes in neighboring wells, or decreasing it by increased resonant tunneling, enables one to design structures with required response time.[265–7]

A set of optical experiments under electric field, only recently used for electro-optical properties,[267a] has yielded profound insight in the problem of superlattice transport.[268] It suffices to mention here that under an applied electric field, the transition from superlattice to Stark ladder states yields a blue shift in the absorption edge at first, as the diminished tunneling probability with electric field raises the superlattice band minimum energy level as seen from Eq. (29). This blue shift is at variance with the usual red shift observed for the single-well QCSE.

b. *Nonlinear Effects*[254,255]

As in any system, nonlinear effects will be widely different in quantum wells depending on whether real or virtual, states are populated.

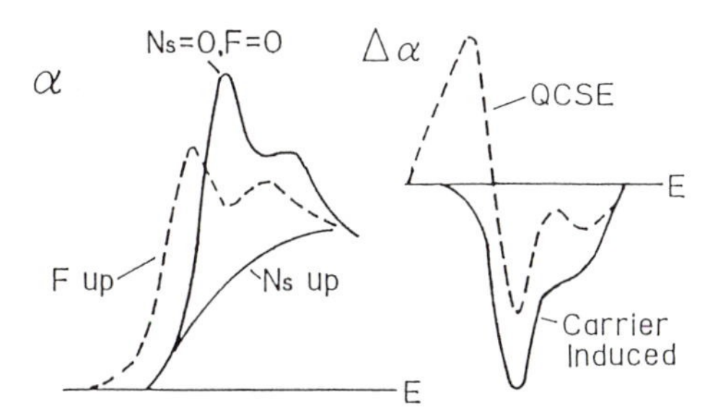

Fig. 49. Schematics of absorption changes under applied electric field due to QCSE ("F up") or to carrier bandfilling ("Ns up") (from Ref. 261).

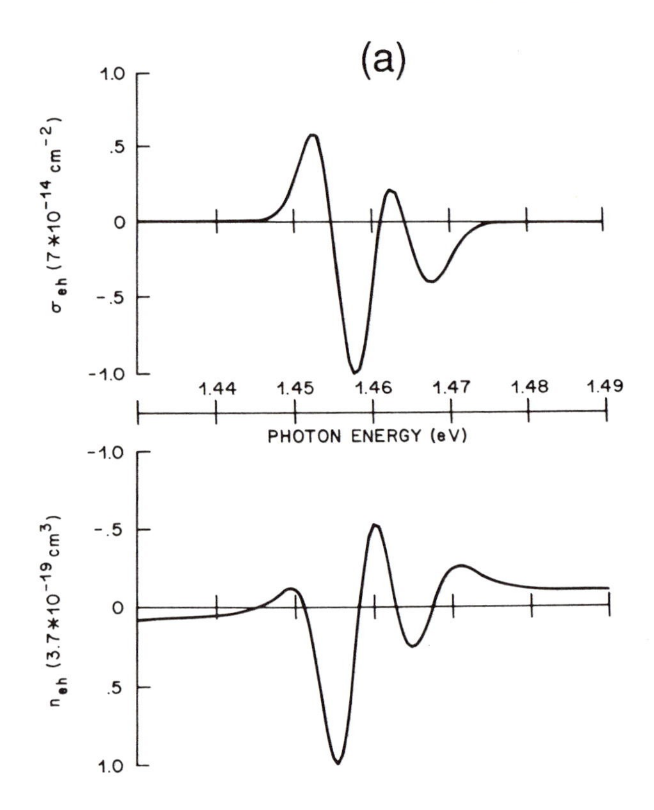

Fig. 50. (a) Nonlinear coefficients σ and η for carrier-induced absorption and refraction index changes, respectively; (b) efficiency per carrier for index of refraction changes as a function of quantum-well thickness (Reprinted with permission from Taylor and Francis Ltd,

The latter case leads to many phenomena that have extremely fast response (instead of being lifetime-limited), but does not yield very efficient systems, since crystal virtual excitations creating nonlinear effects live only for a time $\Delta t \approx \hbar/\Delta E$, where ΔE is the energy detuning between incoming photon energy and crystal eigenenergies. The reader should refer to more specialized reviews or original papers for a description of the various effects in that case.[254]

As can be expected, nonlinear effects with real excitations in quantum wells yield very efficient materials thanks to the large light–matter interaction and reduced density-of-states. The electron-hole pairs created at room temperature by photon absorption, either directly or through exciton dissociation, lead to nonlinear modification of the optical constants through

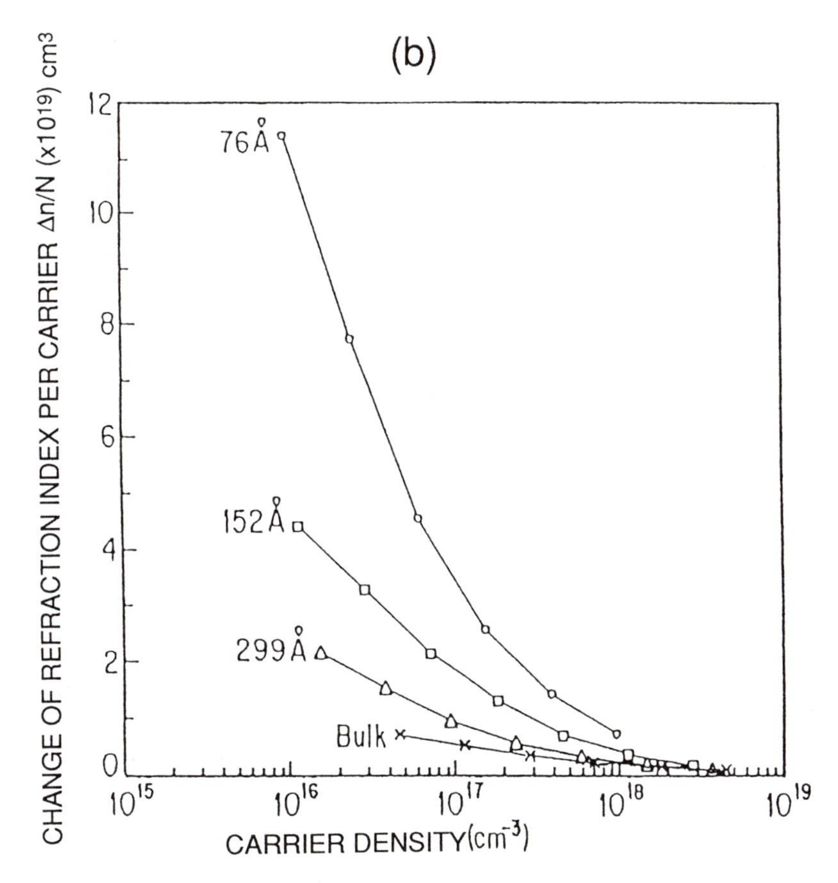

S. Schmitt-Rink, D. S. Chemla and D. A. B. Miller, *Adv. in Phys.* **38,** 89 (1989), *and* from Springer-Verlag, A. Chavez-Pirson, H. M. Gibbs and S. Koch, in "NonLinear Optics of Organics and Semiconductors," (1989).)

several main effects: band-gap renormalization, collision-induced absorption broadening, phase-space filling, electron-hole interaction screening.[254] In the linear (nonsaturated) region of nonlinear effects, the bleaching of excitons can be described by two coefficients, the nonlinear absorption cross-section σ and the nonlinear refractive, as

$$\alpha = \alpha_0 - \sigma N; n = n_0 - \eta N$$

where α_0 and n_0 are the background absorption coefficient and index of refraction, and N is the e-h density. Figure 50a displays typical measurements, which are of course related through the Kramers–Kronig integral. The excellent nonlinear properties of quantum wells are evidenced by degenerate four-wave mixing with reasonable diffraction efficiencies at extremely low pump intensities, such as obtained from a semiconductor laser diode: 5.10^{-5} for a 1.25 μm thick sample at light intensity of 17 W cm^{-2}. This clearly opens the way to full optical signal processing with quantum wells as materials allowing light-by-light beam steering.

Room temperature absorption saturation experiments also evidence superior performance of quantum wells. Figure 50b shows the efficiency per carrier for various quantum-well thickness, as measured in pump and probe experiments.[269] The saturation intensity I_s is defined as

$$\alpha = \alpha_0\left(1 + \frac{I}{I_s}\right)^{-1} \tag{55a}$$

In all nonlinear measurements, one finds a critical (saturation) carrier density $n_{sat} \sim (\pi a_B^2)^{-1}$, the areal density of closely packed excitons. This is quite normal because whatever mechanism is chosen (real space or phase-space filling), this quantity is also (with $L \sim a_B$ in usual quantum wells)

$$n_{sat} \approx \rho_{2D} \times kT\left(\approx\left(\frac{m}{\pi\hbar^2}\right)\left(\frac{\pi^2\hbar^2}{2\,mL^2}\right)\right) \tag{55b}$$

corresponding to phase-space filling of all states up to the energy kT, including those situated at the Rydberg energy from which the exciton states are constructed. The dependence of saturation intensity on well width has been directly traced to the exciton diameter, evidencing a maximum when that diameter is minimal, i.e., at $L \sim 75$Å.[269a]

c. *Electro-optic Applications*[270]

The most popular electro-optic application of quantum wells is the electrooptic light modulator through the QCSE of the light-beam absorption, used for instance in spatial light modulators (see, e.g., the cover picture of this book), a major component of optical computing systems. Compared with other digital devices, quantum wells exhibit both advantages and weaknesses. In the operating modes where no or few carriers are created, like all modulators based on electronic band properties, it has inherent speed, being limited only by RC time constant of the total electric circuit from driving electronics to device electrodes. For a typical $10\mu m \times 10\mu m \times 1\mu m$ (thickness) modulator, the capacitance is ≈ 12fF, which yields a commutation energy $\approx(1/2)CV^2 \approx 150$ fJ at 5V drive.[270] One observes that this is not quite as good as modern, purely electronic digital devices, but at the same time this should be compared with an input/output device of an electronic chip, as the electro-optic modulator is directly connected to the outside world. The comparison is then much better for the QW modulator, because it is well-known that interconnection wires and pads on chips are much larger than individual devices and therefore require high driving currents (and therefore large switching energies). This direct connection to the outside world is one of the advantages of optical interconnects and computing, besides the high degree of parallelism allowed by the 3D geometry of optical systems.[271]

In comparison with other electro-optic materials it should be noted that QWs often operate near the exciton absorption resonance. It could therefore appear that not much is gained as one increases equally both the electro-optic coefficient and the absorption loss[272] (Fig. 51). Although this view is certainly correct and requires great attention in designing integrated optics circuits,[273] one still retains the larger Stark shift of transition energy due to the higher electric fields, which can be applied while retaining oscillator strength in the case of the QCSE, or the very efficient Phase-Space Absorption Quenching (PAQ) effect in the case of free car-

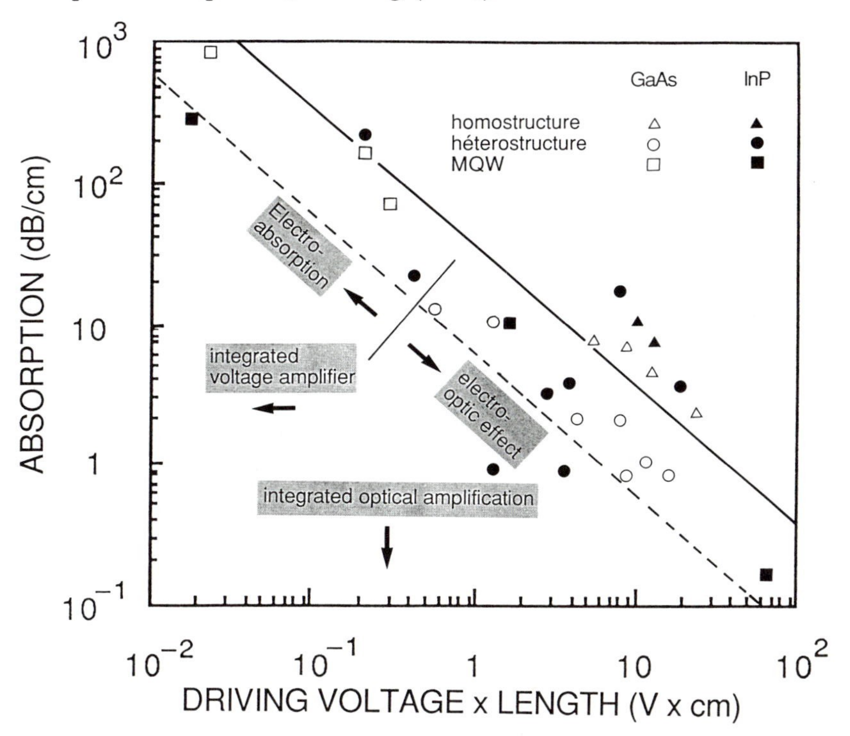

Fig. 51. Comparison of various electro-optic modulators based on changes of index of refraction ("electro-optic") or of absorption coefficient ("electroabsorption"). Note that one figure of merit could be the product of losses times driving voltage times device length, in which case structures are almost equivalent. However, other factors are of importance (see text). (From Ref. 272, IOP Publishing Co. and M. Erman, private communication).

rier-induced absorption changes due to the small density of states of QWs. In any event, the higher electro-optic effect leads to much smaller required interaction lengths. This allows use of shorter integrated optics structures such as small Mach–Zehnder interferometers:[274] this should in turn lead to much higher operating frequencies, as both the RC time constant and propagation/dephasing effects occurring at high frequencies[275] (due to the device size being of the order of the exciting microwave wavelength) are scaled down. It also allows dramatic cost reduction and many-device monolithic integration such as required by integrated-optics switching matrices for fiber network multiplexing.[276] Finally, the very high wavelength selectivity near the resonance might be used for the purpose of wavelength demultiplexing applications.

Of course, it is large overall electro-optic effect that allows efficient modulation in the perpendicular geometry with active thickness ~ a few μm, for which absorption losses play a minor role. Much hope is placed in MQWs modulators to be the fast, high-density Spatial-Light Modulators (SLM) required for optical interconnects and computing systems.[277] Great effort has been devoted to improve the on/off contrast ratio both in transmission and reflection geometries. In the latter case, the use of a multilayer Bragg reflector between the substrate and the active layer allows an efficient double-pass system. By a careful control of the phase of reflected beams, the contrast-ratio has been increased to more than 100, an excellent value allowing many applications.[278] By using two multilayer reflectors on each side of the MQW modulator, a fully integrated Fabry–Perot cavity can be obtained.[279]

d. Nonlinear Effects Applications

So far, the straightforward application of nonlinear properties of QWs has remained a research subject, since the required power levels are still rather high, although much better than those of other nonlinear materials, by several orders of magnitude in the case of infrared intersubband transitions.[280–1] However, by using both the photodetector properties of QWs along with their electro-optic modulation properties, one can design an efficient low-power optical device, thanks to external electrical circuitry, which provides an efficient amplifying feedback. Usually, such systems are made in hybrid form. Here, MQWs can perform both photodetector and modulator functions. A resistor Self-Electro-Optic-Effect Device (SEED) is shown in Fig. 52a as used as a bistable device.[282] The MQWs are imbedded in a p-n junction and serve both as photodetectors and modulators. The incoming photon energy is at the dashed-line position in the absorption spectrum. When no light is shining, all the voltage (10 V) is applied to the structure, resulting in a small absorption. As light starts to

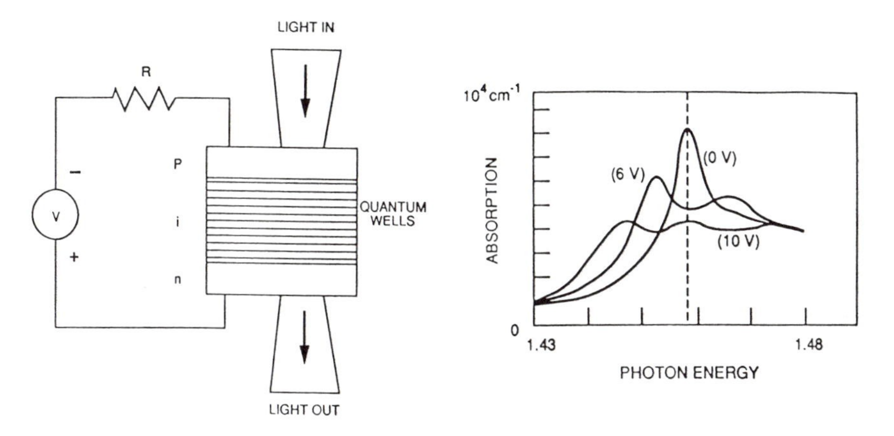

Fig. 52. Schematics of SEED biasing and operation (from Ref. 282).

increase, a voltage drop occurs across the resistor due to the photocurrent, reducing the voltage applied to the MQWs and therefore increasing absorption. Past some critical "switch-down" energy, this process will lead to a runaway, which puts the diode in a high-absorption, high-current, low-applied voltage mode. The critical energy that leads to a nonlinear absorption behavior is of course much smaller than the one required in a fully optical nonlinear effect, since the modification of the absorption edge is provided by an electro-optical effect, which only requires a capacitor charging from an external c.w. electrical source instead of a phase-space filling. The sequence of effects leading to the nonlinear SEED effect can be written as[283]

$$P_{in} \rightarrow P_{abs} \rightarrow N \rightarrow I_{ph} \rightarrow P_e \rightarrow \Delta\mathbf{F} \rightarrow \Delta E \rightarrow \Delta\alpha$$

where P_{in} is the incoming optical power, P_{abs} the absorbed power, N the photocreated current density, I_{ph} the photocurrent, P_e the energy supplied by the electrical generator, $\Delta\mathbf{F}$ the change in electric field applied to the MQWs, ΔE the change in MQW band-edges due to the QCSE, $\Delta\alpha$ the resulting change in absorption coefficient.

The SEED device is bistable because switching to the high-voltage, low-current, low-absorption regime will occur at a lower "switch-up" optical power than at "switch-down." Switching power and speed can be adjusted by the resistor load as the circuit time constant is $\approx RC$, C being the capacitance of the electro-optic device. The optical switching energy is $\approx 1/2\ CV^2$, since switching occurs when the photocreated charge $\approx$ cancels the charge stored in the device acting as a capacitance, i.e., screens out the applied electric field. The switching energy is therefore

$\approx 1.7\ 10^{-7}$ J.cm^{-2} which is $\approx$ 100 times smaller than the energy required to produce an equivalent absorption change in multiple quantum wells through the straight nonlinear absorption (Eq. (55b) or Fig. 50a), and even smaller compared with more usual semiconductor devices.

Many implementations of SEED devices have been demonstrated.[270,284–6] One of the drawbacks of the simple SEED is the lack of intrinsic gain, which requires the device to be operated near the switching threshold so that some amplification is obtained as required to be able to cascade devices. This near-threshold operation also makes the device very sensitive to any fabrication fluctuation, spurious light illumination,[287] etc. To overcome this limitation, the symmetric SEED (S-SEED)[287] was developed. It is another SEED, which provides the load for a given SEED. At a given input power on one of the SEEDs, the other will exhibit the usual bistable behavior as a function of its input power. The interesting property of the device is that if one shines equal powers on the two SEEDs, the devices are in one or the other of two bistable modes, with one diode in the low-absorption mode and the other diode in the high-absorption mode. The diodes do not switch from one state to the other when changing both input powers together, since the ratio of photocurrents does not change. On the other hand, when both powers have been reduced, sending one low-power beam on one of the SEEDs will switch it to the high-absorption mode. Returning to high-input powers on both diodes, a large change in transmitted power has been made, resulting in gain, which, occurring at a different time than switching, is called a "time-sequential gain." This gain is obtained far from a switching threshold, which gives the device a good noise immunity. Many other uses of S-SEED have been demonstrated for logic operations (NAND gate) and memory.[286a] It must be remarked that the S-SEED is a two-beam device, the logic level 0 or 1 being represented by a positive or negative (relative) value of the imbalance of transmitted powers by the two SEEDs. If the "1" of an S-SEED is incident on an S-SEED, it will switch the output to a "0" and vice versa, achieving an inverter function. Although still under active development, the SEED field is already providing impressive figures, with a 2048 S-SEED fully functional matrix, 40 pW holding power per device, and below 1 ns switching speed.[288] SEEDs can easily be integrated with GaAs driving electronics, and preliminary test have demonstrated good lifetimes (>1000 h) for devices grown on Si substrates for monolithic integration with Si IC's.

CHAPTER IV

Electrical Properties of Thin Heterostructures

17. Mobility in Parallel Transport

The flourishing development of modulation-doped heterostructures is based on the extremely high mobility obtained in such structures. As mentioned above, this arises from the spatial separation between charge carriers in the channel and the impurity atoms from which they originate and which remain in the barrier material. It is, however, important to analyze in more detail the various mechanisms limiting the mobility in order to be able to give interface design rules and predict the behavior of the various semiconductor pairs yielding promising interfaces. We shall follow here an analysis first given by Störmer.[289]

Scattering mechanisms are now quite well understood and are measured in *bulk* semiconductors,[290-293] although some higher-order phenomena (such as multiple Coulomb scattering) have never been completely worked out.[293] The scattering mechanisms in the usual perturbative description are decomposed into five contributions.

(1) Optical-phonon scattering (dominant at high temperatures);

(2) Acoustic-phonon scattering due to the deformation potential;

(3) Acoustic-phonon scattering due to the piezoelectric field (III–V and II–VI compounds are piezoelectric due to their lack of inversion symmetry);

(4) Scattering by ionized impurities; and
(5) Scattering by neutral impurities.

The importance of the various mechanisms is shown in Fig. 53 for bulk GaAs as well as experimental results for high-purity VPE GaAs. It is clear that at high temperatures mobility is limited by LO-phonon scattering, very efficient through the Fröhlich mechanism, whereas ionized impurity ($\sim N_D - N_A$) scattering dominates at low temperatures. Two points should be added: (i) In doped bulk GaAs, the mobility depends on the shallow impurity concentration, even at room temperature. (ii) In some optical experiments, ionized impurity scattering can be totally suppressed even at low temperatures ($T \approx 4$ K) owing to the photoneutralization of ionized impurities by photocreated electrons and holes.[294-296] Under such conditions mobilities of $\approx 2 \times 10^6$ cm^2 V^{-1} s^{-1} were observed in *bulk GaAs* by optically measured electron drift velocity[294] and by optically detected cyclotron resonance.[296] Very high hole mobilities were also observed.[296] Such

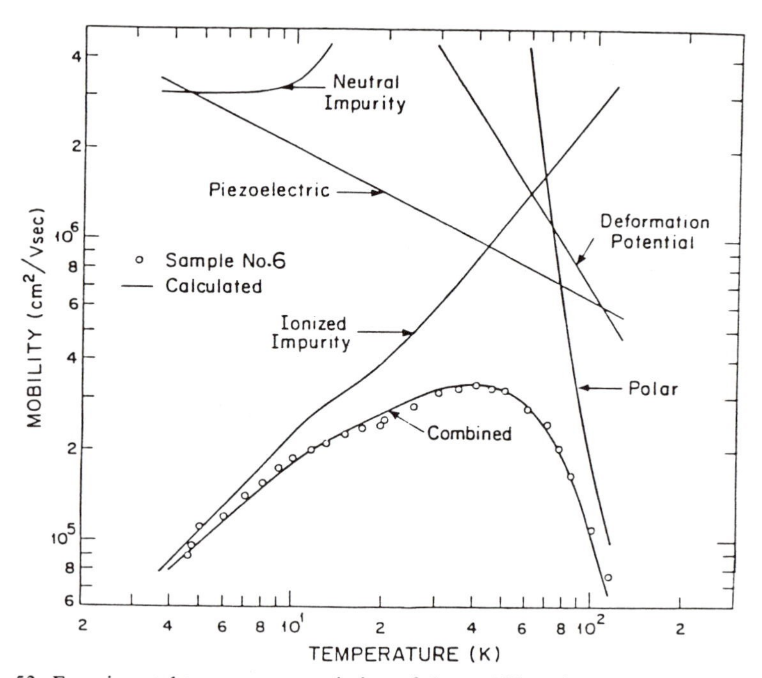

FIG. 53. Experimental temperature variation of the mobility of a high-purity GaAs VPE sample ($N_D = 4.80 \times 10^{13}$ cm^{-3}; $N_A = 2.13 \times 10^{13}$ cm^{-3}) and calculated mobility curves for each scattering process acting separately and for all scattering processes combined (Reprinted with permission from Elsevier Sequoia S. A., G. E. Stillman and C. M. Wolfe, *Thin Solid Films* **31,** 69 (1976).)

experiments give direct evidence of the dominant limiting effect of ionized impurities in bulk material at low temperatures, and show that the suppression of this scattering mechanism indeed leads to mobilities similar to those observed in the best MD 2D samples.

In heterojunctions or quantum wells, the same five mechanisms apply for carriers in the channel, as well as some additional ones.[289]

(6) Scattering by GaAlAs phonons

(7) Scattering by ionized or neutral impurities located in the barrier material (spatially separated from the channel carriers)

(8) Scattering by alloy disorder, either in the barrier material such as in the GaAlAs/GaAs case, in the channel when the channel material is an alloy, as in the case of InP/GaInAs or in both, as in the GaInAs/AlInAs case

(9) Surface phonon scattering, as new propagating surface modes exist at interfaces. However, since the materials are usually similar in density and dielectric functions, these phonons modes never create large scattering probabilities and will not be considered any further

(10) Interface roughness scattering

(11) Intersubband scattering between the quantized levels in the channel

These different mechanisms have been analyzed in great detail by various authors,[297–305] and we will only review their main conclusions pertaining to the GaAlAs/GaAs selectively doped heterointerface, unless otherwise specified.

Mechanisms (1)–(3): The various phonon scattering mechanisms do not change significantly for the quantized channel carriers when compared to the bulk situation. The transition rates from a subband state $|n, \mathbf{k}\rangle$ to another subband state $|n', \mathbf{k}'\rangle$ have been calculated for various phonon scattering mechanisms by Price.[298,300] In short, the usual 3D momentum conservation in the k direction is replaced by overlap integrals $F_{n,n'}(q)$:

$$F_{n,n'}(q) = \iint dz\, dz' \chi_n(z')\chi_{n'}(z')e^{-q(z-z')}\chi_n(z)\chi_{n'}(z) \tag{56}$$

where $\chi_n(z)$ and $\chi_n(z')$ are the envelope wave functions of the $|n, \mathbf{k}\rangle$ and $|n', \mathbf{k}'\rangle$ states and $q = |\mathbf{k} - \mathbf{k}'|$. A complete calculation by Vinter[304] shows a mobility reduction lower than 25% at 77 K. One therefore expects high-temperature (> 80 K) mobilities of 2D carriers to be comparable to those of 3D electrons when phonon scattering is dominant. The limit of the high-temperature mobility of 2D systems is therefore that of high-purity bulk material if all other scattering mechanisms are small (ionized impurities suppressed by modulation doping and low residual channel doping, low interface roughness, etc.). Actual observations[306] show that this is indeed the case and therefore that the LO-phonon scattering rate is similar in 2D and in the bulk. The advantage here is that such values are obtained for highly conductive channels as compared to the low conductivity (low n) of high-purity 3D GaAs.

(4) and (5) ionized and neutral impurities in the GaAs channel: Usually the residual doping is quite small ($\sim 10^{15}$ cm^{-3}) and does not influence the room-temperature mobility. In the case of intentional doping, an impurity contribution to the mobility is observed.[307,308] At low temperatures and low densities, when all other causes of scattering have been reduced ($\mu > 2-3$ 10^5 cm^2 V^{-1} s^{-1}), the limiting factor is still the uncontrolled channel

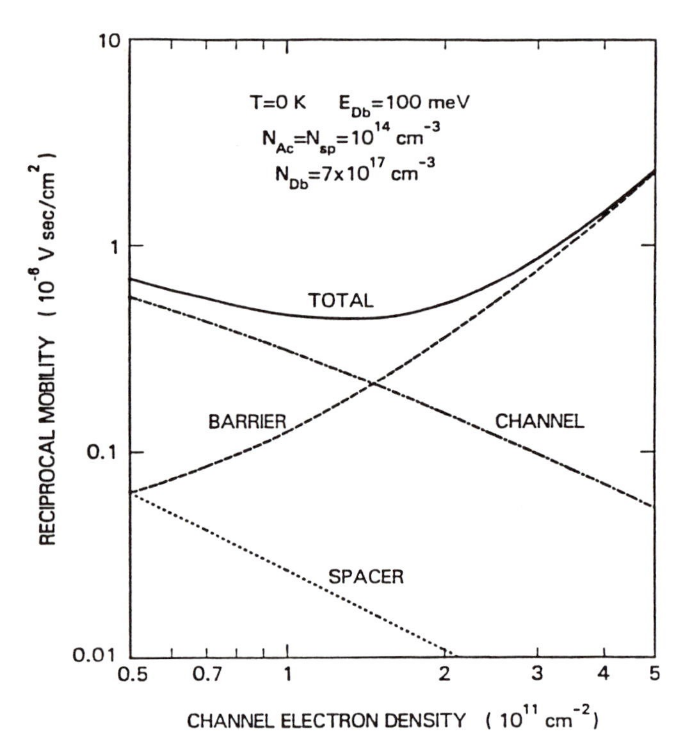

FIG. 54. Calculated low-temperature reciprocal mobility versus channel electron density N_S for a GaAs–Ga$_{1-x}$Al$_x$As heterojunction with 7×10^{17} cm^{-3} donors in the barrier with binding energy $E_{Db} = 100$ meV, a heterojunction barrier height $V_0 = 300$ meV, an acceptor doping level $N_{Ac} = 10^{14}$ cm^{-3} in the GaAs (for which the density of depletion charges, N_d, is 0.46×10^{11} cm^{-2}), and a residual density of charges in the spacer layer also equal to 10^{14} cm^{-3}. The spacer layer thickness d_{sp} is determined for each value of N_S. The three sources of scattering, from the barrier doping itself, from the residual doping in the spacer layer, and from the acceptors in the GaAs, are separately shown (from Stern).[151]

impurity doping.[151] Progress through the years is well evidenced in the sum-up figure (no. 1) in the review by Mendez.[308a]

(6) Scattering by GaAlAs phonons does not play a significant role at any temperature. At low temperatures all phonon mechanisms are suppressed; at high temperatures the GaAlAs phonons can be neglected, as the carrier wave function penetration in the barrier is negligible.

(7) Scattering of channel carriers due to the Coulomb interaction with barrier impurities is an important mechanism of scattering due to the high doping density of the barrier (Fig. 54). Such a mechanism has been calculated in detail and will not be reproduced here.[299] The most remarkable factor appearing in the scattering time is the form factor of the Coulomb interaction matrix element:

$$F(q,z) = \int dz' \,|\chi(z')|^2 \exp(-q|z - z'|) \tag{57}$$

where z is the impurity position, $q = 2k \sin(\theta/2)$ is the scattering vector of the electron with wave vector k, $\chi(z)$ is the confined electron wave function. As expected, the interaction decreases with increasing impurity channel separation thanks to this form factor. This is directly evidenced in front- and back-gating experiments which change the electron wave function penetration in the barrier (see Fig. 60). Another important effect originates from the form factor: as those electrons being scattered are near E_F at low temperatures, one needs to evaluate Eq. (57) for $q = 2k_F \sin(\theta/2)$. When k_F increases with the channel density N_S, this form factor remains significant only for the smaller scattering angles. Such small-angle scattering events, even though efficient in terms of collision time, can be expected to be less efficient for momentum relaxation time (i.e., mobility increases) because of the factor $1 - \cos\theta$ in the momentum-loss integral.

We therefore have the main ingredients of barrier-impurity-limited mobility: it increases both with impurity-channel separation and with channel density. As these two factors vary in opposite directions with undoped spacer thickness for a given alloy doping, one expects a maximum mobility at some value of the spacer. Assuming now a fixed spacer, one has to change the doping density of vary channel density. In such as case there is also some optimal value of N_S (Fig. 55). Such tendencies have been observed experimentally[310] (Fig. 56).

The low-temperature behavior of various high-purity samples shows the delicate balance between the various impurity and phonon scattering mechanisms. The temperature dependence of the mobility switches from positive slope to negative slope when the sample mobility increases, i.e., when the impurity-related scattering rate decreases (Fig. 57). This effect was shown by Lin *et al.*[264] to be due to the balance between impurity-lim-

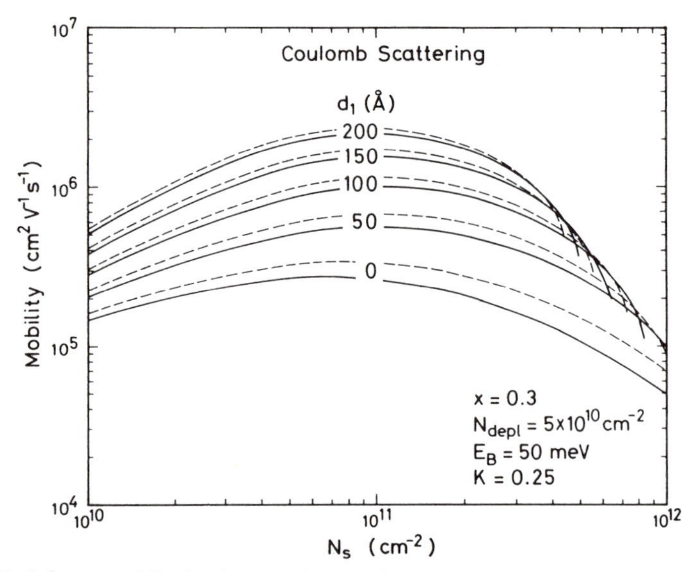

FIG. 55. Influence of Coulomb scattering as a function of channel carrier density N_S and spacer-layer thickness d_1. The increase in N_S is determined by a change in the doping concentration of the GaAlAs barrier. As long as N_S is smaller than N_{dep}, the main effect of increasing N_S is to reduce the scattering rate because of increased electron velocity and channel-impurity screening. Above that value, Coulomb scattering by remote donors in the GaAlAs barrier takes over and decreases the mobility (Reprinted with permission from *J. Phys. Soc. Jpn.* **51**, 3900, T. Ando (1982).)

ited mobility (from the barrier or in the channel) (positive slope) and acoustic-phonon-limited mobility (negative slope). It is remarkable that for the purest samples studied, even though the main scattering mechanism is due to impurities, the temperature dependence arises from the smaller, but strongly temperature-dependent, acoustic phonon mechanisms. Several authors have used this determination of the acoustic phonon scattering rate to evaluate the various phonon scattering mechanisms.[161,300,305,310–313] The form factor in the scattering probability leads to opposite variations of acoustic deformation potential and piezoelectric scattering rates with varying channel electron density N_S. In order to obtain a good fit of the observed decrease of the phonon-limited mobility with N_S using generally accepted phonon coupling parameters, Vinter[313] carried out a calculation involving accurate wave functions and screening of the electron–phonon scattering interaction.

(8) Alloy disorder scattering is due to the statistical composition fluctuations which are unavoidable even in perfectly grown but fundamentally disordered alloys.[314,315] Such fluctuations give rise to a random fluctuating

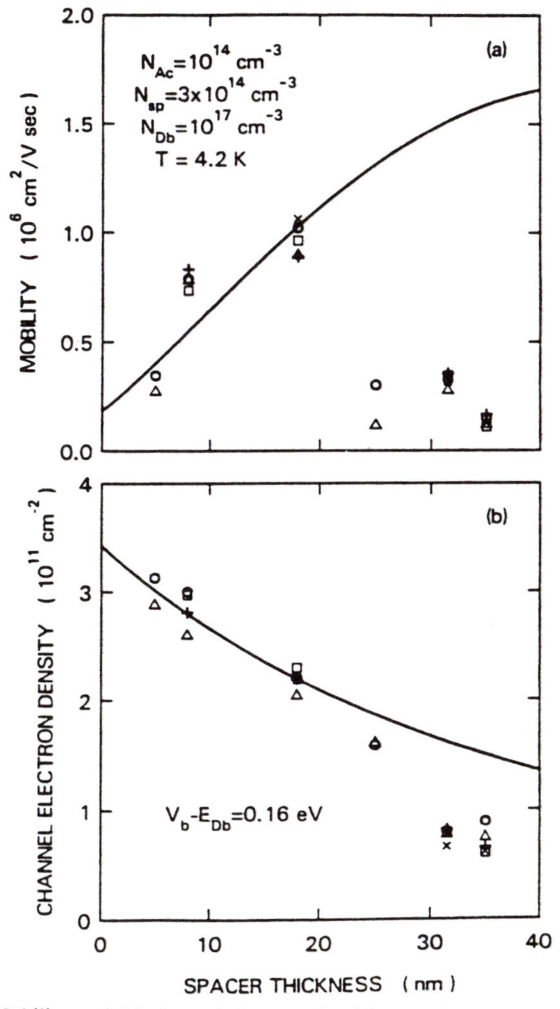

FIG. 56. (a) Mobility and (b) channel electron densities as a function of undoped GaAlAs spacer thickness. Fitting curves are calculated after the method described by Stern. N_{Ac} and N_{sp} are residual acceptor concentrations in the GaAs and GaAlAs layers, respectively, N_{Db} donor concentration in the doped GaAlAs barrier, V_b barrier height, and E_{Db}, donor energy in the barrier material (from Heiblum *et al.*[161]).

potential, well known to limit the mobility in bulk alloy semiconductors. For the GaAs/GaAlAs heterojunction case, there is a weighting factor to the "bulk" alloy mobility given by the channel wave function penetration in the barrier.[299] As this penetration is typically a few percent, reaching such values as maximum N_S, this mechanism is only important at high N_S.

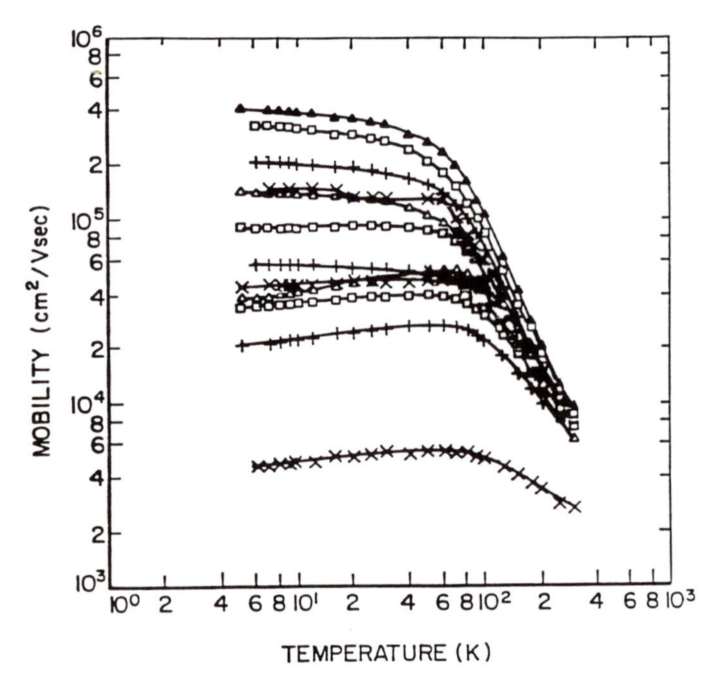

FIG. 57. Temperature dependence of electron mobility in a series of GaAlAs/GaAs MD heterostructures (from Lin[311]).

In the case of an alloy material channel, such as InP/GaInAs, the alloy potential is undiminished by such a factor and sets a rather low maximum mobility to be expected from the 2D electrons.[316,317]

(9) Scattering by interface roughness.

The exact topology of the interfaces is usually unknown. In the case of transport properties, it is modeled by a Gaussian correlation function of the surface position:

$$\langle \Delta(\mathbf{r}) \cdot \Delta(\mathbf{r}') \rangle = \Delta^2 \exp(-|\mathbf{r} - \mathbf{r}'|^2/\lambda^2) \tag{58}$$

where $\Delta(\mathbf{r})$ is the average displacement of the surface height at position $\mathbf{r}$ and λ represents the lateral decay rate of the fluctuations of the interface. Such changes in the interface position can be modeled to act as a spatial variation of the position of the band discontinuity at the interface. One can calculate the transition probability due to such a variable potential by a perturbative approach to the otherwise perfectly plane interface used in the calculations of Section 8. Ando[299] found for the relaxation time $\tau_{IR}(k)$ due to interface roughness

$$\frac{1}{\tau_{\mathrm{IR}}(k)} = \frac{2\pi}{\hbar} \sum_{q} \pi \left(\frac{\lambda\, \Delta F_{\mathrm{eff}}}{\varepsilon(q)} \right)^2 \exp\left(-\frac{q^2\lambda^2}{4} \right) \times (1 - \cos\theta)\, \delta(\varepsilon_k - \varepsilon_{k-q}) \tag{59}$$

where $\varepsilon(q)$ is the static dielectric function of the 2D electron gas and the interface potential effect is represented by F_{eff}, the electric field that confines the channel electrons:

$$F_{\mathrm{eff}} = \int dz\, |\chi(z)|^2 \frac{\partial V(z)}{\partial z} = \frac{e^2}{\varepsilon_0 \varepsilon_R} \left(\frac{1}{2} N_S + N_{\mathrm{dep}} \right) \tag{60}$$

The result of the integration over q is shown in Fig. 58, assuming a mean displacement of the interface $\Delta = 4$ Å and a lateral correlation length $\lambda = 15$ Å. The effect of such an interaction should be observable, at least at high electron densities for extreme-purity samples. It has, however, not been systematically studied. It has only been indirectly shown, as samples grown outside an optimal temperature of 600–700°C, dependent on the growth parameters, show poor mobilities.[318] The high- and low-temperature growth ranges have been correlated by other methods (x rays, TEM, optical spectroscopy) to interface roughness (see the discussion in Section 12), which should then be the mobility-limiting factor if all other scattering causes remain the same.

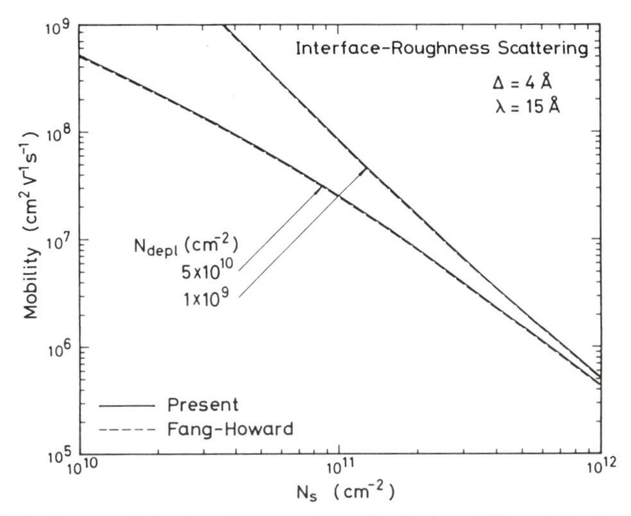

FIG. 58. Calculation of the interface-roughness-limited mobility. Δ is the amplitude of interface position fluctuations and λ the lateral correlation length of these fluctuations (Reprinted with permission from *J. Phys. Soc. Jpn.* **51**, 3893, T. Ando (1982).)

(10) Intersubband scattering occurs at high densities when higher-lying quantum levels ($n = 2, 3, \ldots$) can be populated.[303] The phase space for the final states in scattering events is then larger, increasing the scattering probability and hence diminishing the mobility. Through a study of back-gated Hall samples, Störmer *et al.*[319] were able to show such a decrease in mobility with an increase in channel electron density, and correlate this effect with the population of the $n = 2$ quantum level through the appearance of a double period in Shubnikov–de Haas oscillations (Fig. 59). The effect was also observed by Englert *et al.*,[320] who used a magnetic field parallel to the layer to change the band separation and show the change in sample resistance when the number of populated subbands changes.

The various scattering mechanisms in heterostructures therefore appear rather well understood. Figure 60 shows the development over the years in mobilities obtained in GaAs/AlGaAs heterojunctions.[320a] The best reported mobility to date is in GaAs/AlGaAs and is 11.7×10^6 cm^2 V^{-1} s^{-1} at 0.35 K.[320a] For such mobilities to be explained theoretically, it seems necessary[320b] to go beyond the standard approximation in which the impurity positions in a given plane are considered completely uncorrelated.

We have limited our examples to the case of electrons in the GaAs/GaAlAs system, but a similar analysis of the various scattering mechanisms leading to similar results can be performed on other systems, as has been exemplified on holes in the GaAs/GaAlAs system.[308a,321,322] Several teams have successfully operated a *p*-type FET based on this structure, complementary to the *n*-type structure.[323–325]

Besides the *single* GaAlAs/GaAs heterointerface, several other systems have been considered. Very early, Mori and Ando[326,327] calculated the parallel mobility in modulation-doped superlattices and showed the importance of intersubband scattering. The case of modulation-doped single wells has been considered by Inoue and Sakaki.[328] For rather wide wells, this situation is equivalent to the double heterointerface situation, but gradually changes to a new situation when the well thickness is decreased. The advantage of this QW structure is the higher electron transfer that can be obtained, about twice when compared to the single interface, which leads to better device characteristics due to the better conductivity. Until recently, this structure could not be grown with good equivalent interfaces (GaAs grown on GaAlAs was bad), but recent progress in growth techniques allowed good symmetric, structures to be grown.[329,330] On the other hand, Sakaki used the different mobilities of the asymmetric interfaces (under bias) to design the velocity-modulation transistor (VMT)[331,332]: a large change in channel conductivity is controlled by the gate potential, which confines the carriers on one or the other of the two interfaces of the quantum well, changing their mobility through the deformation of the confined wave function and therefore their interaction with remote ionized impurities in the barrier material (Fig. 61). As one controls the channel

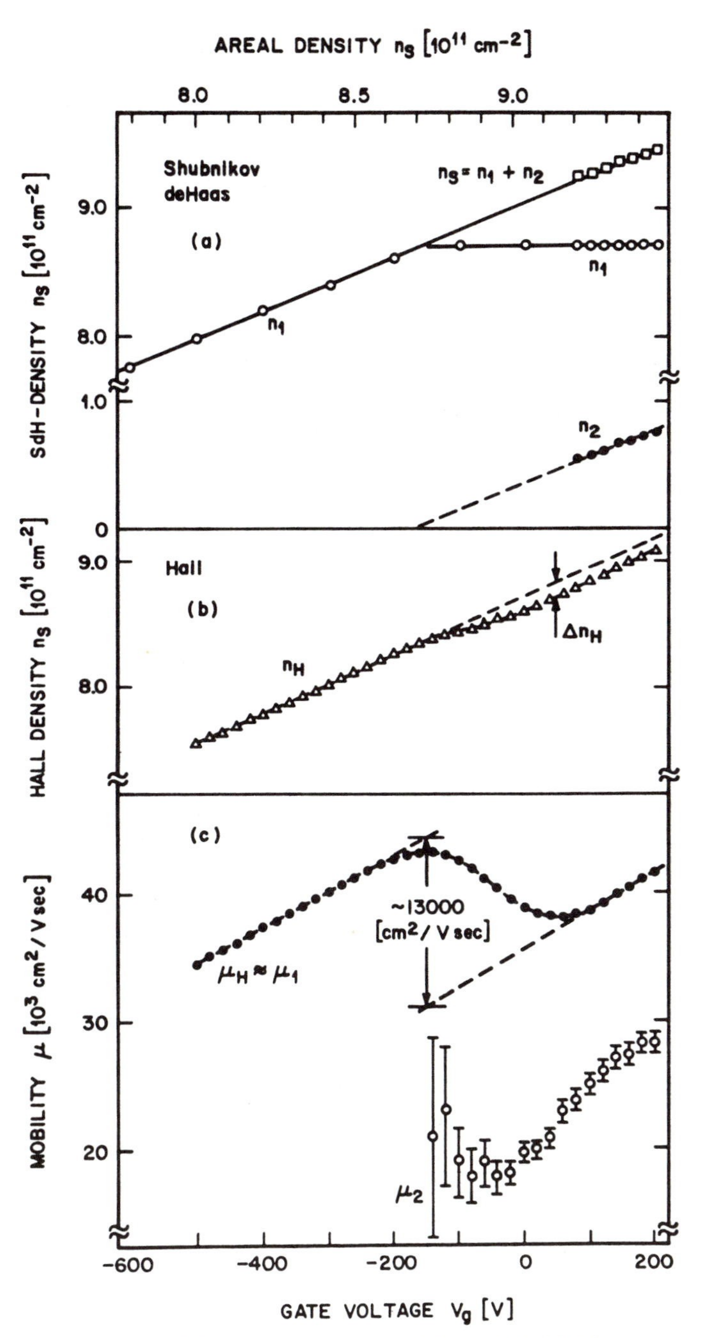

FIG. 59. Onset of intersubband scattering in a GaAs/GaAlAs MD heterostructure. The backside-gate voltage dependence of electron density [(a), (b)] and mobility (c) shows the correlation between the decrease in mobility and onset of upper-subband populations. The Shubnikov–de Haas measurement (a) yields the densities of the two subbands from the two oscillation periods observed. It agrees with the Hall measurement of electron density (b) (Reprinted with permission from Störmer *et al.*[319], © 1982, Pergamon Press plc.)

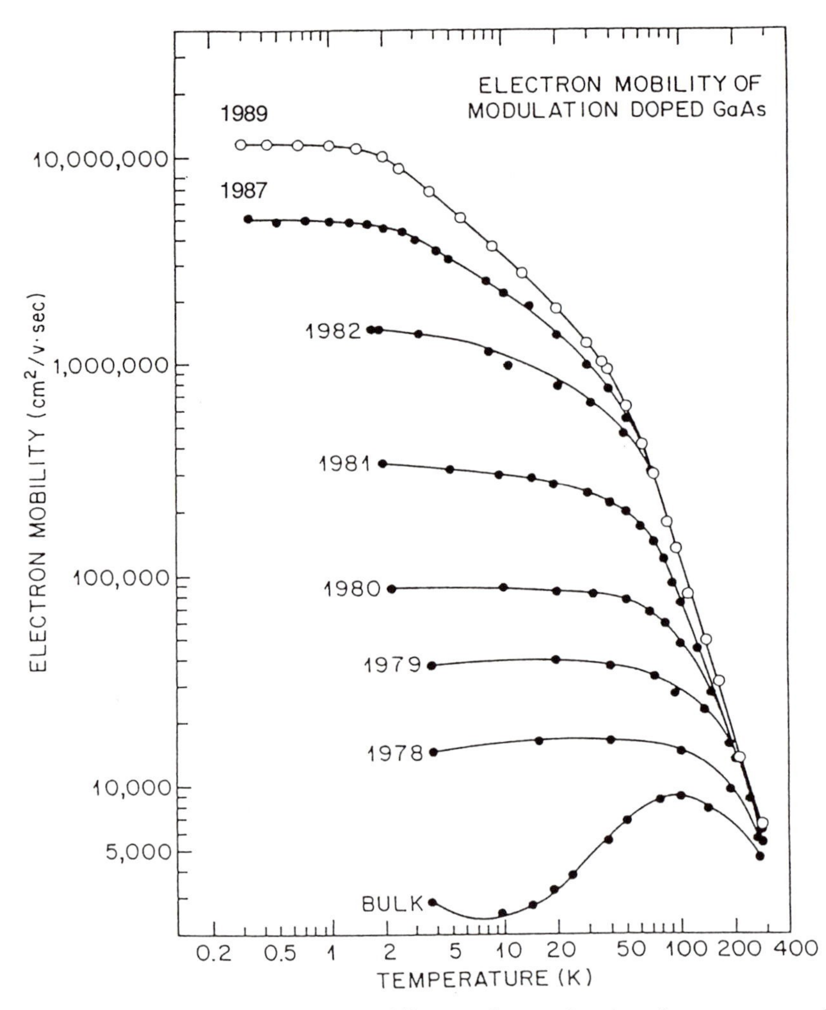

FIG. 60. The mobility of the highest mobility samples as a function of temperature and of year of fabrication (from Pfeiffer *et al.*[320a]).

conductivity by changing the carrier velocity without changing the carrier density, the switching time of the VMT should not be limited by charging-time effects and might reach the subpicosecond range.

18. Hot Electron Effects in Parallel Transport[333]

Since many of the applications of digital heterostructure ICs call for high-speed devices, it is highly desirable to know the high-field properties of such heterostructures. A number of theoretical calculations have been carried out to evaluate the various relaxation rates in 2D systems and the resulting high-field properties [297,334–340] (Fig. 62). The main result of these calculations using different methods is that the 2D energy relaxation rates

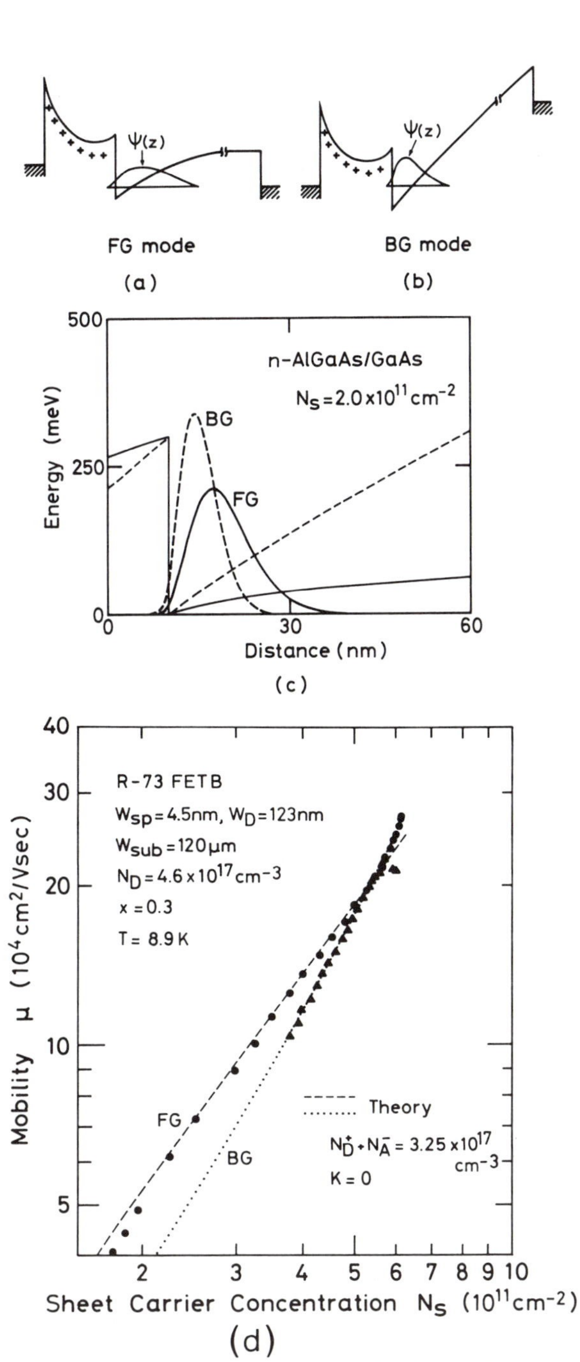

FIG. 61. Control of electron mobility by gate-controlled deformation of the electron wave function. (a) and (b), Schematics of the band diagram under reverse-bias conditions, for front-gate (FG) and back-gate (BG) gating modes, respectively. (c) Calculated electron wave functions in both modes for $N_S = 2.0 \times 10^{11}$ cm^{-2}. (d) Calculated and measured mobilites for both modes. Note the lower mobility in the back-gate mode, for a given N_S, due to increased penetration in the barrier material and thus increased interaction with ionized barrier donors (after Hirakawa *et al.*[332]).

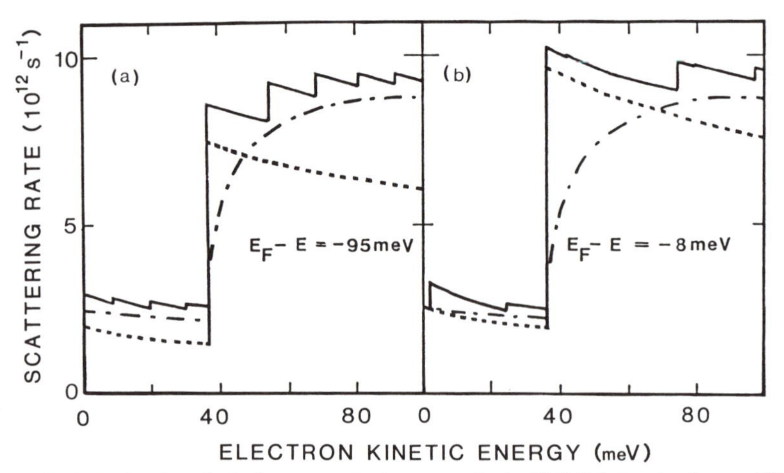

FIG. 62. Calculated optical phonon scattering rates via the Fröhlich mechanism at 300 K. (—·—) represents the 3D scattering rate, (---) includes only the intraband scattering probability ($n = 1$ to $n = 1$ confined state) for scattering out of 2D electrons from the lowest ($n = 1$) subband, while (——) includes inter- and intraband scattering in a GaAs/GaAlAs heterostructure. (a) $N_S = 4 \times 10^{10}$ cm^{-2}; (b) $N_S = 6.2 \times 10^{11}$ cm^{-2}. The abruptness of the onset of phonon emission at $E \sim 36$ meV is characteristic of the square 2D DOS. Note the comparable scattering rates in 2D and 3D except near onset (after Vinter[304]).

should be comparable to the 3D rates. Ridley has predicted that, due to the peculiar 2D momentum, energy relaxation and intersubband scattering, intrinsic negative differential resistance (NDR) could occur.[337–339] Hess and his collaborators[340] have in addition predicted and shown that carrier heating in a heterostructure gives rise to a new mechanism of NDR by real-space transfer over the potential barriers. Such concepts of real-space transfer were applied by Kastalsky *et al.* to design a number of new high-frequency devices.[29–31]

Room-temperature hot-electron characteristics have been studied by a number of groups using Hall measurements under pulsed applied electric field in order to avoid lattice heating.[341–344] Velocities significantly higher than in bulk GaAs were obtained at 300 K and increased even more at 77 K (Fig. 63). However, the peak velocity as observed by microwave methods[344a] has been found to be lower in heterojunction structures than in undoped bulk GaAs both at room temperature and at 77 K. Those observations are also corroborated by recent Monte Carlo simulations of hot electron transport.[344b]

Several additional effects can occur for the hot-electron regime at *low temperatures:* Schubert and Ploog[344] observed a decrease in conducting electron density by Hall measurements at 77 K in the GaAlAs/GaAs

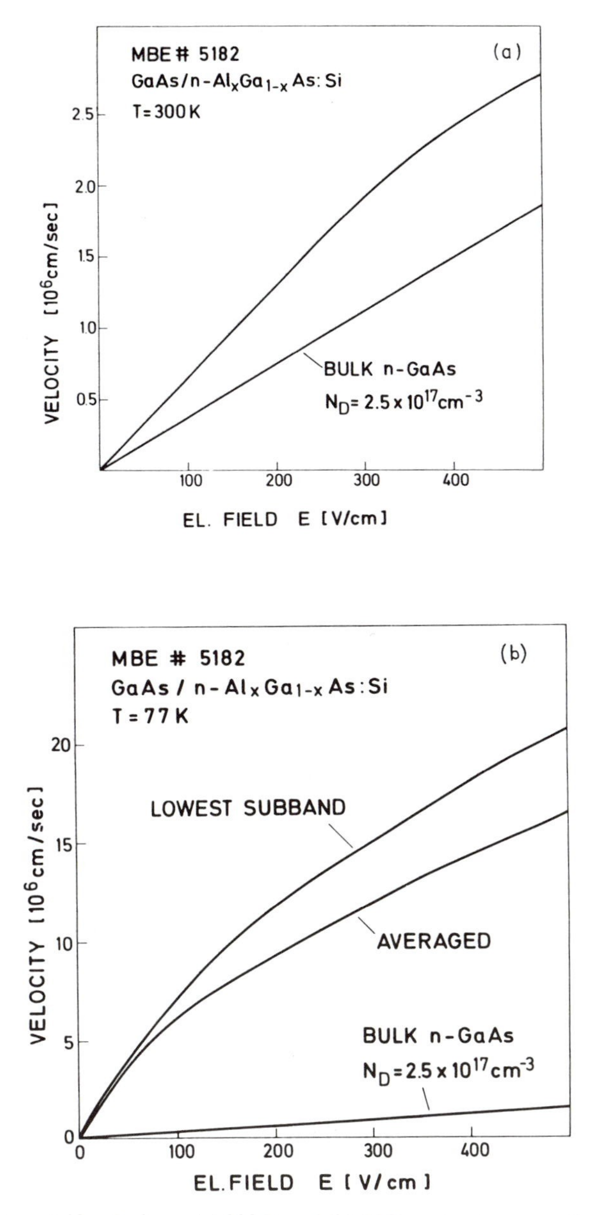

FIG. 63. Electron drift velocity at (a) 300 K and (b) 77 K under strong applied electric field for bulk or modulation-doped GaAs/GaAlAs heterostructures. The 77 K curves displayed correspond to the fraction of electrons in the 2D lowest subband of the channel and to the whole averaged electron gas. Note the large increase in velocity between the bulk and 2D electrons at 77K (from Schubert and Ploog[344]).

interface. They explain this effect by the scattering of electrons into the higher-lying confined subband level E_2 where they have a low mobility, and by the trapping of hot electrons in localized states situated in the barrier material near the interface. At liquid helium temperatures, the electron density tends to increase in both the barrier and channel due to impact ionization of neutral Si donors in the barrier material in the hot-electron regime.

In the warm electron regime the parameter β, which describes the lowest order deviation from Ohm's law, $\mu = \mu_0(1 + \beta E^2)$, has been measured and used to determine energy relaxation rates[344c] and to measure the effective deformation potential D for electron–phonon scattering.[344d] A value of $D = 11$ eV, larger than the commonly accepted bulk value of 7 eV, was found in agreement with measurements based on the temperature dependence of the low field mobility.[309,312,313]

Energy relaxation rates were measured optically at low temperatures[345,346]: as in the 3D case, carrier heating is deduced from the line-shape analysis of the photoluminescence line (high-energy slope $\sim\exp(-h\nu/kT_{\text{eff}})$, where T_{eff} is the carrier effective temperature), the carrier heating being produced by an electric field applied to the illuminated area. Energy relaxation rates are deduced from power-balance equations, which equate the energy loss to the lattice with the energy gained per carrier from the applied field:

$$P = \left(\frac{dE}{dt}\right)_{T_{\text{eff}}} = e\tau E^2$$

These rates have been measured both for the electron and hole gas in the GaAlAs interface. For electrons, carrier heating can be detected for fields as low as 0.3 V/cm. Such an efficient electron heating, due to the very high mobility, has also been observed by the Hall effect[344,347] and damping of the Shubnikov–de Haas oscillations[344,347,348] (Fig. 64). Comparing electron and hole relaxation rates, Shah *et al.* found a scattering rate 25 times larger for holes than for electrons.[346] This difference, which cannot be explained by 2D or coupling effects, has been attributed to the accumulation of hot phonons, well above the thermal number, which interact predominantly with electrons.[224a]

19. Perpendicular Transport

As mentioned in the introduction, the hope for new effects in perpendicular transport gave impetus to the development of superlattices and heterostructures. The semiclassical equations of free motion for electrons in an energy band $E(\mathbf{k})$ (infinite solid) with an electric field $\mathbf{F}$ are

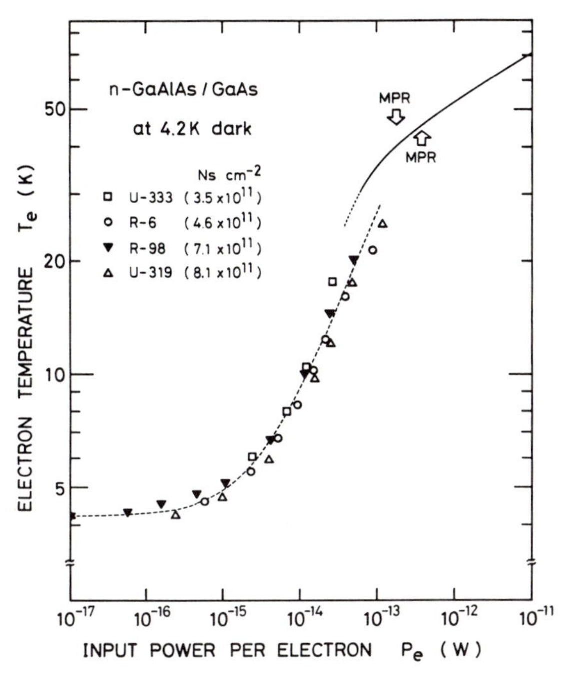

FIG. 64. Electron heating as deduced from the damping of the Shubnikov–de Haas oscillation (——) represents the results of Shah *et al.* deduced from the analysis of the luminescence line shape (from Sakaki *et al.*[347]).

$$v(\mathbf{k}) = \frac{1}{\hbar}\frac{\partial E(\mathbf{k})}{\partial k}; \qquad \hbar\frac{d\mathbf{k}}{dt} = e\mathbf{F} \tag{61}$$

In a steady applied field,

$$\mathbf{k}(t) = \mathbf{k}(0) - e\mathbf{F}t/\hbar \tag{62}$$

For electrons in a band, **k** therefore changes linearly with time. The energy of the electrons also changes according to the dispersion curve $E(\mathbf{k})$, and so does $v(\mathbf{k})$. In the reduced Brillouin zone scheme, once the electron reaches a zone boundary point k_{ZB}, it is Bragg reflected in the opposite direction; i.e., it appears at the $-k_{\mathrm{ZB}}$ point. Thus $v(\mathbf{k})$ is an oscillatory function of time with a period equal to the time needed for k to cross the Brillouin zone, $T = (2\pi/d)(eF/\hbar)^{-1}$, where d is the lattice periodicity (Fig. 65). The motion in real space would have the same frequency, and a very fast oscillator called a Bloch oscillator could be achieved.[3,349] However, the period has to be shorter than the collision time, which is currently impossible when d is an atomic lattice constant ($T = 10^{-11}$ s for $F = 10$ kV cm^{-1}

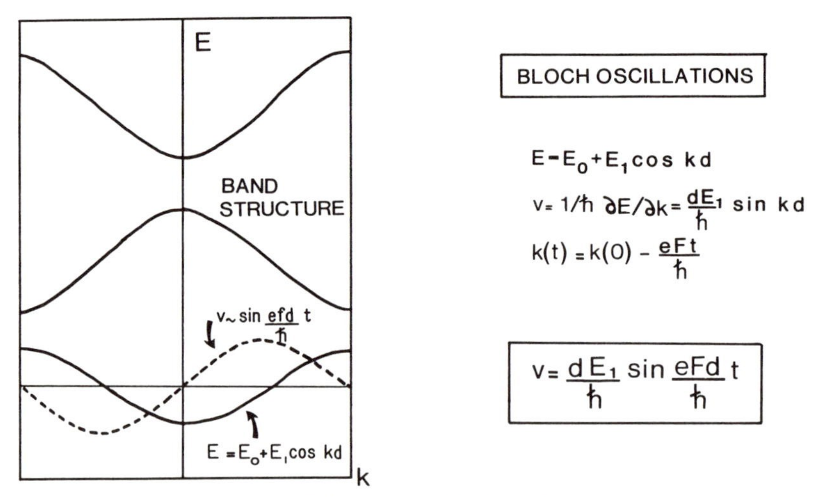

FIG. 65. Schematics of Bloch oscillation.

and $d = 3.5$ Å), but should become possible when the lattice constant d is that of a superlattice, about 10 to 50 times larger.

The existence of Bloch oscillations was, however, challenged quite early:[352] the main argument is due to the fundamental modification of the band structure in an electric field, which allows interband transitions at Brillouin zone boundaries rather than Bragg reflections. We refer the reader to two recent discussions on the validity and conditions for observation of Bloch oscillations.[349,353]

Taking collisions into account, Esaki and Tsu[3,350] calculated the drift velocity in an *infinite superlattice* using a classical method.[351] The velocity increment in a time interval dt is, from Eqs. (61) and (62)

$$dv_z = \frac{eF}{\hbar^2}\frac{\partial^2 E}{\partial k_z^2}\,dt \tag{63}$$

The *average drift velocity* imposed by collisions occurring with a frequency τ^{-1} is

$$v_{\mathrm{d}} = \int_0^\infty e^{-t/\tau}\,dv_z = \int_0^\infty \frac{eF}{\hbar^2}\frac{\partial^2 E}{\partial k^2}\,e^{-t/\tau}\,dt \tag{64}$$

As k is changing with time, $\partial^2 E/\partial k^2$ is a function of time, and one requires the knowledge of $E(k)$ to proceed further. Assuming a sinusoidal dependence of E on k, $E = E_0 + 2E_1 \cos kd$, one finds

$$v_{\mathrm{d}} = \frac{\pi\hbar}{m_{\mathrm{SL}}d}\,\frac{\zeta^2}{1+\pi^2\zeta^2} \tag{65}$$

where $\zeta = eF\tau d/\pi\hbar$ and $1/m_{\mathrm{SL}} = (1/\hbar^2)(\partial^2 E/\partial k^2)$.

The v_d versus F curve has a maximum for $\pi\zeta = 1$ and exhibits an NDR beyond this value. The condition to be fulfilled on τ to achieve NDR is about 6 times easier than that required to achieve Bloch oscillations.

Effects in finite (i.e., a low number of barriers and wells) heterostructures were also considered very early by Tsu and Esaki using a multibarrier tunneling model.[120] A pioneering theoretical analysis of the $I-V$ characteristics was provided by Kazarinov and Suris as early as 1972.[355a] In addition to being very tractable with few-interface problems, such a formalism allows one to treat the case of *intersubband* electron tunneling transfer which occurs at high electric fields, an effect not easily described in the formalism of Bloch transport [Eqs. (61) and (62)]. The negative differential resistance observed in superlattices by Esaki *et al.*[13,354,355] was actually explained by resonant electron transfer between adjacent wells due to coincident *ground and excited states.* The formalism also allows one to take into account the effect of unequal layer thickness and/or interface disorder (caption of Fig. 1) which, in the Bloch oscillator formalism, would lead to untractable scattering events, as they destroy the coherence of the superlattice wave function [of the type described by Eq. (28)]. The electron motion is then most easily described in a hopping model between localized states. The transition between the two types of transport, superlattice or hopping, is discussed by Calecki *et al.*[355b] and compared to experiments. A discussion of the theoretical foundations of quantum transport in heterostructures can be found in Barker.[355c]

The resonant transmission of single and double heterostructures is the subject of renewed interest due to recent advances in growth control.[356–358] As shown schematically in Fig. 66 a double barrier structure can lead to a region of negative differential resistance: The level in the well descends when a voltage is applied to the two contacts; when the level in the well descends below the conduction band of the emitter, the electrons have a very low probability of being transmitted so that the current decreases with increasing voltage.

The simplest theory to describe the effect quantitatively was first given by Tsu and Esaki.[120] The emitter and collector contacts are considered as reservoirs of electrons distributed according to Fermi–Dirac statistics with a difference in Fermi level equal to the applied voltage. The current from left to right is then given by the sum of the current density of each state over occupied states multiplied by their transmission probability through the structure. Since the transmission only depends on the motion perpendicular to the interfaces, it is possible to group together the states that have the same energy for perpendicular motion but different kinetic energy for parallel motion:

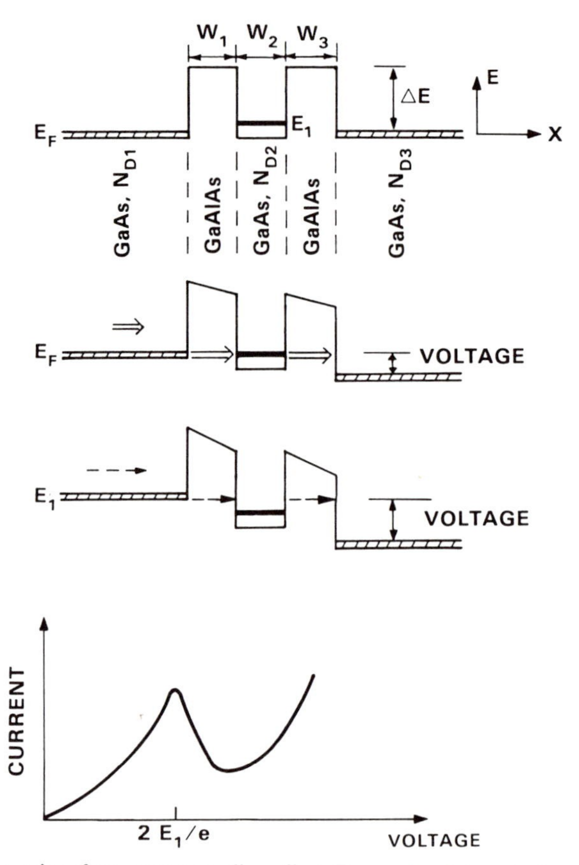

FIG. 66. Schematics of resonant tunneling effect. Energy levels in a single-well double-barrier heterostructure (top three drawings) under bias increasing from the top. The electron energy is indicated as a function of position. Parameters are $N_{D1} = N_{D3} = 10^{18}$ cm^{-3}, $N_{D2} = 10^{17}$ cm^{-3} and $W_1 = W_2 = W_3 = 50$ Å. Resonant transmission occurs for $V = 2E/e$, when electrons tunnel resonantly into the $n = 1$ well state from the left electrode (from Sollner *et al.*[25]).

$$\mathbf{N}(E_z) = \int_0^\infty dE_\| \frac{D(E_\|)}{1 + \exp\left(\dfrac{E_z + E_\| - E_F}{kT}\right)}$$

$$= \frac{kTm^*}{\pi\hbar^2} \log\left[1 + \exp\left(\frac{E_F - E_z}{kT}\right)\right] \tag{65a}$$

as found in Section 8d. This function is often referred to as the "supply function." The current density from left to right can then be written

$$j^{\rightarrow} = e \sum_{k_z > 0} \frac{\hbar k_z}{m^*} T^{\rightarrow}(E_z)\mathbf{N}(E_z)$$
$$= \frac{e}{2\pi\hbar} \int_{E_c}^{\infty} T^{\rightarrow}(E_z)\mathbf{N}(E_z)\, dE_z, \quad (65b)$$

where E_c is the conduction band of the emitter. A similar expression holds for the current from right to left, so that the net current is

$$j = \frac{e}{2\pi\hbar} \int_{E_c}^{\infty} T^{\rightarrow}(E_z)[\mathbf{N}_e(E_z) - \mathbf{N}_c(E_z)]\, dE_z \quad (65c)$$

where $\mathbf{N}_e$ and $\mathbf{N}_c$ are the supply functions of emitter and collector, respectively. For small applied voltages V, one finds, by insertion of the supply functions and taking right and left transmissions equal,

$$j = \frac{em^*kT}{2\pi^2\hbar^3} \int_{E_c}^{\infty} T(E) \log \left[\frac{1 + \exp\left(\dfrac{E_F - E}{kT}\right)}{1 + \exp\left(\dfrac{E_F - E - eV}{kT}\right)} \right] dE \quad (65d)$$

For large applied voltages, $eV \gg (E_F - E_c)$ and $eV \gg kT$, the current from right to left can be neglected, the denominator in the argument of the logarithm of (65d) becomes equal to one, and the current is dependent only on the voltage through the variation of the transmission as the field across the double barrier increases. The rapid decrease in current then occurs when a strong resonance in the transmission disappears when the level in the quantum well crosses the emitter conduction band edge E_c.

The expressions derived here have been found to describe quite well the I-V characteristics in the voltage regions where strong resonant tunneling occurs, i.e., the peaks in current. The prediction of the simple model of a peak in current at an applied voltage of $2E_1$, where E_1 is the energy of the bound state in the well at zero applied voltage, was verified by Chang *et al.*[355] in 1974. Systematic studies of current-voltage characteristics in $Al_xGa_{1-x}As/GaAs$ double barrier structures have been performed by Tsuchiya *et al.*,[356a] and many other groups have reported excellent peak-to-valley ratios in the same system, e.g., Paulus *et al.*[356b] and Diamond *et al.*,[356c] with maximum ratios of up to 3.9 at room temperature and more than 14 at 77 K. In the same system resonant tunneling of holes has also been observed by Mendez *et al.*[356d] Much larger peak-to-valley ratios up to 30 at room temperature have been reported in double barriers based on the AlInAs/GaInAs system.[356e,f,g]

The valley current is always found to be larger than predicted by the resonant tunneling theory. This indicates that incoherent processes by which an electron in the emitter tunnels into the well assisted by some interaction, optical phonon, acoustic phonon, impurities, alloy disorder, or interface roughness. The process whereby an electron enters the well by emitting an optical phonon has recently been observed,[357a,b] and several theoretical discussions of the incoherent processes have been published.[357c–e]. It has been shown that the major part of the discrepancy between the experimental peak-to-valley ratios and those found in the coherent theory can be accounted for by the intrinsic (phonon and alloy disorder assisted) incoherent processes.[357f] In the resonant peak one can expect contributions to the current both from coherent resonant tunneling and incoherent tunneling in which an electron suffers a scattering process in one of the barriers or in the well. The incoherent processes broaden the *transmission* resonance and reduce the peak transmission;[357g] however, it has been shown[357h,i,j] that the peak current, which is essentially the integral over the whole transmission resonance (Eqs. 65c–d) does not depend on the partition into the coherent and incoherent contributions, *provided that the width of the broadened transmission resonance is smaller than structures in the supply function around the resonance energy.*

The negative differential conductivity implies that the double barrier diode is unstable in the NDR region. The instability can lead to qualitatively different behavior depending mainly on the external circuit connected to the diode. If the diode is inserted in a resonating circuit or a cavity, oscillations may be generated at a frequency characteristic of the resonator; this effect is exploited in very high frequency oscillators as described in Section 24. In other circumstances bistability may occur, in which the diode for a given polarization can be either in a high-current state in which the level in the well carries a resonant tunneling current or in a low-current state in which the current is only carried by incoherent processes. The bistability can be extrinsic owing to the external circuit, but recently is has also been shown that the bistability can be intrinsic: In the high-current state the well level contains many electrons, the charge of which tends to push the level upward in energy, thus stabilizing the high-current state; in the low-current state the well level is almost empty, which stabilizes the low-current state.[358a,b]

The subject of resonant tunneling structures is still in rapid evolution and several corrections to the simple theory have been studied, e.g., by Wigner function methods and by including the self-consistent influence of the charge distribution on the shape of the potential. We can refer only to a few specialized articles on these theories.[359]

A great number of optical methods have been applied to the study of tunneling. Time-resolved photoluminescence measurements on double quantum wells in an applied electric field[359a] have shown a rapid increase in tunneling rate from one well to the other when the final electron level decreases below one optical phonon energy below the initial level as expected from theory.[359b] Similar experimental studies[359c,d] and theories[359e] have revealed that holes can tunnel much faster than expected from their heavy mass. Transport properties in superlattices have also been studied by time-resolved and CW photoluminescence methods.[359f] Finally, it should be mentioned that optical methods have also been applied to double barrier diodes to determine the tunneling escape time of electrons in the well[359g] and to estimate the amount of charge that is stored in the well when the diode is in resonance.[359h,i,j]

In superlattices optical techniques have also been applied to monitor perpendicular transport: Photoexcited carriers moving in conduction and valence bands are trapped in deliberately introduced enlarged quantum wells, which act as probes of spatial transport.[360,360a]

20. Quantum Transport[361]

As in studies of 3D electrons, transport measurements under strong magnetic fields (in the so-called quantum regime, where $\omega_c \tau \gg 1$, τ being the carrier collision time) provide a vast amount of information about the parameters of the 2D electron gas. They have also opened the large new area of quantum Hall effects, a major advance in solid-state physics. We therefore devote a detailed description to such studies, broken into three parts: the effect of magnetic fields on 2D electrons, the Shubnikov–de Haas effect, and the quantum Hall effect.

a. *Effect of a Magnetic Field on 2D Electrons*

In the 3D quantum-mechanical problem of electrons in a magnetic field B_z,[361] the motion in the $x-y$ plane is described by Landau levels. The wavefunction is given by (in the Landau gauge of the vector potential $A = [0, xB, 0]$)

$$\psi_{nk}(\mathbf{r}) = (L_y L_z)^{-1/2} \varphi_n(x - X_k) e^{ik_y y} e^{ik_z z} \tag{66}$$

where L_x, L_y, and L_z represent the dimensions of a 3D crystal, the φ_n functions are normalized wave functions of a harmonic oscillator with the quantum number n centered at point $X_k = -r_c^2 k_y$, r_c being the classical cyclotron radius of the lowest oscillator ($n = 0$) orbit,[362]

$$r_c = \sqrt{\hbar / eB}$$

The oscillator quantum number n can take the values 0, 1, 2, ... The energy eigenstates are

$$E_{nk} = (n + \tfrac{1}{2})\hbar\omega_c + g^*\mu_B B_z + E_z \tag{67}$$

where ω_c is the cyclotron frequency eB/m^*, $g^*\mu_B B_z$ is the spin magnetic energy, g^* being the Landé factor, and E_z is the energy associated with the z motion of the carrier.

From Eq. (67), it is easily shown that the quantum states in k space are located on cylinders with their symmetry axes along the z direction (Fig. 67a). In the z direction the usual quasi-continuum free-particle DOS has the value $L_z/2\pi$. For the $x-y$ motion, states are characterized by the cyclotron energy $(n + \frac{1}{2})\hbar\omega_c$, located on circles with radii $k_x^2 + k_y^2 = (2m^*/\hbar^2)(n + \frac{1}{2})\hbar\omega_c$. The degeneracy of each single-spin Landau level (i.e., the number of states on each circle) can be found from the number of possible cyclotron orbits in the crystal. One has to ascertain that the center of the quantum state is within sample boundaries, i.e., $0 < X_k < L_x$; this can be transformed into

$$0 < k_y < m\omega_c L_x/\hbar \tag{68}$$

From the density of states in the k_y direction $L_y/2\pi$, this means that the number of possible states in the range $[0, m\omega_c L_x/\hbar]$ is $L_x L_y m\omega_c/2\pi\hbar$; i.e., the DOS per unit area is $m\omega_c/2\pi\hbar = eB/h$. Comparing this with the number of states in zero field contained within the energy separation between two Landau states, i.e.,

$$(\hbar\omega_c)[m^*/(2\pi\hbar^2)] = m\omega_c/2\pi\hbar = eB/h \tag{69}$$

we find the same value! Thus the average density of states in a quantizing magnetic field is unaffected. Instead of having a 2D continuum of states, these states are all collapsed in a single degenerate Landau state. For a 3D electron system, the occupied states within a given Fermi energy E_F are contained in a 3D k sphere of radius $k_F = (3E_F/4\pi)^{1/3}$ if no magnetic field is applied. When a quantizing field exists, all the states situated within the sphere on the allowed state cylinders are occupied. The density of states, given by dN/dE, is shown in Fig. 67b. It shows a divergence typical of 1D systems, each time a new cyclotron state comes into the Fermi sphere. For that new Landau state, the number of states is given by the degeneracy of each k Landau state times the density of states for the z motion, i.e.,

$$\rho(E) = \frac{dN}{dE} = \left(L_y L_z \frac{eB}{h}\right)\left(\frac{L_z}{2\pi}\frac{dk_z}{dE}\right) \tag{70}$$

which diverges as the cylinder of allowed states is then tangent to the Fermi sphere, yielding numerous new states for a small change in magnetic field ($\hbar\omega_c$) or charge carrier density (change of E_F). In real systems, broadening

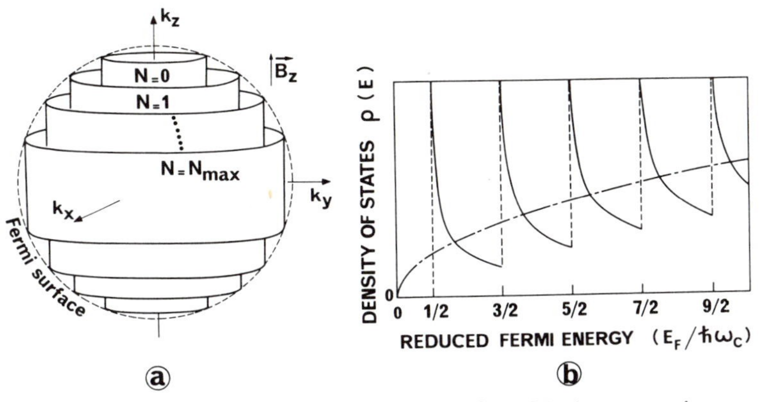

FIG. 67. Allowed states and density of states (DOS) for a 3D electron gas in a magnetic field B_z. (a) Momentum-space occupied states: Allowed states are characterized by the relation $E = (n + \frac{1}{2})\hbar\omega_c + E_z$. Such a relation defines cylinders of axis k_z and radii $k_x^2 + k_y^2 = 2m^*/\hbar^2(n + \frac{1}{2}\hbar\omega_c)$. (b) Density of states dN/dE. The change in Landau state degeneracy is smooth, whereas the z-motion DOS diverges each time a new Landau state enters the Fermi sphere, which is reflected in the total DOS. The 3D DOS is shown for comparison.

will wipe out the divergence, but the periodic behavior of the DOS is retained. The period is given by the change in the number of cyclotron states in the Fermi sphere, determined by

$$(n + \tfrac{1}{2})\hbar\omega_c = E_F$$

For a fixed number of carriers, it can be shown that the DOS at the Fermi energy oscillates with the magnetic field. As many physical quantities depend on the DOS at the Fermi energy, they will exhibit oscillations with the magnetic field. Such effects such as Shubnikov–de Haas [oscillatory magnetoresistance due to an increase in the scattering rate whenever $\rho(E)$ diverges], de Haas–van Alphen (magnetic susceptibility), etc., have long been observed in 3D systems and have been widely used to analyze the electronic properties of metals and semiconductors.[361]

In 2D heterostructures, with a magnetic field perpendicular to the layer plane, the same Landau quantization occurs (Fig. 68). However, the effect is even more dramatic as the z motion of carriers is also frozen by the confining potential leading to a "completely confined quantum limit" system. The energy-level structure is made up of a ladder of cyclotron levels for each confined state, each level having a singular DOS (Dirac-like function) with a degeneracy of eB/h. As in 3D, this degeneracy is equal to the number of 2D states contained in the energy spacing between two consecutive Landau levels. For real systems, disorder (random impurities, alloy fluctuations, interface roughness, . . .) will broaden this singular

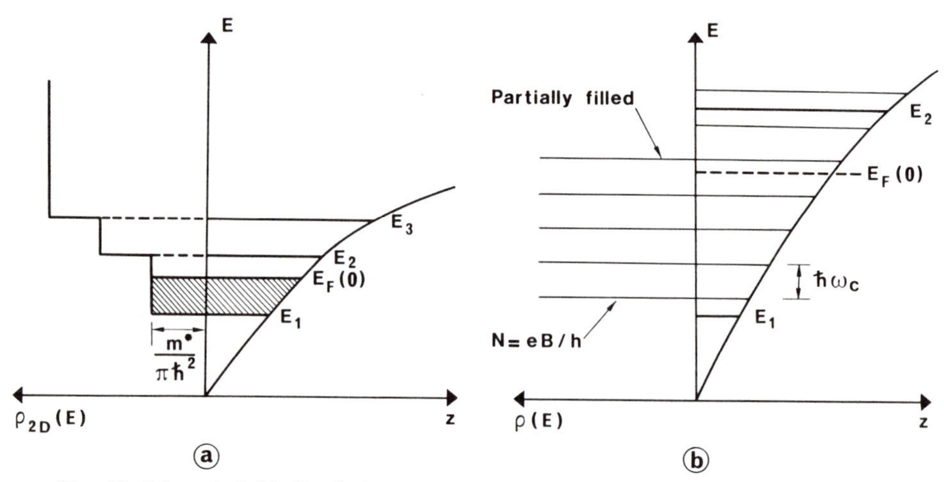

FIG. 68. Magnetic field effect in 2D systems. (a) Energy levels and DOS of a heterojunction without magnetic field. (b) Energy levels and DOS of a single quasi-2D level in a magnetic field [scaled up compared to (a)]. Electrons occupy Landau levels up to some last partially filled Landau level.

DOS (Fig. 69). The states in the tails of the levels are localized in space and will be shown below to play an important role in the existence of the quantum Hall effect.

The oscillatory behavior of several quantities has been calculated[363,364]. as in 3D, the Fermi energy oscillates. When a Landau state is not completely filled, the Fermi level lies in that state and has therefore a smooth variation with the magnetic field or electron density. However, when the last occupied Landau state is filled, the next electron must lie in the next Landau state and the Fermi level jumps there. The result of the calculation of E_F including a Gaussian broadening of the Landau states is shown in Fig. 70a. Other quantities have been calculated, such as the magnetization (de Haas–van Alphen effect) (Fig. 70b), the specific heat (Fig. 70c), the thermoelectric power (Fig. 70d), etc. The peculiar shape of the specific heat curve (Fig. 70c) is due to the existence of inter- or intra-Landau state thermal excitations. At finite temperatures and low enough fields, where $\hbar\omega_c \approx kT$, intersubband excitations can occur and show up as sharp peaks whenever the Fermi level lies in between two Landau states. Such effects have been observed in heat-pulse experiments by Gornik *et al.*[365] The typical oscillatory behavior of the thermoelectric power has also been demonstrated by Obloh *et al.*[366]

Many spectacular effects have been observed in the quantum regime of 2D heterointerface systems. The 2D cyclotron resonance has been demonstrated for confined electrons and holes in various heterostructures (GaAlAs/GaAs,[367] GaInAs/InP,[368] GaSb/InAs[369]) and has allowed the determination of electron and hole[93] masses as well as polaron[368,369] and screening effects[370] in 2D. In particularly pure samples, a specific oscillation of the cyclotron-resonance linewidth has been observed, which has been related to the oscillatory character of the scattering probability with the filling factor[371] or to the softening of the 2D magnetoplasmon mode.[372]

The de Haas-van Alphen measurement of the oscillatory magnetic susceptibility allows a more direct determination of the 2D DOS as compared to Shubnikov-de Haas magnetoresistance measurements (which involve carrier scattering) and cyclotron resonance (which yields a combined DOS of initial and final states). Such measurements are extremely difficult, as the total number of 2D electrons to be measured is very small, as compared to a 3D case. Nevertheless, using a 272-layer sample, Störmer *et al.*[373] were able to measure the 2D electron gas magnetic susceptibility. Switching from a SQUID detection to a torsional balance magnetometer recently allowed a 100-fold gain in sensitivity.[374]

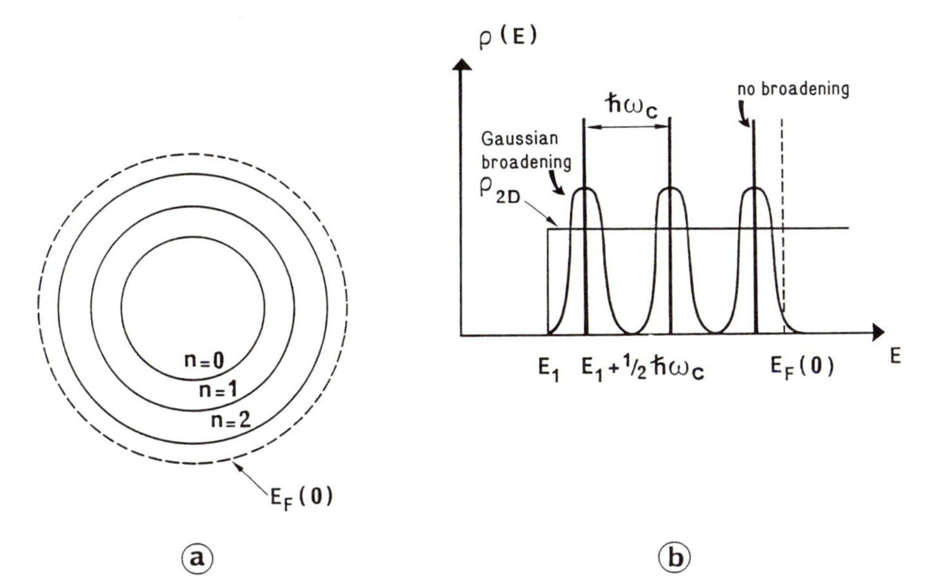

FIG. 69. Fermi level in a 2D system. (a) In k_x-k_y plane. Landau states are all filled up to some fractionally occupied state where the Fermi level lies. (---) represents the Fermi disk, which contains all allowed states when no magnetic field is present. (b) Energy representation: The Landau states are broadened, which smoothes out the transition of the Fermi level from the last fully occupied Landau state to the next empty one when adding an electron or changing magnetic field.

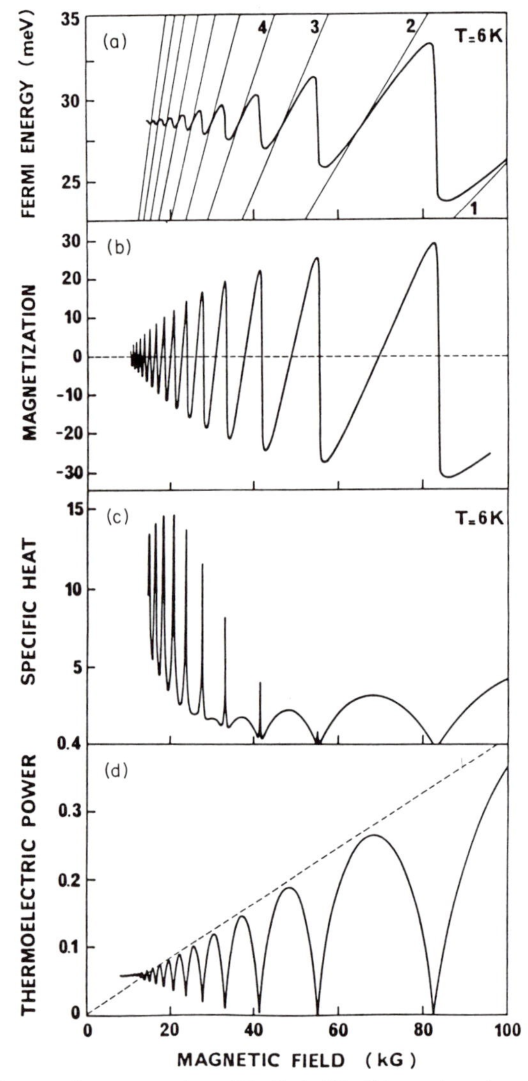

FIG. 70. Oscillatory phenomena in a 2D GaAs/GaAlAs system in a magnetic field. (a) Fermi level; (b) magnetization; (c) specific heat; (d) thermoelectric power. A Gaussian broadening of 0.5 meV is assumed (Reprinted with permission, after Zawadski and Lassnig[364], © 1984, Pergamon Press plc.)

When the magnetic field B is *parallel* to the heterostructure layers, the z motion in the confining potential is only slightly perturbed by the applied magnetic fields. Conversely, the usual cyclotron motion is inhibited by the confining potential. Therefore, the above description of 2D or 3D states collapsing into degenerate Landau level is invalid and the only effect of $B_{\parallel}$ is to increase the separation between low-lying confined quantum levels.[375] As mentioned above (Section 17), this effect was used by Englert *et al.*[320] to study the onset of intersubband scattering. However, in an extremely high parallel magnetic field, a new oscillatory effect sets in when the cyclotron orbit becomes of the order of, or smaller than, the confined wave function. This effect has been observed in QW structures and gives rise to a new form of SdH oscillation.[376]

We now concentrate on the most widely used magnetic field techniques in physics and assessment of 2D heterostructure systems, the Shubnikov–de Haas effect and the quantum Hall effect.

b. The Shubnikov–de Haas Measurements

The Shubnikov–de Haas effect, i.e., the oscillations of the longitudinal resistance in a quantizing field ($\omega_c\tau > 1$, $\hbar\omega_c > kT$) has long been a premium technique to study 2D systems. In Si MOSFETs, Fowler *et al.*[377] showed that the oscillation observed with changing electron number (by varying the gate voltage) has a constant period, which proves that each Landau level has the same number of states in 2D. This would not be the case in 3D due to the k_z motion, and can provide a signature for the 2D character of the electronic system. Another specific effect is the directional dependence of the SdH effect: only the perpendicular component of the B field confines the $x-y$ motion of carriers and determines the SdH oscillation period, which thus changes as $\cos\theta$ in 2D. This was already observed in early papers on modulation doping[131,132] (Fig. 71).

From Eq. (69) one can deduce the carrier density from the period of the SdH oscillation between two adjacent Landau levels $\Delta(1/B)$:

$$N_S = (e/h)/\Delta(1/B) \tag{71}$$

These measured values are usually in excellent agreement with those determined by Hall measurements (see Fig. 59), provided that no parallel conductance occurs in the GaAlAs barrier. The cyclotron mass does not enter the value of the oscillation period because the mass factor of ω_c is cancelled by the mass factor entering the determinations of E_F from the 2D density. However, the temperature dependence of the SdH oscillation amplitude allows one to extract an effective mass. Ando *et al.*[100] have calculated the low-field ($\omega_c\tau \lesssim 1$) oscillatory conductivity as

$$\sigma_{xx} = \frac{N_S e^2 \tau_f}{m^*} \frac{1}{1+(\omega_c\tau_f)^2}\left[1 - \frac{2(\omega_c\tau_f)^2}{1+(\omega_c\tau_f)^2}\frac{2\pi^2 k_B T}{\hbar\omega_c}\right.$$
$$\left. \times \cosh\left(\frac{2\pi^2 k_B T}{\hbar\omega_c}\right)\cos\left(\frac{2\pi E_F}{\hbar\omega_c}\right)\exp\left(-\frac{\pi}{\omega_c\tau_f}\right)\right] \tag{72}$$

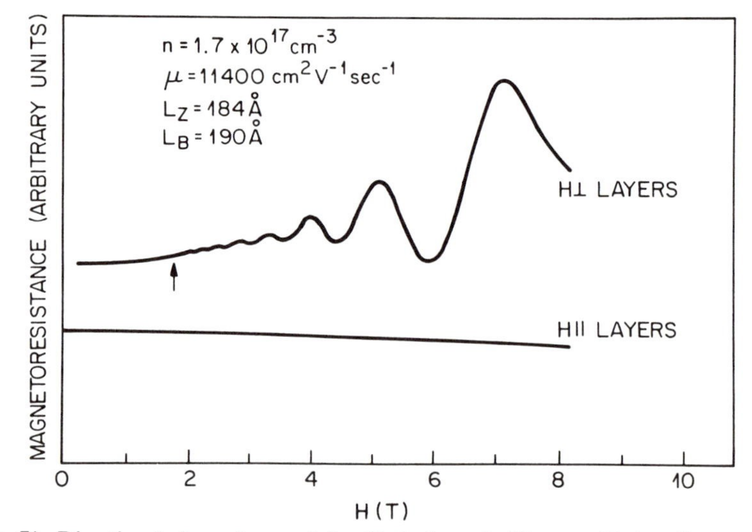

FIG. 71. Directional dependence of the Shubnikov–de Haas oscillation (from Störmer *et al.*[131] IOP Publishing Co.)

where τ_f is the scattering time corresponding to the dephasing of the Landau state.

From the temperature and magnetic field dependences of the oscillation amplitude, it is thus possible to extract m^* and τ_f. It must however be remembered that this τ_f is quite different from that deduced from Hall mobility measurements, as the small-angle collisions can play a much more important role in SdH oscillations, depending on the scattering mechanism. Harrang *et al.*[378] carried out a detailed comparison of both determinations.

In a number of cases, the spin splitting of the quantized levels has been observed[379,380] (see Fig. 73 below). The effective g^* value, defined as the distance between two spin-split states observed in the SdH measurements, is strongly enhanced as compared to the 3D value of -0.44.[381,382] This has been explained in terms of the *electron–electron correlation energy,* which depends strongly on the spin of occupied electronic states in the partially filled Landau states. g factors up to 5 in GaAs/GaAlAs have been measured,[379,383] whereas *direct* spin-resonance measurements[383] reveal *an uncorrelated spin splitting* with $g^* \simeq 0.2$. In this latter case, the g factor is at variance with the 3D value because of the lifting of the Kramers degeneracy of the conduction band by the confining electric field.

When two or more confining levels are occupied, the structure of the SdH oscillations becomes more complex.[131] If two levels are occupied, two oscillations will occur with two different periods due to the different densi-

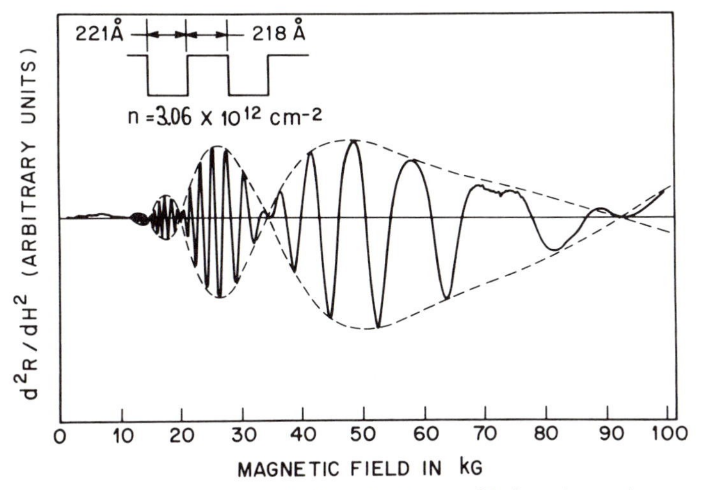

FIG. 72. Interference effect in the Shubnikov–de Haas oscillations due to the occupancy of two confined subbands. The second derivative of SdH oscillations is shown. After data reduction, calculations show that the two subbands are separated by 8.6 meV (from Störmer *et al.*[131] IOP Publishing Co.)

ties in the two levels [Eq. (71)]. This effect is shown in Fig. 72 and is very useful in ascertaining the number of occupied subband levels. It is also used to differentiate the conducting channels in TEGFET-like structures when parallel conductance in the GaAlAs is present.[384]

c. *Quantum Hall Effect*[385,386]

When observed at high magnetic field, at low temperatures and in high-purity samples, the SdH effect and the Hall effect exhibit a very marked departure from the usual behavior, a linear change of Hall voltage and smooth oscillations of the longitudinal magnetoresistance with magnetic field (Fig. 73).[387,388] Zeros of the longitudinal resistance are observed, corresponding to well-defined plateaus of the Hall resistance. Also remarkable is that these features exist over a wide range of sample parameters (electron density, mobility, temperature, . . .) and are not dependent on the exact shape of the sample.[386] Although first reported in Si-MOSFET samples,[387] this effect, the quantum Hall effect (QHE), has since seen an enormous development in the GaAlAs/GaAs system, the main reason being the lighter electron mass ($\approx 0.07 m_0$ instead of $0.19 m_0$ in Si), which increases by the same amount the cyclotron frequency for a given magnetic field, rendering the QHE obtainable at lower magnetic fields. The higher mobility of GaAlAs/GaAs heterostructures also leads to better resolved plateaus.

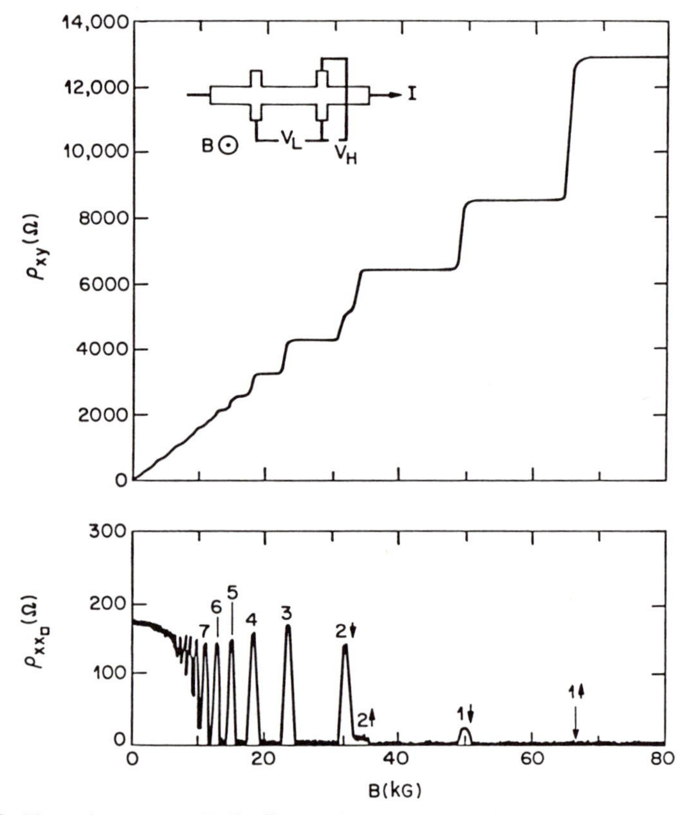

FIG. 73. Normal quantum Hall effect (NQHE) observed in the Hall resistance ρ_{xy} and parallel resistance ρ_{xx} of a selectively doped GaAs/GaAlAs interface at 50 mK. From the low magnetic field, where ρ_{xy} and ρ_{xx} display a typical "classic" behavior, the NQHE behavior develops from ~ 10 kG. Note the large $n = 1$ state spin splitting due to the strong electronic correlation (from Paalanen *et al.*[397]).

A standard Hall bar geometry can be used (see inset of Fig. 73). The current I is imposed while the magnetic field, perpendicular to the layer plane, is swept. The Hall resistance $\rho_{xy} = V_{\rm H}/I$ and the longitudinal resistance $\rho_{xx} = gV_L/I$, (where g is a geometric factor depending on the exact geometry) are measured. As usual, the resistivity tensor $\bar{\bar{\rho}}$ is related to the conductivity tensor $\bar{\bar{\sigma}}$ by $\bar{\bar{\rho}} = \bar{\bar{\sigma}}^{-1}$, i.e.

$$\rho_{xx} = \frac{\sigma_{xx}}{\sigma_{xy}^2 + \sigma_{xx}^2}; \qquad \rho_{xy} = \frac{\sigma_{xy}}{\sigma_{xx}^2 + \sigma_{xy}^2} \tag{73}$$

with $\sigma_{yx} = -\sigma_{xy}$; $\rho_{yx} = -\rho_{xy}$; $\sigma_{xx} = \sigma_{yy}$; $\rho_{xx} = \rho_{yy}$.

The *classical Drude model* can give a useful physical insight in the problem.[385] When no collisions are present, an electron moving classically in crossed electric (F_y) and magnetic field (B_z) describes a cycloid in the $x-y$ plane. The equation of motion is

$$m\frac{d^2\mathbf{r}_\perp}{dt^2} = -e\mathbf{F} - e\frac{d\mathbf{r}_\perp}{dt}\times\mathbf{B} \tag{74}$$

with the solution

$$\frac{dx}{dt} = \frac{F}{B}(\cos\omega_c t - 1); \qquad \frac{dy}{dt} = \frac{F}{B}\sin\omega_c t \tag{75}$$

where initial conditions $v_\perp = 0$ have been chosen. The *time-averaged motion* occurs in a direction *perpendicular* to the electric field, i.e., to the potential drop, and occurs with a *constant drift velocity F/B*. In that case, $\sigma_{xx} = \sigma_{yy} = 0$; $\sigma_{xy} = -\sigma_{yx} = N_S e/B$. The Hall voltage is given by $\rho_{xy} = B/N_S e$. There is no power dissipation in the absence of scattering and the movement of electrons is perpendicular to the electric field.

In presence of collisions Eq. (74) can be simply modified by adding a phenomenological friction term $m\boldsymbol{v}_\perp/\tau$, where τ is the collision time. The time-averaged motion now becomes

$$\left\langle\frac{dx}{dt}\right\rangle = \frac{F}{B}\frac{\omega_c^2\tau^2}{1+\omega_c^2\tau^2}; \qquad \left\langle\frac{dy}{dt}\right\rangle = \frac{F}{B}\frac{\omega_c\tau}{1+\omega_c^2\tau^2} \tag{76}$$

from which one deduces

$$\sigma_{xx} = \frac{N_S e}{B}\frac{\omega_c^2\tau^2}{1+\omega_c^2\tau^2}; \qquad \sigma_{xy} = -\frac{N_S e}{B} - \frac{\sigma_{xx}}{\omega_c\tau} \tag{77}$$

Quantum mechanically, electron motion (in reasonably low electric fields) occurs in Landau levels, i.e., closed cyclotron orbit. The electric field superimposes over this cyclotron motion a drift motion which is given by the same expression as in Eq. (74), but where σ_{xx} has now its quantum-mechanically computed value. The 2D DOS leads to a peculiar situation when the Fermi level is located between two Landau levels numbered i and $i+1$ (Fig. 69b). In such an occasion, no quasi-elastic scattering can occur at low temperatures; all states below the Fermi level are occupied, and an electron requires an energy $\hbar\omega_c$ (neglecting broadening) to be scattered to the next empty Landau state. In that case, $\sigma_{xx} = 0$ and σ_{xy} is given by the *classical collisionless value!* From the density of states *per Landau level,* eB/h, we deduce $N_S = ieB/h$ and therefore

$$\sigma_{xy} = i\frac{e^2}{h}; \qquad \rho_{xy} = \frac{1}{i}\frac{h}{e^2} \tag{78}$$

The Hall resistivity takes *quantized* values $(1/i)(h/e^2)$ whenever the Fermi level lies in between filled Landau levels. The remarkable feature of Eq. (78) is the fundamental nature of the parameters involved. The particular semiconductor does not even play a role. When compared to the observed SdH and QHE curves, the predicted values $\rho_{xx} = 0$ and Eq. (78) are extremely well verified;[386] resistivities as low as 10^{-10} $\Omega/\square$, equivalent to 10^{-16} Ωcm in 3D, have been measured. This value is three orders of magnitude lower than any other nonsuperconducting material. The accuracy of the corresponding plateau in ρ_{xy} is one part in 10^7. Such a high precision is of fundamental physical importance and can be used to calculate the fine-structure constant $\alpha = e^2/4\pi\varepsilon_0\hbar c$. The QHE resistance $R = (1/i)R_K$ from samples with different origin has been measured to be the same within a relative uncertainty of $<5 \times 10^{-8}$. This is more precise[390,390a] than the earlier SI reference resistors, so that the QHE was adopted[390b] in 1990 as a new resistance standard, i.e., by definition $R_K = 25812.807$ Ω.

There is, however, a major difficulty in the explanation just given above. It cannot explain the existence of a *finite width* for the QHE plateaus and for the zero longitudinal resistance dips: if there are no states between the successive conducting Landau levels, the Fermi level jumps from the last-occupied Landau level to the next higher-lying one. The Fermi level never lies in between conducting Landau levels as the magnetic field is swept and quasi-elastic scattering is always present. Therefore, one has to invoke the existence of localized, i.e., non-current-carrying, states in the tails of the current-conducting Landau levels. The existence of such localized states is well justified within our present understanding of disordered systems, the disorder here being due to random distribution of defects, impurities, or to the random interface topology. When varying the magnetic field or the number of charge carriers, the Fermi energy will either lie in delocalized, current-carrying states where quasi-elastic scatterings are possible, with $\rho_{xx} \neq 0$, or in localized states, in which case the lower-lying current-conducting charges will require a finite energy to be scattered into an empty conducting state. In such a case $\rho_{xx} = 0$ at low temperatures and the Hall resistance ρ_{xy} retains a constant value due to the constant number of current-conducting carriers while the Fermi energy is swept through localized states.

The new, astonishing phenomenon is the value of ρ_{xy}, exactly equal to $(1/i)h/e^2$, as if all electrons were in conducting states, independent of the fact that a fraction of them are in localized states, which crucially depends on sample disorder and therefore should vary from sample to sample. Several explanations have been given to explain this amazing result: it was shown by calculating the current carried by delocalized carriers in the presence of disorder that their speed is modified in order to exactly com-

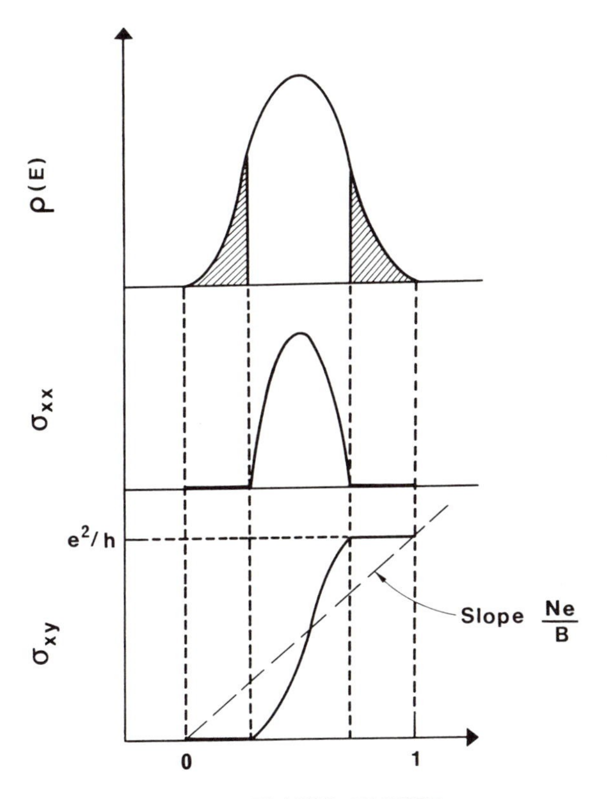

FIG. 74. Density of states (top), longitudinal (middle), and Hall (bottom) conductivities for a single Landau state as a function of the filling factor $\mu = N/d$, where N is the electron density and d the degeneracy of the Landau state. The shaded areas indicate the localized states. Note that the Hall conductivity of the filled Landau state is the classical, collisionless value e^2/h, independent of the fraction of localized states (Reprinted with permission, after Aoki and Ando,[391] © 1981, Pergamon Press plc.)

pensate for the lack of conduction of the localized electrons[391,392] (Fig. 74). A classical image is that of an obstacle in a pipe carrying a fluid: Around the obstacle, the fluid will flow faster than in the rest of the pipe, in order to conserve a constant fluid flow along the pipe. It is however clear that such an important feature of QHE must be due to first-principles arguments, which were outlined by Laughlin.[393] He showed the accurate quantization of QHE to be due to two effects:

(1) gauge invariance of the interaction of light with matter; and
(2) the existence of a mobility gap.

From these two assumptions, Laughlin was able to demonstrate that, whenever the Fermi level lies within a mobility gap, $\rho_{xx} = 0$ and $\rho_{xy} = h/ie^2$. Detailed discussions of the significance of gauge invariance were given by Laughlin,[393] Aoki,[394] and Hajdu.[395]

We therefore now have a satisfying explanation of the QHE: Plateaus are due to Fermi levels situated in localized states due to disorder. Well-defined values of the QHE resistance independent of sample and detailed experimental conditions are due to the adjustment of conducting carriers to compensate for localized electrons. The detailed shape of the observed features, however, depends on the sample parameters, and has opened the way to numerous fundamental studies of 2D systems. The localized fraction of the DOS determines the width of the plateaus. These have been observed with up to a 95% width with 5% transitions. The width of the plateaus has been correlated with sample mobility. High-precision measurements of $\rho_{xy}(T)$ have demonstrated its dependence upon the residual $\rho_{xx}(T)$ (Eq. 77).[396–398] The temperature and voltage dependence of ρ_{xx} gives information about the transport mechanism in localized states.[397] The breakdown of the QHE as a function of applied voltage has also been studied and explained by various heating mechanisms.[399,400] The influence of sample shape and contact interconnections on the sample has been studied in a number of fascinating experiments.[401]

Going to lower temperatures and in the extreme quantum limit ($\hbar\omega_c > E_F$), novel correlation effects have been observed in the 2D electron gas as the Coulomb interaction between electrons exceeds their kinetic energy, which is almost completely frozen by the magnetic field and the heterojunction confining potential.[402] The extreme quantum limit is characterized by the filling factor $\nu = N_S/d \approx 1$, where d is the degeneracy of Landau levels. In that situation, all electrons are in at most one or two Landau levels. The signals observed under such a situation are shown in Figs. 75 and 76, for highly perfect samples. At even ν, the Fermi level is between Landau states of different n, whereas, at odd ν, it resides between the spin levels of a given Landau level. The spin splitting in GaAs being much smaller than the cyclotron splitting $\hbar\omega_c$, the QHE is better observed for even values of ν (top curve of Fig. 75a). However, new plateaus appear at the lower temperatures for *fractional* values of ν.[403–405] Many rational values p/q have been observed. Until 1987 all values found had q odd, but recently a plateau has been observed[405a] for $p/q = 5/2$. These plateaus can be very well defined, with the $\nu = \frac{1}{3}$ plateau defined to better than one part in 10^4.

Such a new effect, named the fractional quantum Hall effect after its resemblance to the integer quantization of the normal QHE (NQHE) described above, cannot be explained in the framework of NQHE. It must, however, rely on a similar type of explanation; i.e., it requires the Fermi

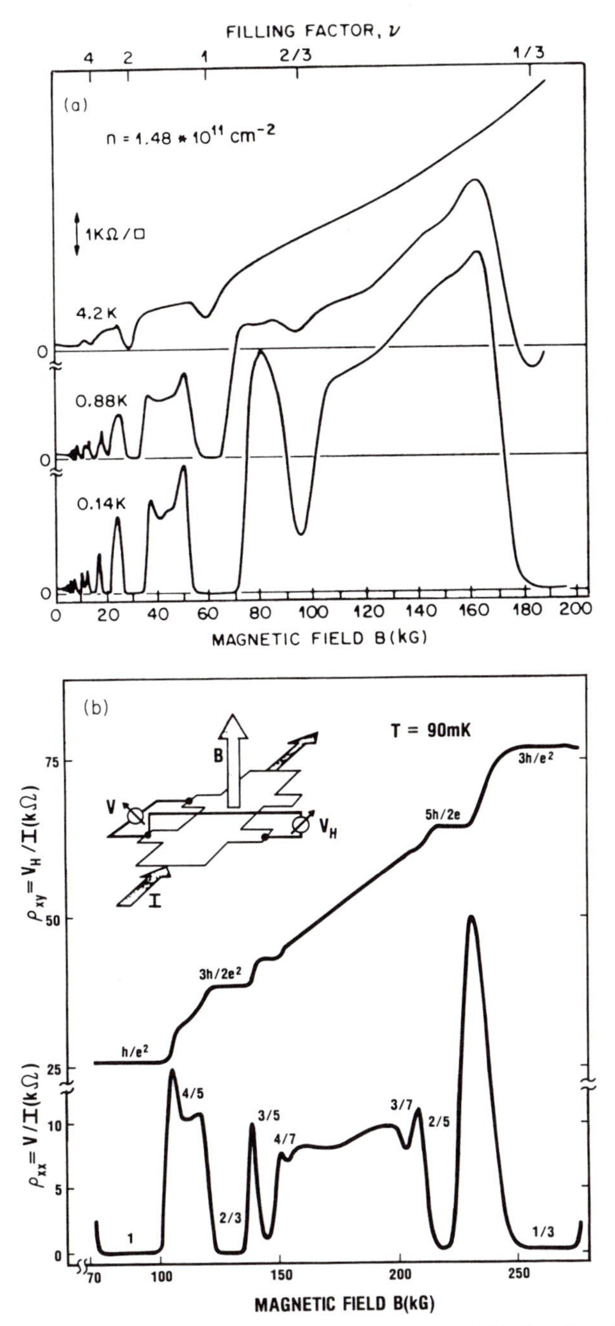

FIG. 75. Normal (NQHE) and fractional (FQHE) quantum Hall effect for a GaAs/GaAlAs sample. (a) At 4.2 K, only the NQHE is observed, with a small $n = 3$ dip in the SdH curve, as is expected from odd values of the filling factor (see text). When lowering the temperature, dips develop at fractional values of the filling factor (from Tsui *et al.*[389]). (b) Observed SdH and quantum Hall effect at 90 mK (courtesy H. L. Störmer, AT&T Bell Laboratories).

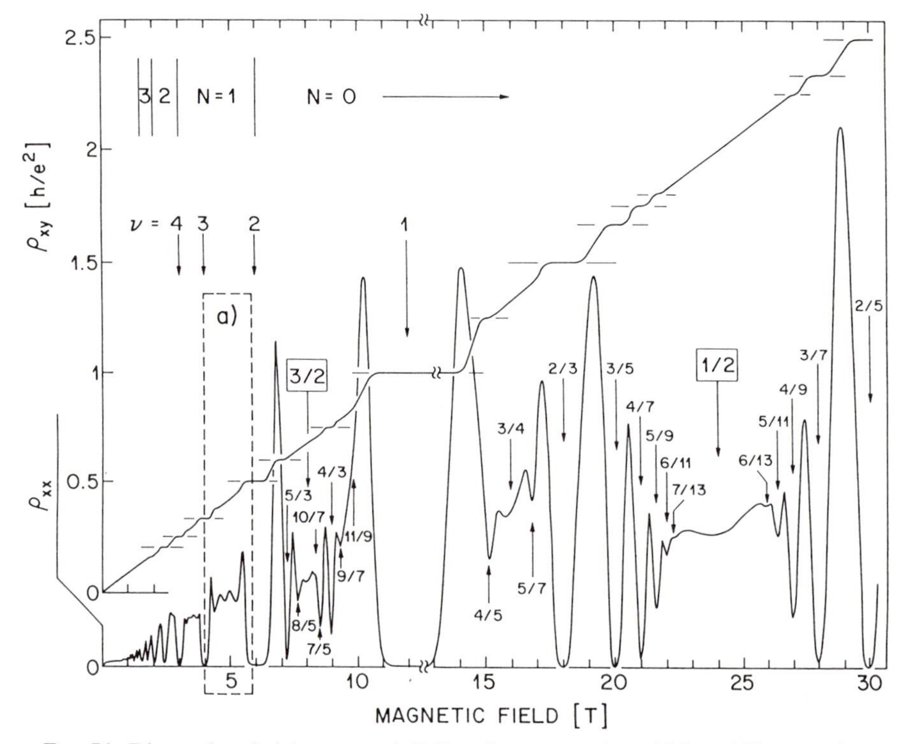

FIG. 76. Diagonal resistivity ρ_{xx} and Hall resistance ρ_{xy} in a high-mobility sample at temperatures below 150 mK. The high field ρ_{xx} trace is reduced in amplitude by a factor 2.5. Filling factor ν and Landau levels indicated (from Willett *et al.*,[405a]).

level to lie for fractional ν's in a true gap or in a mobility gap. The long-predicted Wigner solidification of an electron gas could explain such anomalous behavior if the solid would preferentially form at given fractional values of ν. However, all calculations give smooth variations of the cohesive energy of the solid Wigner crystal on the filling factor.[406] Experiments would also reveal the pinning of the solid at the existing potential fluctuations due to disorder, and yield nonlinear current–voltage characteristics, which have never been observed. A numerical calculation for a finite system of 4, 5, and 6 electrons confined in a box in a magnetic field has shown that minima in the total energy could exist a fractional values of ν, significantly lower than that of a Wigner crystal which is therefore *not* the ground state of the system.[407] Laughlin has proposed a new many-particle wave function for the description of the ground state, which gives states at filling factors of $1/q$, where q is odd.[408] The elementary excitations from this ground state are situated above a gap $\approx 0.03e^2/r_c \approx 5$ K at

150 kG and have e/q charge. These quantum states form an incompressible fluid with no low-lying excitations, implying a flow with no resistance at $T = 0$. This model, therefore, explains satisfactorily at present the phenomena observed for $\nu = 1/q$ and, due to particle–hole symmetry, those observed at $\nu = 1 - 1/q = (q - 1)/q$. Some extensions of the theory have been proposed to explain the other observed fractional values of ν.[409] The field of FQHE is still very vigorous and aims to attain a fuller basic understanding as well as to explore all its implications: statistics, phase condensations, crystallization, etc. The quantum Hall effect has become a whole new discipline in solid state physics. Two books have appeared[410,411] and a guide to the extensive literature can be found as a Resource Letter.[412]

The specific features of the NQHE and FQHE have been observed in several systems other than GaAs/GaAlAs, such as GaInAs/InP,[413] GaSb/InAs.[414] They were also observed for holes in the GaAs/GaAlAs system. A general trend is that the NQHE is better observed (wider plateaus) in samples with some degree of disorder, i.e., nonoptimum mobilities, whereas the FQHE requires samples of the utmost purity.[415] A converging argument is provided by frequency-dependent measurements: Long *et al.*[416,417] were able to switch from NQHE to FQHE with increasing frequency, which diminishes the length scale of transport and therefore localization effects. The very different origins of NQHE and FQHE are emphasized by the experimental facts, the former requiring disorder and the latter being based on the intrinsic properties of the 2D electron gas in a magnetic field.

CHAPTER V

Applications of Quantum Semiconductor Structures

21. Electronic Devices Based on Parallel Transport

The possibility of producing near-perfect barriers and ultrathin structures has given rise to the invention and demonstration of many new electronic semiconductor devices with higher performances in standard applications or with new functionalities. So far, one has been limited by either material problems or electrostatic behavior: Classically, potential barriers are due to metal-semiconductor association (Schottky voltage barriers), insulator-semiconductor association (potential barrier due to band-gap difference), or *p-n* junctions, where the potential barrier originates in the electrostatic buildup in the depletion zone. The two former types of potential barriers, although abrupt, are associated with poor interface properties, in particular mobility, due to the imperfect crystal growth of such material pairs; with the exception of some Si cases (such as $CoSi_2/Si$, where the Si properties are, however, still degraded), these materials are grown amorphous over Si with interface morphology disorder degrading the Si transport properties[419]. *p-n* junctions do not adversely affect material properties, but the potential barrier extends over the finite extension of the depletion zone, determined by doping levels, which in turn are limited by metallurgical considerations. This restricts the barrier minimum length to about 50 nm. Semiconductor heterojunctions with their perfect interface properties allow abrupt potential barriers—and at the same time even improved transport properties—thanks to the carrier transfer into an undoped region as described in Section 8.

One classification of these devices is obtained by considering the principal current flow through the device: In this section we consider devices with transport parallel to the interfaces; in the next section we describe devices based on perpendicular transport.

New implementations of the *field effect transistor* (FET) are the main representatives of the first kind. In FETs the current between two contacts, source and drain, is controlled by the voltage of a third electrode, the gate, which is electrically isolated from the source-drain current. The function of such a device is simple in principle. In a logic circuit it functions as a switch: depending on the gate voltage the connection between source and drain is broken or closed. In analogue circuits a small time-varying voltage signal on the gate creates a time-varying current between source and drain, and since the gate current is ideally a pure displacement current, a very small input power can be amplified.

The most important "classical" transistor type (with respect to production volume) based on III-V compound semiconductors is the GaAs metal epitaxial semiconductor field effect transistor (MESFET). In a MESFET the source-drain current is carried in a thin highly doped epitaxial semiconductor layer. The current is controlled by a gate which forms a Schottky barrier on the semiconductor and therefore—depending on the applied gate voltage—depletes more or less the semiconductor layer of electrons under the gate.

A new heterostructure type of FET has been developed over the last few years: the two-dimensional electron gas field effect transistor (TEGFET) (also called high electron mobility transistor (HEMT), modulation doped field effect transistor (MODFET), or selectively doped heterojunction transistor (SDHT) depending on manufacturer). It has features in common with both MESFETs and metal-oxide-silicon field effect transistors (MOSFETs).

The structure is based on the heterojunction between AlGaAs and GaAs as described in Section 8. Its essential structure is shown in Fig. 77 and consists of a semi-insulating substrate on which is first grown a buffer layer of nonintentionally doped GaAs; on top of this is grown a thin (~50 nm) layer of $Al_xGa_{1-x}As$, part of which is rather heavily ($\sim 10^{18}$ cm^{-3}) *n*-type doped. The gate metal forms a Schottky barrier to the AlGaAs and by making the ternary layer thin enough, the gate can completely deplete the AlGaAs layer of electrons. Then the density of electrons on the GaAs side of the heterojunction is controlled by the voltage applied to the gate, so that the current between the source and the drain

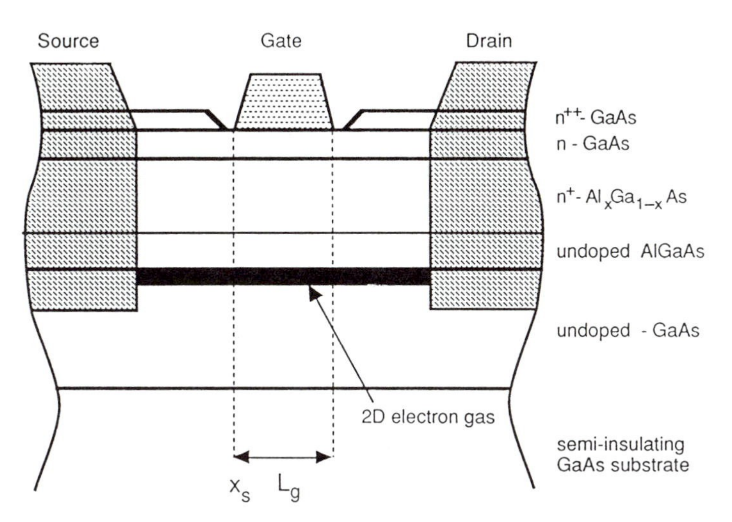

FIG. 77. Schematic structure of a two-dimensional electron gas field effect transistor (TEGFET). The source and drain contacts are diffused to make contact with the channel.

contacts can be controlled by the gate voltage.

The impact of the heterostructure is that the potential is now abrupt (as compared to the Schottky potential drop at the metal-semiconductor interface in the MESFET), which leads to a more vigorous control of the channel-charge density by an externally applied voltage.

a. Simplest Model of FET

In the simplest description the charge of the electrons in the channel varies linearly with the gate voltage as in a capacitor:

$$n_s Z L_g = C(V_g - V_{\text{th}}) \tag{78}$$

where n_s is the electron density per surface area, L_g is the gate length, Z is the gate width, $C \cong \varepsilon_r \varepsilon_0 Z L_g / d_{\text{AlGaAs}}$, and V_{th} is a threshold voltage. In normal transistor operation the source-drain voltage is so large that almost all the electrons under the gate move with their saturation velocity v_s, independent of drain voltage, so that the current between source and drain is given by

$$I_{\text{DS}} \cong n_s Z e v_s \tag{79}$$

so that I_{DS} is also roughly linear in gate voltage. A very important parameter is the intrinsic transconductance:

$$g_{m0} = \frac{\partial I_{\text{DS}}}{\partial V_g} \tag{80}$$

which in this simple model is given by (78) and (79) as

$$g_{m0} \cong Zev_s \frac{dn_s}{dV_g} = C\frac{v_s}{L_g} \tag{81}$$

or

$$\frac{g_{m0}}{C} = \frac{v_s}{L_g} = \frac{1}{\tau} \tag{82}$$

where τ is the transit time for an electron under the gate. It is obvious that for fast operation it is necessary to reduce as much as possible the gate length and to find structures and materials that have a high average velocity under the gate.

For reference we present a slightly better model in which the velocity of the electrons in the channel as a function of electric field has the dependence shown in Fig. 78. Below a critical field E_c the velocity of the electrons is proportional to the field $v = \mu E$ and above is a constant saturation velocity v_s. Furthermore, it is assumed that the voltage in the channel varies along the channel, $V_{\text{ch}} = V(x)$, but that the density of electrons follows the law described by (78). The current is then given by

$$I = Zen_s(x)v(x) = \frac{C}{L_g}[V_g - V_{\text{th}} - V(x)]\mu\frac{dV}{dx} \tag{83}$$

as long as $dV/dx \leq E_c$. Since the current is conserved we can integrate the equation by separation:

$$I(x - x_s) = \mu\frac{C}{L_g}\int_{V(x_s)}^{V} [V_g - V_{\text{th}} - V']\,dV' \tag{84}$$

where x_s is the position of the entrance to the channel on the source side. If we define the saturation current to be the current for which the field on the drain side of the channel $x_s + L_g$ just reaches E_c, we find the implicit equation for I_{sat}:

$$I_{\text{sat}} = g_{m0}\{\sqrt{(V_g - V_{\text{th}} - R_s I_{\text{sat}})^2 + E_c^2 L_g^2} - E_c L_g\} \tag{85}$$

where

$$g_{m0} = \frac{C\mu E_c}{L_g} = \frac{Cv_s}{L_g} \tag{86}$$

Here R_s is the resistance of the materials between the source contact and the entrance to the gate at x_s, and the source voltage is at 0.

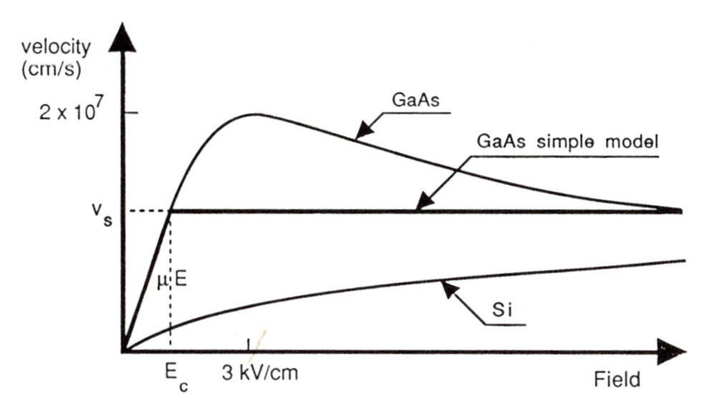

FIG. 78. Schematic velocity field characteristic for electrons in GaAs and Si and the simplified model applied in the text.

It is instructive to study two limits of (85):

$$I_{\text{sat}} \cong \frac{g_{m0}}{1 + R_s g_{m0}} (V_g - V_{\text{th}} - E_c L_g) \quad \text{for } E_c L_g \ll V_g - V_{\text{th}} - R_s I_{\text{sat}} \tag{87}$$

i.e., a linear behavior as in (78) but with an extrinsic transconductance $g_m = g_{m0}/(1 + R_s g_{m0})$ for a short highly conductive channel; and

$$I_{\text{sat}} \cong \frac{g_{m0}}{2\, E_c L_g} (V_g - V_{\text{th}})^2 \quad \text{for } E_c L_g \gg V_g - V_{\text{th}} - R_s I_{\text{sat}} \tag{88}$$

i.e., a quadratic dependence on V_g for a long channel near pinch-off.

This rather crude model, which is a standard model also for MOSFETs, describes quite well the basic characteristics of a normally functioning TEGFET.[420] Note that when the intrinsic transconductance becomes very large, the performance of the transistor depends strongly on the access resistance R_s (Eq. (87)), so that it is increasingly important to reduce this parameter.

For a TEGFET one can use the same basic small-signal equivalent circuit model as for other field effect transistors shown in Fig. 79. The drain current is represented by a current generator which is controlled by the voltage across the gate capacitor. Circuit theory leads to the expression for the current gain:

$$h_{21} = \frac{I_{\text{D}}}{I_g} = \frac{g_{m0}}{j\omega C} = \frac{f_{\text{T}}}{jf} \tag{89}$$

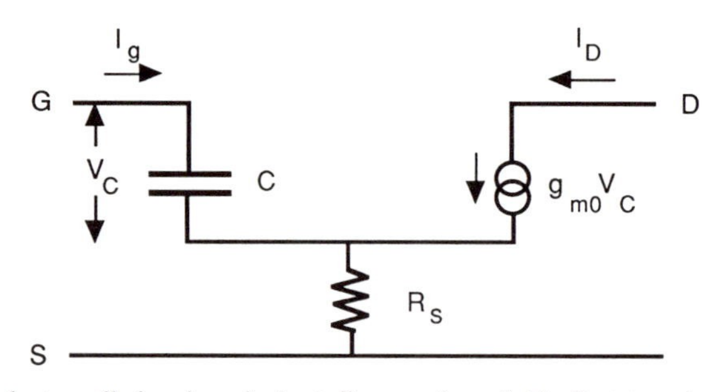

FIG. 79. Simplest small-signal equivalent diagram for a field effect transistor. C is the gate to source capacitance, g_{m0} is the intrinsic transconductance, and R_s is the access resistance between source and gate entrance.

where the cut-off frequency $f_T = g_{m0}/2\pi C$. This again shows the importance of keeping g_{m0}/C as large as possible for obtaining gain at high frequencies. The elementary circuit model is too simplified to give a finite maximum available power gain, which falls off with frequency as $(f_{max}/f)^2$, where f_{max} depends on various other parasitic elements in the equivalent circuit.[421] (see Sect. 21c). Several of the parameters mentioned are improved in a heterojunction FET relative to a MESFET: The mobility is higher, increasing g_{m0}, and the smaller fringe capacitance, which enhances the effective gate length over the geometrical one, increases the cutoff frequency f_T; also the higher mobility reduces the access resistance R_s, which increases f_{max}.

b. More Refined Models

Charge Transfer

More refined models for the dependence of the charge in the channel on the gate voltage have been reported. The fullest treatment is obtained by solving the self-consistent problem of the Schrödinger and Poisson equations along the lines described in Section 8c. The main difference is in the boundary conditions, which for the gate is that the conduction band is $eV = \Phi + eV_g$, where Φ is the Schottky barrier of the gate-semiconductor interface and V_g is the applied gate voltage. Inside the semiconductor, the Fermi level is constant. The self-consistent solution then yields the charge n_s and its distribution in the semiconductor for a given voltage,[143] and by taking the derivative $e\, dn_s/dV_g$ one obtains the gate-channel capacitance C. Results from such calculations are shown in Figs. 80a and 80b. One sees that below pinch-off ($V_g < -2V$ for 3468) there are no electrons in the system; with increasing gate voltage the electrons begin to go into the channel in GaAs (curve a) and the density increases almost linearly with V_g

as in the simple model (78); around $V_g = -1.5\text{V}$, however, the channel density tends to saturate because the additional electrons now either neutralize donors in the AlGaAs barrier (curve d) or enter the AlGaAs as free electrons (curve b). This means that it becomes less and less possible to control the density of useful electrons, that is to say, those that are in the high-mobility channel in GaAs, which has an important influence on capacitance and transconductance as can be seen from Fig. 80b. By definition, the derivative for all electrons in the system (curve a) gives the gate capacitance. Note that it is actually not very constant: from pinch-off it rises rapidly, tends to flatten out a little around -1.5V and then increases again, above all because of the electrons that neutralize the donors in

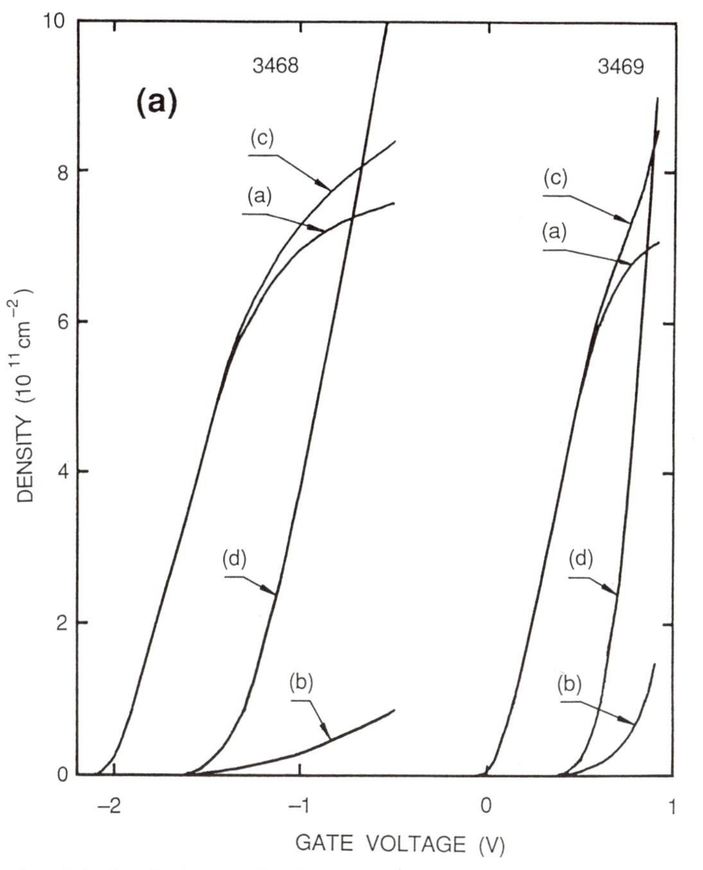

FIG. 80a. Calculated electron density versus gate voltage for the transistors shown in Fig. 22b. (a) free-electron density in the GaAs channel; (b) free-electron density in AlGaAs; (c) total density of free electrons; (d) density of neutralized donors in AlGaAs.

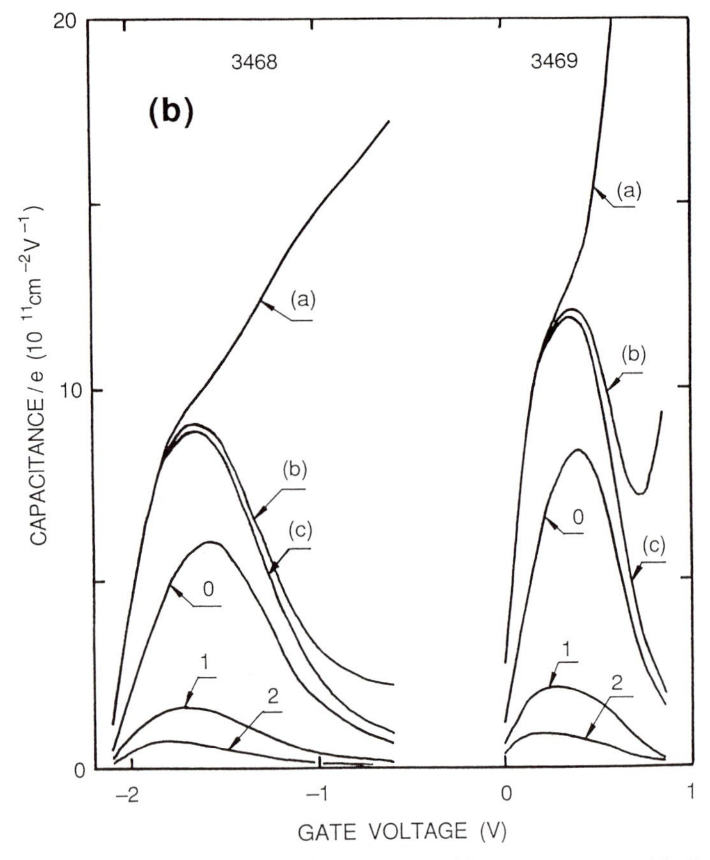

FIG. 80b. Calculated capacitance of (a) all electrons; (b) all free electrons; (c) all electrons in the channel in GaAs; and (0, 1, 2) electrons in the lowest three channel subbands (Vinter[143]).

AlGaAs. The curves corresponding to the channel electrons (c) and all the free electrons (b) show a different behavior: an increase, a maximum around −1.5V, followed by a decrease. Since current from source to drain can be carried only by free electrons, one can understand from these curves that qualitatively the transconductance must show a behavior similar to curve (b). Therefore, the transistor cannot function efficiently beyond the voltage for maximum transconductance, which is seen to be due to the fact that the density of electrons in the channel cannot surpass 7 or 8×10^{11} cm^{-2} because the barrier between the AlGaAs and the GaAs is less than ~300 meV.

Velocity Effects

In order to obtain a high speed and a high f_T it was seen that the time spent by an electron under the gate should be as short as possible (Eqs. 82 and 89). Compared with Si devices two factors favor GaAs: Even though the source-drain fields under the gate are large and bring the electrons far out of the ohmic region, the average velocity is seen from Fig. 78 to be considerably larger than in Si. Comparing a MESFET structure with a modulation doped structure, the higher mobility in the undoped channel of a heterostructure transistor further increases the maximum frequencies for current gain[421a].

Another effect which is important when gate lengths shrink below 1 μm is the overshoot phenomenon: The maximum velocity shown in Fig. 78 is obtained in the stationary regime and is essentially determined by the transfer of hot electrons in the Γ-valley of the GaAs conduction band to the higher-lying L- and X-valleys; the transfer is only possible for electrons which have sufficient energy for the transfer; however, the electrons that have not reached this threshold may have a velocity several times the stationary velocity and may therefore in short structures pass the gate before they are transferred. The light mass of Γ-valley electrons in GaAs again favors that material over Si in that respect and the high mobility of heterostructure FETs is even more advantageous[421a] as can be seen from the fact that TEGFET performances increase substantially when the temperature is decreased.

c. Performance Analysis

Microwave Devices

In analog applications three factors are of prime importance: power gain, noise, and total microwave output power. The limited useful gate-voltage excursion demonstrated in Figs. 80a and 80b prevents the use of a simple heterojunction transistor for power amplification. A hybrid multiple channel MODFET, in which the active zone is a number of modulation-doped quantum wells, has been proposed and an output power of 1 W at 30 GHz has been obtained[422]. However, the most interesting application of heterojunction transistors is for low-noise very-high-frequency amplification. The high cutoff frequencies f_T are necessary to obtain power gain since the maximum available power gain can be expressed as[422a]

$$f_{\max} \cong \frac{f_T}{2(G_0(R_g + R_s) + 2\pi f_T C_{dg} R_g)^{1/2}} \tag{89a}$$

where $G_0 = \partial I_D / \partial V_{DS}$ is the output conductance, R_g is the resistance of the metal gate, and C_{dg} is a parasitic capacitance between gate and drain. From the expression it is clear that reducing parasitic resistances and capacitances is crucial for increasing the maximum frequency for power gain. The microwave noise performance as measured by the Fukui formula for the noise figure[422b]

$$\mathrm{NF} \cong 10 \log\left[1 + K_f\left(\frac{f}{f_T}\right)(g_{m0}(R_g + R_s))^{1/2}\right] \tag{89b}$$

where K_f is an empirical constant, is also seen to depend crucially on f_T and the access source resistance R_s. Both of these are consistently improved in a heterostructure compared with a MESFET.

For more detailed accounts on the functioning and performances of AlGaAs/GaAs TEGFETs we refer to Chapters 3 and 4 in Refs. 20 and 21. A few recent results are worth mentioning, however: Technological improvements and especially the reduction of gate length to submicron sizes have pushed the frequencies of operation and gain gradually higher and the noise figures to previously unimaginable values. Amplification at 94 GHz has been demonstrated in a 0.25 μm transistor,[423] and noise figures down to 2.7 dB at 60 GHz have been reported.[424] A 0.1 μm gate TEGFET has shown[425] an extrapolated unity current gain cutoff frequency $f_T = 110$ GHz.

It can be seen in Figs. 80a and 80b that the transconductance levels off when the electrons added by an increase in gate voltage start entering the AlGaAs barrier material. To enhance the transconductance and performance it is therefore advantageous to increase the barrier height. This has led to two other families of TEGFETs: the AlInAs/GaInAs heterojunction field effect transistors in which the ternary alloys are lattice-matched to the InP substrate; and the pseudomorphic TEGFETs, in which a thin epitaxial strained layer (~20 nm) of GaInAs with an In content of 15–25% is grown between the GaAs buffer and the AlGaAs barrier of a normal GaAs-based TEGFET. In both cases one profits from higher velocities in the GaInAs channel and of a higher confinement barrier, which increases the current carrying capability and transconductance of the transistor. The former type was first fabricated by Pearsall *et al.*[426] and Chen *et al.*[427] already in 1982, and recently very high frequency transistors with 0.1 μm gate length have been reported[428] with current gain cutoff frequencies f_T of 170 GHz. In the pseudomorphic structures current gain cut-off frequencies above 120 GHz have been obtained in 0.2 μm gate length devices[429] and amplification has been reported at 94 GHz.[430]

Figure 80c shows the dependence of the cutoff frequency f_T as a function of gate length and materials. The general trend, for each material pair, is an increase in cutoff frequency inversely proportional to the gate length as expected from Eq. (82) except for very small lengths, when the geometri-

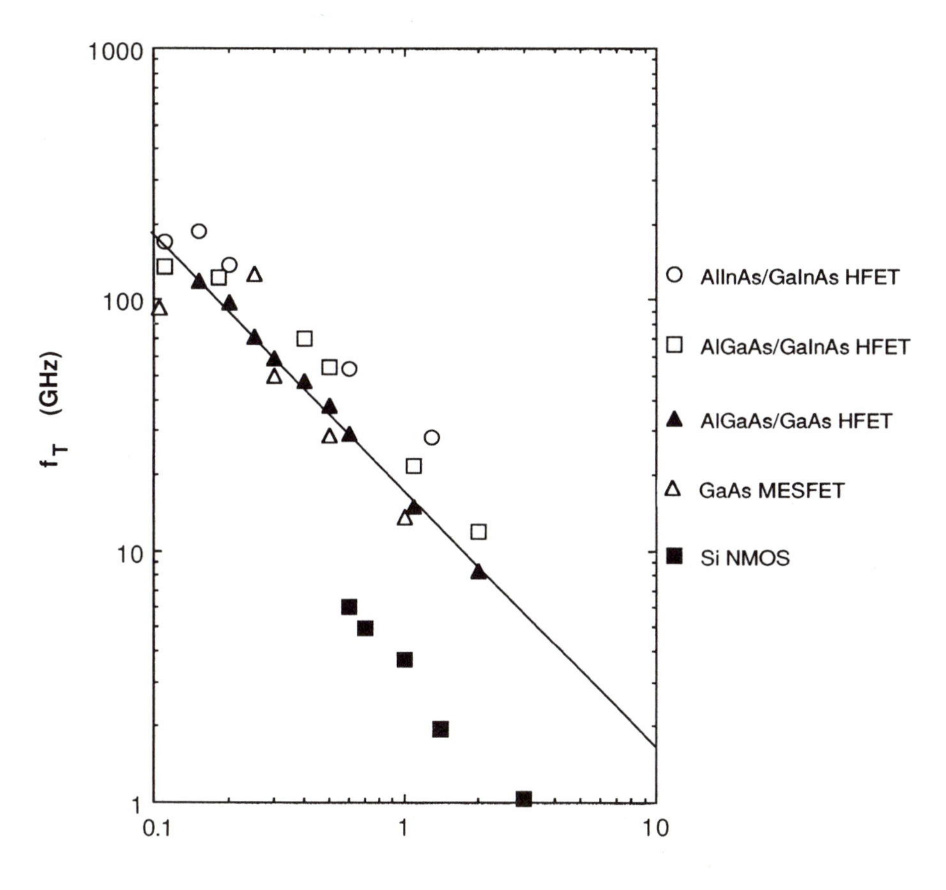

FIG. 80c. Unity current gain frequency f_T versus gate length for several families of FETs.

cal gate length becomes smaller than the effective gate length owing to fringe capacitances in the gate-source capacitance.

Digital Devices

For digital applications the main issues are the switching delay and the associated power consumption: In order to charge a capacitance on the input of a stage it is necessary to furnish a charge $C\Delta V$, where ΔV is the voltage swing to change the state of the stage. To do this rapidly a high driving current must be delivered, which increases power dissipation. A figure of merit is therefore the power-delay product per stage which, however, depends intimately on the circuit described. In a very simplified general way the advantages of GaAs and heterojunction transistors again lie in their high g_{m0}/C ratio or their high f_T which allows a high current in the "on" state to charge a relatively small capacitance assuring high speed. Another way of describing this advantage is to note that since the gate-channel separation can be very short owing to the abrupt barriers, the intrinsic capacitance is relatively large, so that the contribution of parasitic capacitances in the device and in the circuit is less important. Furthermore, the high mobility of GaAs and even more of heterojunction transistors allows for a smaller logic voltage swing for a given logical noise margin, so that power dissipation can be kept low.

Naturally these general considerations will be modified by the exact applications and the state of availability, and it is often the case that the best Si technology can deliver circuits comparable with GaAs in speed. Those better (advanced bipolar) Si circuits require a much more complex technology and are not available on the market. For this reason the fast circuitry of the CRAY III computer is fabricated in GaAs technology[430a].

The lower logic voltage permitted by the high mobility can only be exploited at the cost of a very stringent control of threshold voltages of all devices on a chip. Figure 81 shows the distribution of HEMT threshold voltages observed across a full three-inch-diameter wafer[430b]; the measured standard deviation of 11 mV corresponds to a thickness control of better than 1 nm across the whole wafer and a similar control of doping in the AlGaAs layer.

The level of integration in GaAs circuits has increased tremendously; the most recent announcement claims process technologies that permit one million transistors on a chip[430c]. While this density is surpassed by the most advanced Si bipolar complementary MOS circuits, the latter have a longer gate delay at comparable integration and require a much more complex process technology.

In the simplest small-scale integrated circuits the fastest switching delays obtained are below 10 ps per gate at room temperature[430d] and 5.8 ps at 77 K.[430e] Several examples of large-scale integrated circuits in HEMT technology can be found in Ref. 430a. For a comparison of advantages and disadvantages of GaAs versus Si technologies for integrated digital circuits we refer to the paper by Solomon.[430f] For a detailed account of MESFET and TEGFET technology we also refer to two very recent volumes in the Semiconductor and Semimetals series.[430g]

d. Variants of Heterojunction FETs

A variant of the heterojunction field effect transistors is the semiconductor insulator semiconductor field effect transistor (SISFET, or MIS-like

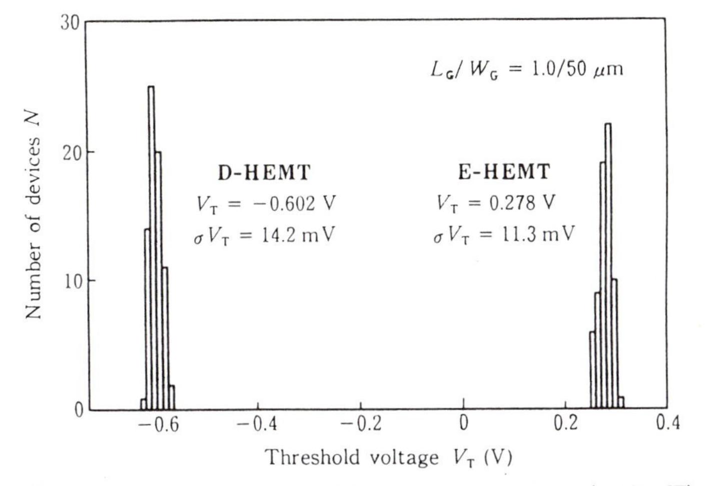

FIG. 81. Histograms of measured threshold voltages of normally on (D-HEMT) and normally off HEMTs (E-HEMT) across a full three-inch-diameter wafer (Reprinted with permission from IEEE, from Abe *et al.*,[430b]).

FET). It consists of a highly doped GaAs gate on top of an undoped barrier of AlGaAs, which separates the gate from the channel in undoped GaAs. This structure resembles very much the Si MOSFET, only the AlGaAs creating a barrier of about 300 meV is supposed to replace the oxide insulator, which creates a barrier of more then 3 eV. For this reason the SISFET is expected to work only at low temperatures to limit leakage by thermoemission. On the other hand, the SISFET is interesting for applications involving complementary circuits because both *p*- and *n*-channel transistors can be fabricated on the same chip and also because of the much smaller fluctuation in threshold voltage owing to the undoped barrier. Results on SISFETs were first reported by Solomon *et al.*[431] and Matsumoto *et al.*[432] Figure 82 recapitulates the conduction band diagram of several recent field effect devices.

The relatively low heterojunction barriers make it possible for electrons in the channel to be accelerated by the source-drain field to energies high enough to permit a real-space transfer into the barrier material or across the barrier. This effect is exploited in several structures: in real-space transfer diodes[433] in which electrons in one channel are heated so much that they transfer to a cooler channel, thus reducing the current for an increasing electric field, leading to a negative differential resistance; in three- and four-terminal devices such as the NERFET[434] (negative resis-

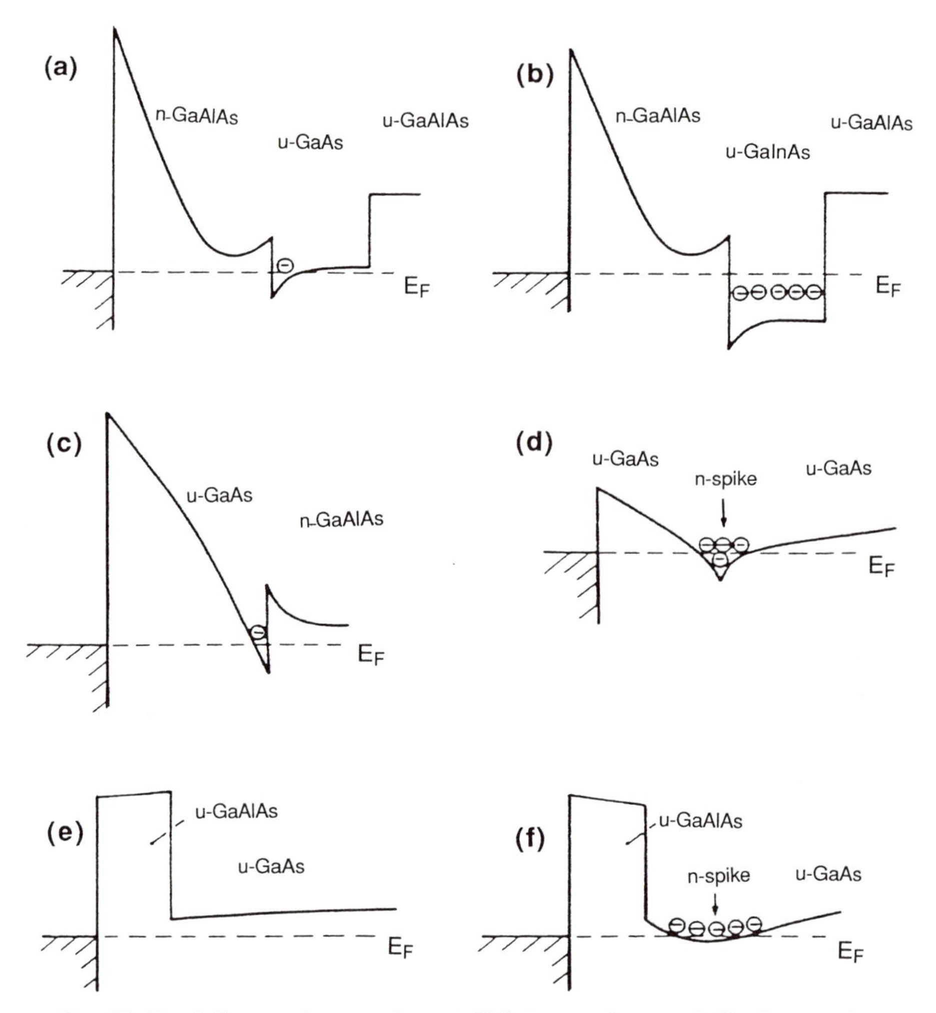

FIG. 82. Band diagrams for several recent high-transconductance field effect transistors. (a) AlGaAs-buffer layer TEGFET; (b) pseudomorphic TEGFET; (c) inverted TEGFET; (d) δ-doped MESFET; (e) SISFET; (f) channel-doped SISFET (Ref. 432a).

tance FET) in which electrons heated in the quantum-well channel of a heterojunction FET are collected in the gate or backgate electrode. The unusual properties of these and several other real-space transfer structures are well described in the review of Luryi,[435] to which we refer the reader for details.

22. Electronic Devices Based on Perpendicular Transport

Historically, the concept of a field-effect transistor was already patented in the 1920s and 1930s[435a], but the first semiconductors that actually worked as electronic devices were based on perpendicular transport across barriers created by bipolar doping as in the p-n-diodes and bipolar Ge and Si transistors. Although the MOS field effect transistors in the meantime have become the simplest, most widely employed structures for large-scale integrated circuits, devices based on bipolar vertical transport still survive even though their technology is considerably more complex.

It was seen in the previous section that an important limitation for high-frequency operation of field effect devices is the time the charge carriers take to traverse the active region under the gate, so that the gate length must be reduced as much as technologically possible. Much shorter active regions can be obtained by epitaxial growth, so that it can be advantageous to utilize transport perpendicular to heterojunction layers to increase speed. A further advantage of perpendicular transport is that the current is essentially proportional to the area of the device, so that large currents can be handled for high-frequency power amplification and also for high speed driving of large fan-out circuits.

In the field of devices based on perpendicular transport, the advantage of using heterostructures will also be due to shorter distances allowed by abrupt potential barriers, and on higher functionalities made possible by complex association of layers (see the following passages on RHETs, tunneling transistors, etc.).

a. *Two-Terminal Devices*

The simplest heterojunction devices are two-terminal and the most actively studied is the double barrier diode based on the resonant tunneling effect described in Section 19. The double barrier diode has a characteristic that shows a region of negative differential conductivity, so that if it is polarized to that region, the system is potentially unstable. If one mounts the diode in a resonating waveguide system the diode can oscillate at a frequency determined by the external circuit. In Fig. 83 we demonstrate how a negative differential resistance can lead to oscillations: the diode, very simply represented by a negative resistance in the small signal equivalent circuit, is connected to the load resistor via a resonating LC circuit. The Laplace-transformed voltage response to a current fluctuation is then given by

$$V(s) = \left(\frac{1}{R_L} - \frac{1}{R} + \frac{1}{sL} + sC \right)^{-1} I(s) \tag{90}$$

which may be written

$$V(s) = \frac{s}{2j\omega_0 C}\left(\frac{1}{(s-\gamma-j\omega_0)} - \frac{1}{(s-\gamma+j\omega_0)}\right)I(s) \qquad (91)$$

with $\gamma = (1/R - 1/R_L)/2C$ and $\omega_0{}^2 = 1/LC - \gamma^2$. In the time domain the response to a current impulse $I(t) = \alpha\delta(t)$ is therefore

$$V(t) = \alpha\frac{e^{\gamma t}}{C}\left(\frac{\gamma}{\omega_0}\sin\omega_0 t + \cos\omega_0 t\right) \qquad (92)$$

This shows that if $R < R_L$ then $\gamma > 0$ and any little current fluctuation will provoke an oscillation to start with exponentially increasing amplitude. Of course the amplitude will stabilize when nonlinearities increase the average dynamic resistance of the diode.

These microwave diodes operate in a completely different way than the transferred electron or Gunn devices[421]: In the latter the velocity of a fixed number of electrons is reduced by an increasing field (see Fig. 78) once the field is larger than the field for maximum velocity, whereas in the former the number of electrons participating in the current is reduced with increasing applied voltage.

The double-barrier diode resembles much more the classical tunnel or Esaki diode[421], which is an extremely highly doped *p-n* junction. For low applied voltages electrons can tunnel across the gap of the semiconductor, and with increasing voltage that process becomes blocked so that the current is reduced. For the performance of resonant tunneling diodes the theory developed for Esaki diodes can be taken over. The main difference between the two types is that the double-barrier diode has fairly low doping so that the depleted layer between the barriers and the collector can be rather long and the capacitance of the diode can be much smaller. This makes a very high frequency of oscillation possible before effects intrinsic to the diode limit speed. Thus it has been shown[436] that the negative differential resistance persists to 2.5 THz, i.e., a response time below 10^{-13} s. In a real setup the maximum obtainable frequency is therefore mainly determined by parasitic access resistance, lead inductance, and impedance matching between diode and waveguide. If one adds the diode capacitance C_d, the access resistance R_s, and the lead inductance L_s between emitter and barriers to the diode equivalent circuit as in Fig. 83 it is simple to derive the real part of the output impedance of the diode in its resonator seen by the load as:

$$R_{out} = R_s + \frac{-R}{1 + (\omega R C_d)^2} \qquad (92a)$$

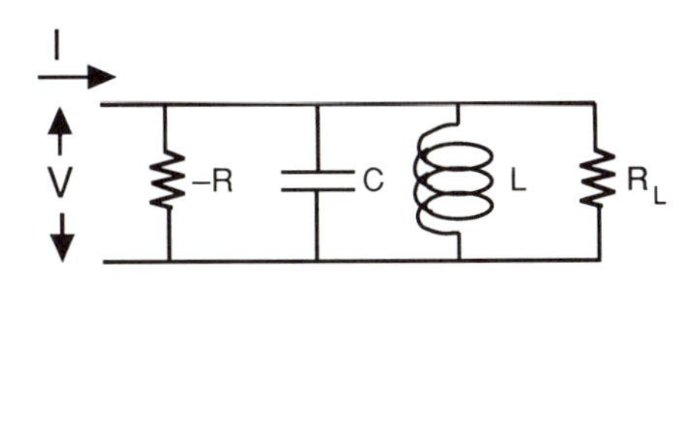

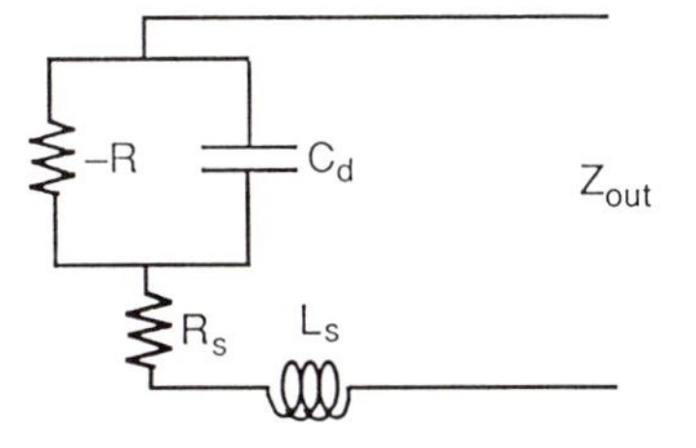

FIG. 83. Simplified small-signal equivalent circuit of a negative differential resistance diode, R, connected to a load, R_L, via a resonating LC-circuit (top). Equivalent circuit of diode with parasitic access resistance and diode capacitance (bottom).

To sustain oscillations this resistance must be negative, so that a cutoff frequency exists given by:

$$f_r = \frac{1}{2\pi RC_d}\sqrt{\frac{R}{R_s} - 1} \tag{92b}$$

so that a high-frequency oscillation requires a small capacitance, a small absolute value of differential resistance (i.e., a steep descent after the maximum in the I-V characteristic and a high peak-to-valley ratio), and an even smaller access resistance. These design rules go against the power that can be extracted from the diode. First, a small output impedance is very difficult to adapt to a useful load like a waveguide or an antenna, which requires diodes of very small cross section and very small waveguide resonators connected to ordinary waveguides via a gradually widened transformer waveguide. Second, a large microwave output requires a large current and voltage swing in the active part of the diode characteristic, which is obtained by a high current density and a larger absolute value of the differential resistance. All these effects are obtained in very thin barrier diodes which have a large current density at the expense of a smaller peak-to-valley ratio and a smaller cutoff frequency: If one assumes a constant negative differential resistance between the peak current I_p and voltage V_p and valley current and voltage I_v, V_v, the maximum available power is easily found as

$$P_{\max} = \frac{1}{8}(I_p - I_v)(V_v - V_p) \tag{92c}$$

for a perfectly matched load, which shows that a large difference in peak and valley currents is more important for the microwave power than just a high peak-to-valley ratio.

The development of oscillators based on double-barrier diodes has almost exclusively been pursued by a group at MIT[437]; their highest frequency result (1989) is 420 GHz with an output power of 0.2 μW. The diode employed has AlAs barriers of only 1.1 nm thickness. The strong nonlinearity of the I-V characteristic of double-barrier diodes has also been employed by the MIT group to demonstrate high-frequency mixing and fivefold frequency multiplication with a decent conversion yield[437].

b. Three-Terminal Devices

Heterojunction Bipolar Transistors

Most applications require more than two-terminal devices. The bipolar transistor in which a *p*-type base creates a barrier for electrons diffusing from an *n*-type emitter to an *n*-type collector is the standard vertical transport three-terminal device[421] (see Fig. 84a). The obvious advantage of such a structure is the short base thickness w_{B} that has to be traversed by the electrons, which is counterbalanced, however, by the diffusive motion in the base region. Standard bipolar transistor theory gives the collector current density as

$$j_c = \frac{D_n e n_{p0}}{w_{\mathrm{B}}} \left(e^{eV_{\mathrm{BE}}/kT} - 1\right) \tag{92d}$$

where V_{BE} is the emitter-base voltage and n_{p0} is the equilibrium electron density in the base and the emitter doping is larger than that of the base. This expression can be interpreted as an average diffusion velocity $v_d = 2D_n/w_{\mathrm{B}}$ times the average density of electrons injected into the base. The exponential dependence on the emitter-base voltage V_{BE} gives rise to a very large transconductance especially at large current levels:

$$g_m = \frac{\partial I_c}{\partial V_{\mathrm{BE}}} = \frac{e}{kT} I_c \tag{92e}$$

which is advantageous for digital applications since it reduces the necessary input voltage swing. On the other hand the minority charge stored in the base corresponds to a capacitance C given by

$$C = \frac{\partial Q_{\mathrm{B}}}{\partial V_{\mathrm{BE}}} = \frac{e}{kT} \frac{e n_p}{2} w_{\mathrm{B}} \tag{92f}$$

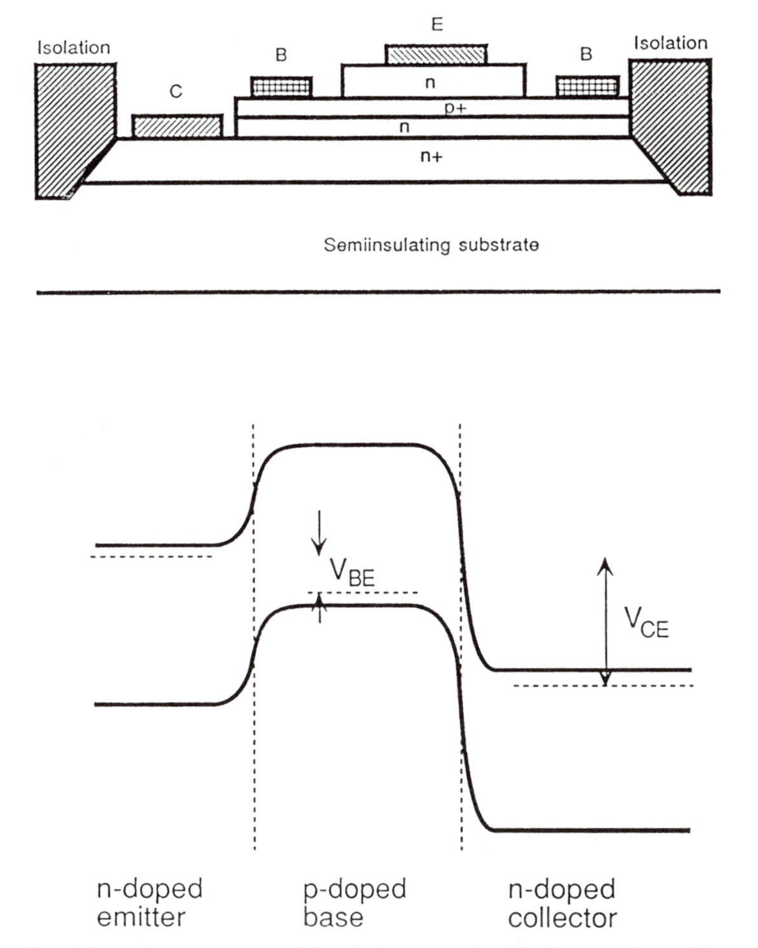

FIG. 84a. Schematic structure and band diagram of a bipolar transistor. The metallic contacts are designated E, B, C, for emitter, base, and collector. Isolation between devices can be obtained, for example, by implantation damage as shown or by etched grooves.

where n_p is the minority carrier density at the base entrance. Again it can be seen that $g_m/C = 2D_n/w_B{}^2$ is an important parameter which describes the inverse of the time it takes for an electron to traverse the base, so that base thickness should be minimized in order to obtain high speed. Since the capacitance has to be charged via the base (and the emitter) it is necessary to increase base doping at the same time, in order to keep the access resistance low. If one wants to draw analogies between FETs and bipolar transistor structures, gate length corresponds to base thickness, source access resistance to emitter access resistance, time under the gate to time in the base, and base resistance to gate resistance. Comparing intrinsic structures, Ladbrooke[437a] thus obtained the ratio of power gain obtainable by the two structures as

$$\frac{G_{\mathrm{BJT}}}{G_{\mathrm{FET}}} = 4\left\{\frac{L_g + X_g/2}{w_{\mathrm{B}} + X_{\mathrm{B}}/2}\right\}^2 \left(\frac{\sigma_{\mathrm{B}}}{\sigma_g}\right)\left(\frac{m_{\mathrm{B}}L_{\mathrm{B}}}{Z_{\mathrm{B}}}\right)\left(\frac{w_{\mathrm{B}}}{L_g}\right)\left(\frac{Z_g}{m_g h}\right) \tag{92g}$$

where the X's are due to gate fringe fields/depletion layers, σ's are gate metal/base resistivities, m's are numbers of gate fingers/base stripes, Z's are gate/base widths, and h the gate thickness. For usual values of the parameters the ratio is of the order of 1, so that intrinsically one cannot designate one technology as superior to the other. For a more detailed comparison, see also the work of Long[437b] and the previously mentioned article by Solomon[430f].

However, the fact that the depletion zones between emitter and base have a macroscopic thickness can to a great extent be eliminated by introducing a heterojunction, which is abrupt on atomic scales between emitter and base: The emitter is made of n-doped AlGaAs and the base of p-type GaAs on an n-type GaAs collector. This is the heterojunction bipolar transistor (HBT). A schematic band-diagram of an HBT is shown in Fig. 84b. Since the base current is predominantly due to the injection of holes from the base into the emitter, the heterojunction barrier for holes reduces strongly the base current, thereby enhancing current gain and/or allowing a higher base doping (even higher than that of the emitter, especially important in GaAs transistors because it is difficult to dope n-type to more than 10^{19} cm^{-3}) with a lower base resistivity. Furthermore, the electron velocity introduced in Eq. (92d) is based on pure diffusion transport. In Si bipolar transistors an increase in velocity is obtained by a gradual doping in the base which creates a pseudo-field drift enhancement of the velocity; in HBTs this enhancement can in addition be obtained by making a base which contains Al in a concentration which decreases from the emitter junction to zero at the collector junction creating an accelerating internal field. Furthermore, the contribution of ballistic electrons traversing the base without any scattering as described in Section 21b also enhances the transit velocity to a much greater extent than in Si for comparable base widths.

For digital applications a simplified expression for the propagation delay in standard circuits has been given by Tang and Solomon[437c] as:

$$t_{\mathrm{pd}} = k_1 r_{\mathrm{B}} C_d + k_2 \tau + k_3 R_{\mathrm{L}} C_{\mathrm{L}} \tag{92h}$$

where the k's are constants. The first term represents the charging time of the diffusion capacitance of Eq. (92f) through the base resistance r_{B}; the second is the transit time through the base; and the third is due to the load resistor R_{L} charging all capacitances on the output side of the tran-

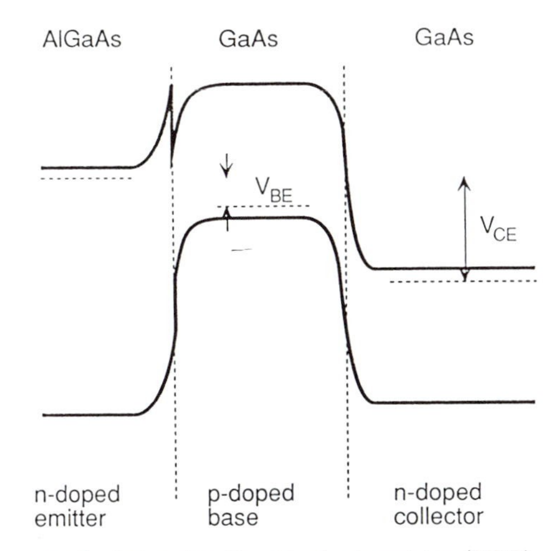

FIG. 84b. Band diagram of a heterojunction bipolar transistor (HBT).

sistor. The first term is roughly proportional to the collector current (Eq. (92f)), the second independent of current, and the third is inversely proportional to the current, since R_L has to be scaled to keep a fixed logic swing, $R_L = \Delta V/I_C$. Qualitatively, the influence of the three terms is depicted in Fig. 84c. The minimum delay time is empirically found for $I_C \cong 100\text{kA/cm}^{-2}$, so that very fast circuits can be obtained because a low voltage swing can be permitted, but at the cost of a high power dissipation, which limits the integration level.

Several schemes to reduce the parasitic resistances have also been implemented: By growing a series of smaller band-gap layers of GaInAs and GaAs on top of the emitter one can greatly reduce the emitter access resistance[438] and by introducing In in the base a pseudomorphic HBT with even lower base resistance has been reported[438a].

From this short description it can be seen that trade-offs in the design of an HBT have to be made and technologically the transistor presents many difficulties. Nevertheless very good performances have been obtained recently: Commutation times of 3.8 ps/gate at room temperature in ring oscillators[439], 1/4 frequency dividers working at 35 GHz[440], power densities of 4.4 W/mm at 10 GHz with 42% power-added efficiency[441].

In InP-based structures, either with an InP emitter and GaInAs base and collector or with an AlInAs emitter, very high-frequency operation has been reported: In the latter system current gain cut-off frequencies of 52 GHz[442] and 80 GHz[443] have been obtained; in the former an extrapolated cutoff frequency of 244 GHz[444] has been found at a temperature of 80 K.

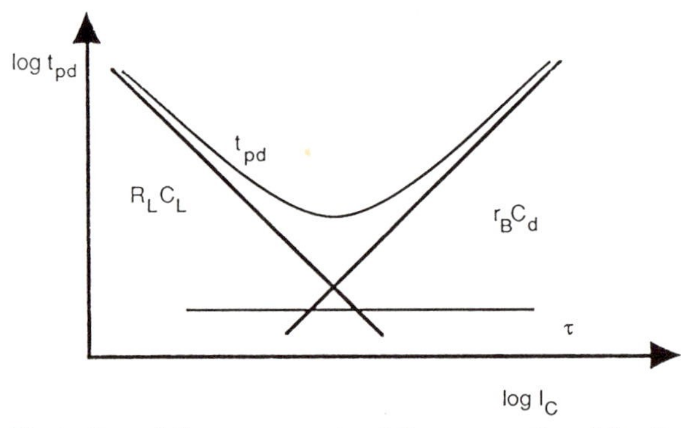

FIG. 84c. Illustration of the components of the propagation delay time as a function of collector current, as seen from Eq. (92h).

For more details on the various HBT structures we refer to Refs. 31, 445, and 446.

Other Vertical Transport Devices

A double-barrier diode in which the well potential is controlled by a third contact has been obtained under the name bipolar quantum resonant tunneling transistor (BiQuaRTT):[447] The well is much too thin to be contacted individually if it is doped *n*-type as the emitter and collector contacts, so the necessary insulation between the well and the outer terminals is realized by doping the well type *p*. In this structure the resonance conditions are controlled directly by the emitter-well voltage and peculiar transistor characteristics showing regions of negative transconductance have been observed.

The introduction of heterojunction barriers in semiconductor structures makes it possible to design vertical transistors, which are unipolar and utilize only the majority carriers. The simplest device of this type is the hot electron transistor (HET) and the tunneling hot electron transistor amplifier (THETA). In Fig. 85 we show the conduction band diagrams of these devices. It can be seen that in both structures the AlGaAs barrier serves to inject electrons, which when they enter the base have a substantial kinetic energy and therefore traverse the base region very rapidly, possibly without being scattered. Potentially this hot electron effect should allow very high-frequency operation. By varying the base-collector voltage it is possible[448] to use the second barrier as an analyzer of the energy distribution of the electrons that arrive at the collector and thereby measure the fraction of the electrons that go ballistically through the base region. For the AlGaAs/GaAs HET and THETA devices, most studies have been made by Yokoyama *et al.*[24,449] and Heiblum *et al.*;[450] although

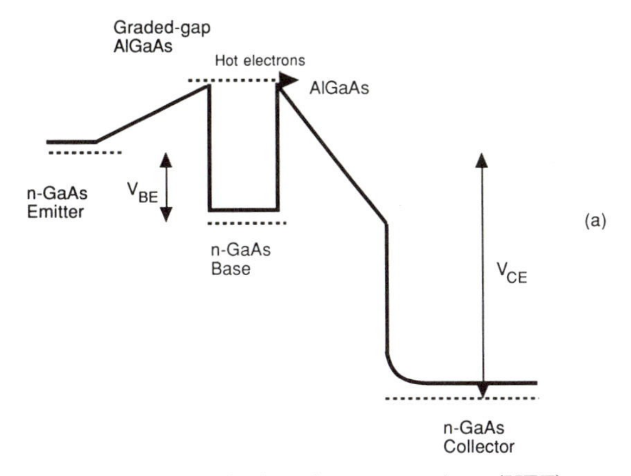

FIG. 85a. Conduction band diagram of a hot electron transistor (HET).

the common emitter current gain of at most nine[451] obtained with a base width of only 30 nm is still too limited to make the device competitive for amplification purposes, it has been shown[451] that 75% of the electrons traverse the base without scattering for the 29 nm GaAs base. Using a pseudomorphic base of GaInAs, Seo *et al.*[452] have increased the current gain to 27 at 77 K, and the system $AlSb_{0.92}As_{0.08}$ (emitter) InAs (base) GaSb (collector) has made room-temperature operation with a current gain of more than 10 possible.[453]

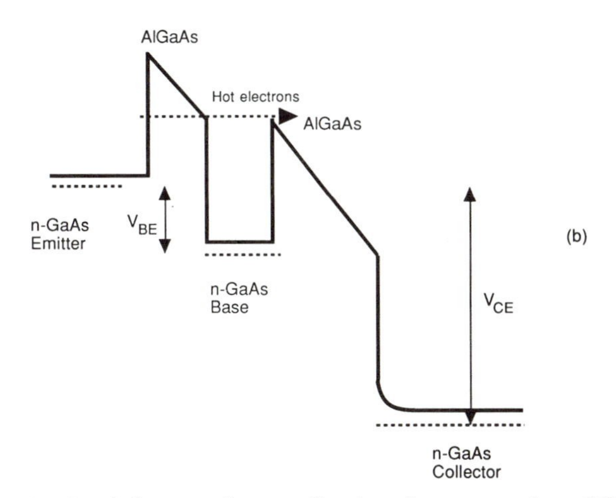

FIG. 85b. Conduction band diagram of a tunneling hot electron transistor (THETA).

A hybrid of the tunneling hot electron transistor and the heterojunction bipolar transistor is the tunneling emitter bipolar transistor,[454] in which a thin layer of AlAs in the emitter in an otherwise all GaAs bipolar transistor creates a barrier that permits only hot electrons to enter the base, and in which the base is made so thin (50 nm) that ballistic transit is possible. Such a transistor has shown current gain up to 40 GHz.[454]

The last vertical transport devices that we shall describe combine the resonant tunneling effect with the hot electron transistor concept: the resonant-tunneling hot electron transistors[455] (RHET) and the resonant-tunneling bipolar transistor[456] (RBT). Figure 86 shows the band diagram of the RHET. It can be seen that the double barrier in the emitter functions as a filter, which only lets electrons at the resonant energy enter the base region; those electrons are then already hot, but more interestingly the negative differential conductivity of the double barrier introduces a region of negative transconductance in the collector current as shown in Fig. 87. Apart from the characteristics of the hot electron transistor, the RHET can be used to create new logic functions as the exclusive NOR: If one adds the logic values of two variables at the base, it can be seen that the collector current will be high if and only if one of the variables is high; if both are low or both are high the collector current will be low. Such circuits (also shown in Fig. 87) have been realized by Yokoyama *et al.*[457] The RBTs differ from the RHET only by the fact that the base is doped *p*-type.

We have not tried to make a comprehensive inventory of all the devices that utilize the possibilities offered by the combination of heterojunctions and the very precise control of layer thicknesses and interface quality. A

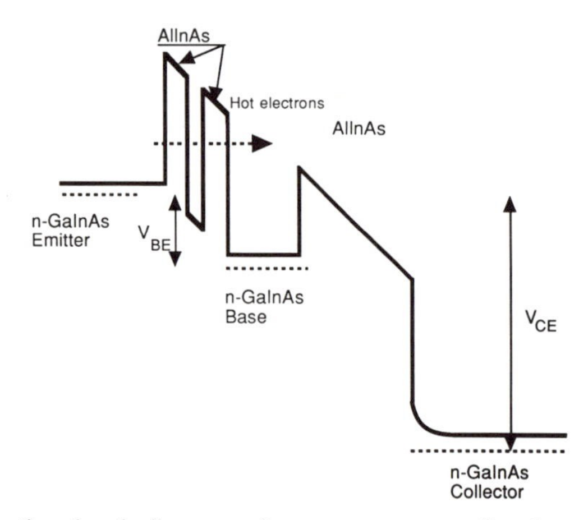

FIG. 86. Conduction band diagram of a resonant tunneling hot electron transistor (RHET).

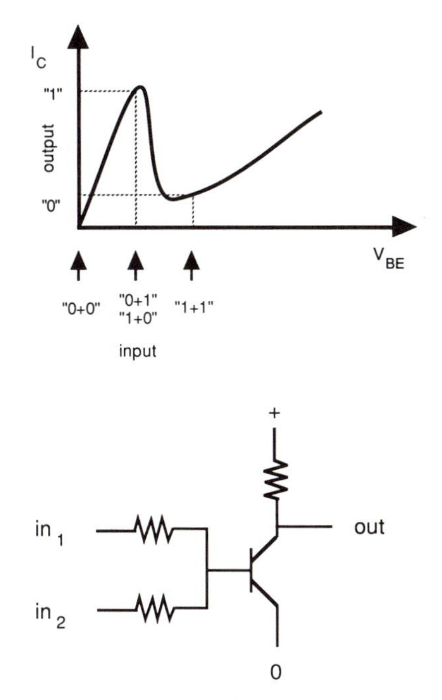

FIG. 87. Schematic collector-current base-voltage characteristic of a resonant tunneling hot electron transistor showing negative transconductance and application of the RHET in an exclusive-NOR circuit.

great number have been proposed; many have been realized and characterized. Several reviews of such devices have appeared recently in which the reader may find further information. References can be found in the selected bibliography in particular the two books edited by Capasso and Magaritondo[439] and Capasso[437] as well as the other chapters in volume 24 of the series "Semiconductors and Semimetals."[20,21,31]

23. Quantum Well Lasers

a. Basic Description[15,16,458–59] *of Laser Action*

A standard double-heterostructure (DH) laser is shown in Fig. 88. This device was truly the first band structure–engineered structure because it simultaneously serves two purposes:

(1) The larger index of refraction of the inner layer guides the optical wave between the two outer layers, resulting in a compression of the optical wave compared with its natural extension in an unbounded active medium.

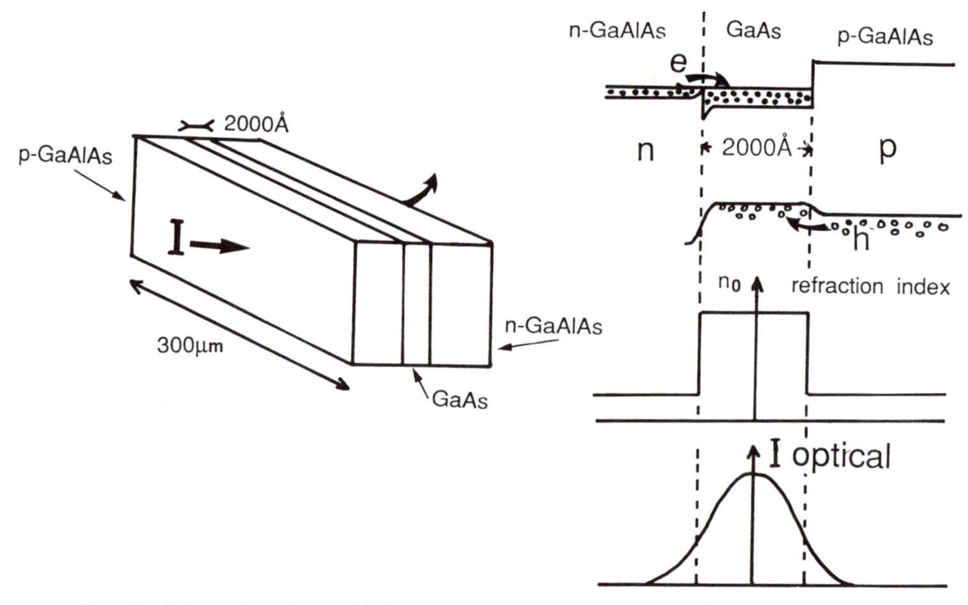

FIG. 88. Schematics of a double heterostructure and the associated profiles of band structure, index of refraction, guided optical wave.

(2) The band discontinuities in the conduction and valence bands confine the electrons and holes within the active layer, resulting in more concentrated carriers as compared with the usual homogeneous situation where carriers diffuse over distances of the order of microns. This thinner active volume has then a smaller number of states to be inverted to reach gain.

These two effects result in a structure that has a more efficient stimulated light emission, since it is easier to reach inversion of carrier population with a given injected current and since concentrated light will more easily induce stimulated radiative transitions. A diminution of $\simeq 100$ of the threshold current density was thus obtained in 1970, which made the semiconductor laser at last usable.[460]

Two relations determine the current threshold: the threshold condition and the gain-current density relation.

The threshold condition states that light intensity I_o is unchanged after one round trip in the laser cavity:

$$I_o R_1 R_2 \exp(2\Gamma g_{\rm th} L) \exp[-2L(\Gamma \alpha_i + (1-\Gamma)\alpha_c)] = I_o \qquad (93)$$

where R_1 and R_2 are the facet reflectivities at each end of the double-heterostructure waveguides, Γ is the optical confinement factor, g_{th} is the volume gain of the active layer at threshold, L the cavity length, α_i the internal loss coefficient of the active layer, and α_c the loss in the confining layers. Equation (93) can be rewritten under the usual form

$$\Gamma g_{th} = \Gamma \alpha_i + (1 - \Gamma)\alpha_c + \frac{1}{2L} \log \frac{1}{R_1 R_2} \tag{94}$$

Typical values for the right-hand side are 40–80 cm^{-1}, with the uncoated-mirror cavity loss playing a major role ($\simeq$30 cm^{-1} for $L = 300$ μm and $R_1 = R_2 = 0.3$).

The optical confinement factor Γ plays an essential role in deriving the design rules for any semiconductor laser. It depicts the overlap of the optically guided wave with the quantum well, according to the formula

$$\Gamma = \int_{-d/2}^{+d/2} |E(z)|^2 \, dz \bigg/ \int_{-\infty}^{+\infty} |E(z)|^2 \, dz \tag{95}$$

It will describe which fraction of the photons of the guided optical wave interacts with the active or passive layer material: When g is the volume gain per unit length of the active material, the amplification of the optical wave is Γg per unit length. If α_i is the loss per unit length in the unbounded active material, $\Gamma\alpha_i$ is the loss per unit length of a guided wave. Conversely, $1 - \Gamma$ is the fraction of the optical wave outside the active material and $(1 - \Gamma)\alpha_c$ will be the loss of the optical wave per unit length if α_c is the absorption coefficient of the confining material at the lasing wavelength. It is from these modifications of g, α_i, α_c, due to Γ for guided waves, that one deduces Eq. (93).

As can be seen from Fig. 89, Γ varies widely for different DH and quantum-well structures. For very thin layers Γ is vanishingly small, as d^2. This originates from the fact that when the optical waveguide is below some critical value ($\simeq$1000 Å for the GaAs/GaAlAs case), the optical wave is less confined in the heterostructure (i.e., it leaks into the confining layers). To overcome this effect while retaining quantum wells as active layers, one uses the separate confinement heterostructure (SCH) scheme where optical confinement is provided by a set of optical confinement layers, while carrier confinement in a quantum well occurs in another imbedded layer.[461]

Standard wave propagation in a stratified medium yields the value of Γ. Figure 90 gives the result of the calculation[462] of the confinement factor Γ per angstrom as a function of total active material thickness for the most widely used structures: single or multiple or SCH quantum wells (which have sometimes[463] been called "modified" MQW's in the case of MQW's

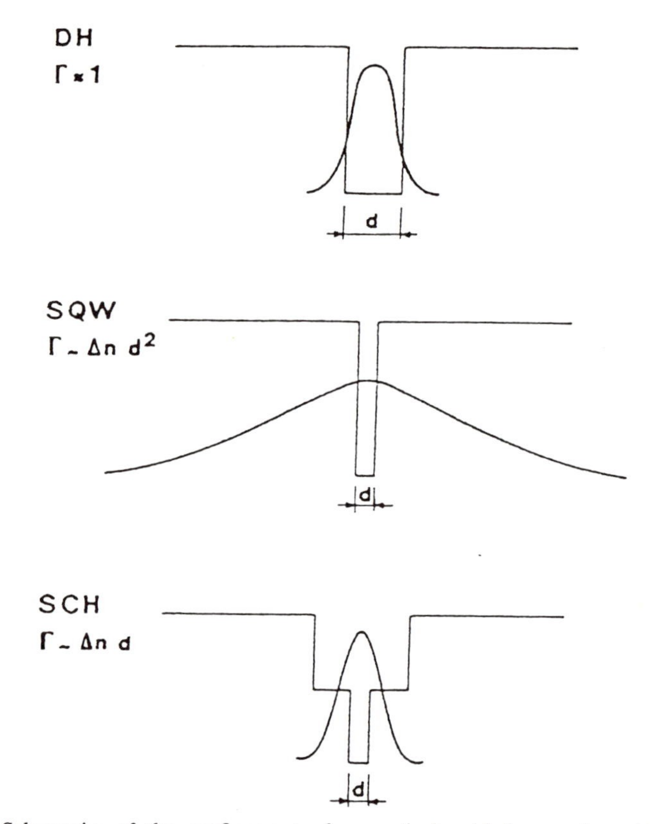

FIG. 89. Schematics of the confinement of an optical guided wave for wide, narrow, separate-confinement waveguides.

imbedded in a SCH). It also includes the ratio r of the active material to barrier material thicknesses in the quantum well regions for MQW's. The standard double-heterostructure laser case corresponds to the MQW curve with $r = 0$. One remarks the maximum "squeezing" of the optical wave-function between the two confining layers occurring at $\simeq$1000 Å for GaAs layers imbedded between $Ga_{0.54}Al_{0.46}As$ layers.

The gain equation is calculated from the radiative recombination rate as obtained from standard perturbation theory in solids. The spectral gain curve can be written as[15,16,458–9]

$$g(E) = \frac{he^2}{2n_r\varepsilon_0 m_0^2 cE} \int M^2 \rho_j(E)(f_c(E) - f_v(E))\, dE \tag{96}$$

where n_r is the refraction index, M the interband matrix element, $\rho_j(E)$ is the joint DOS between the valence and conduction bands and, for a given

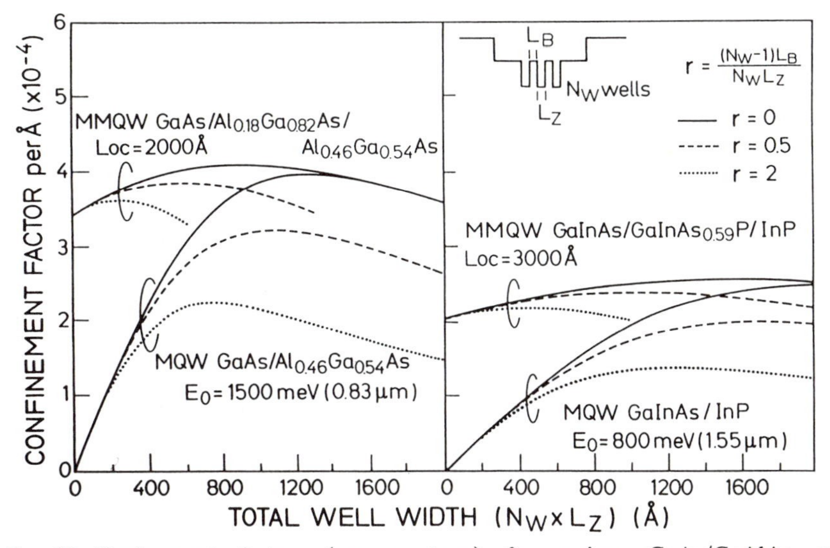

FIG. 90. Confinement factors (per angstrom) for various GaAs/GaAlAs and InGaAs/InGaAsP/InP (lattice-matched) structures as a function of total well material. MQW structures correspond to multiple quantum wells, and different ratios r of well and barrier thicknesses are indicated. MMQW structures correspond to separate confinement structures. Single quantum wells and DHs correspond to cases with $r = 0$. GRIN–SCH values for Γ are very similar to those of SCH with corresponding cavity widths.[462]

current injection, f_c and f_v are the values of the Fermi–Dirac distribution functions in the conduction and valence bands, respectively, at energies determined by the vertical optical transition with energy E.

The net gain condition $f_c(E) - f_v(E) > 1$, called the *inversion condition*, was derived by Bernard and Durrafourg to be[464]

$$E_{fc} - E_{fv} > E \tag{97}$$

where E_{fc} and E_{fv} are the quasi-Fermi levels dependent on injection for electrons and holes, respectively.

From Eq. (96) it is easy to derive the gain-current relation. One injects some value N for the electron and hole densities, calculates the quasi-Fermi levels and then the spectral gain curve from the transition rate. The relation between the carrier densities and injection current is obtained either by assuming a fixed recombination time τ (in which case $J = eNd/\tau$, where d is the layer thickness), or by calculating that time directly for each concentration from the spontaneous recombination rate. In any event, one obtains gain curves, the maximum of which determines the lasing threshold. Figure 91 illustrates the buildup of gain for various carrier densities injected in double heterostructures or quantum wells. The re-

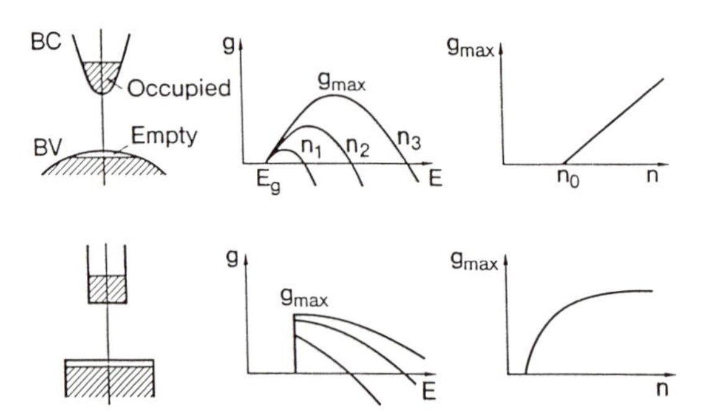

FIG. 91. Schematics of the gain formation in DH lasers (top) and QW lasers (bottom).[465]

markable simplifying feature is that in 3D, the gain-current relation, i.e., the variation of the volume gain curve maximum versus injection current is a straight line to an excellent approximation,[15,16] which can be written as

$$g_{\max} = g = A(J - J_o) \tag{98}$$

A is therefore the differential gain and J_o the so-called transparency current. J_o represents the injected current needed to reach the carrier inversion: up to J_o, carriers do not produce any gain at any energy. Above J_o, carriers start to be effective for gain and laser action. It can be inferred that J_o will decrease with active layer thickness.

From Eq. (94) and (98), one can deduce the current threshold, which becomes, when including an internal quantum efficiency η representing the fraction of carrier recombination that is radiative and therefore participates in gain,

$$J_{\text{th}} = \frac{J_O}{\eta} + \frac{\alpha_i}{\eta A} + \frac{(1-\Gamma)\alpha_c}{\Gamma\eta A} + \left(\frac{1}{\eta\Gamma A}\right)\left(\frac{1}{2L}\right)\log\left(\frac{1}{R_1R_2}\right) \tag{99}$$

For good GaAs/GaAlAs material, η can often be taken as unity, whereas this is far from true for many other material pairs. The variation of threshold current with DH active layer thickness (Fig. 92) can then be easily analyzed: Above 1000 Å, J_o increases with increasing active layer thickness, like the number of quantum states to be inverted; A decreases because the same current density populates fewer states per unit volume; and Γ increases only slightly. As a result, J_{th} increases with d in that region. Below the critical thickness for Γ, the last term in (99) increases dramatically with decreasing active layer thickness as Γ diminishes very fast (as d^2) whereas J_o decreases only as d, and A increases only as d^{-1}.

As is evident from Fig. 92, many quantum well lasers can perform much better than DH's. They also display a number of additional advantages,

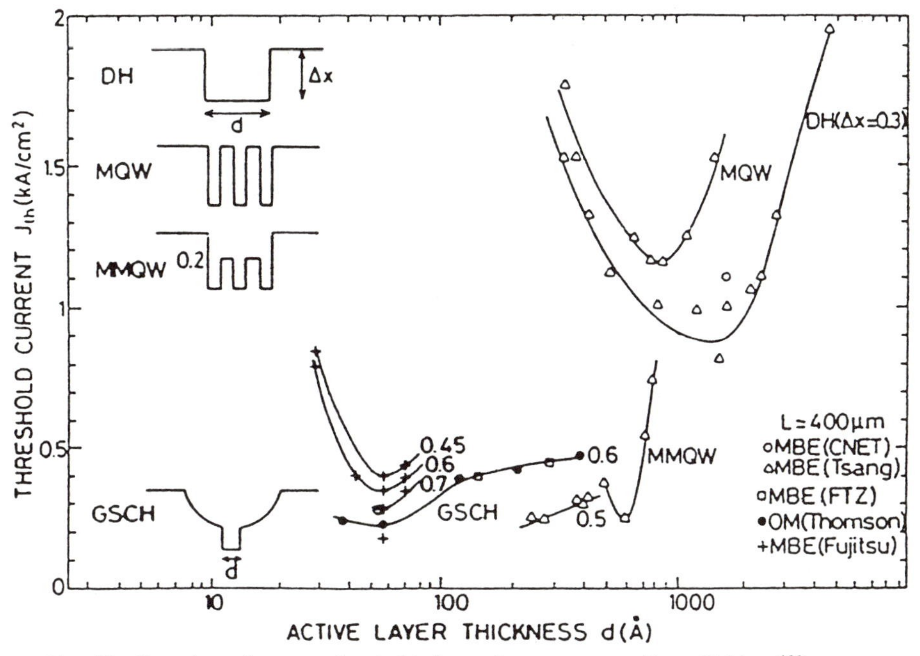

FIG. 92. Overview of current thresholds for various structures (from Noblanc[466]).

which are explained from a more detailed analysis, which we outline below for the various structures used (Fig. 93); all have their advantages and inconveniences.

b. Single Quantum Well (SQW) Laser Operation

The use of a very thin active layer has several consequences:

(1) Energy levels in the conduction and valence band become quantized. Therefore, lasing will occur at energies determined by the band gap and the confining energies. Several other effects contribute to the precise energy level position, such as band gap renormalization and Coulomb effects, which will not be discussed here any further.[467]

(2) Due to wavefunction and energy quantizations, the density-of-states becomes 2D. The buildup of gain for three different injected carrier densities in a quantum well is schematically depicted in Fig. 91. As can be seen, carriers are more "efficient" than in 3D since added carriers will contribute to gain at its peak (the bottom of the 2D-band), whereas in 3D added carriers move the peak gain away from the bottom of the band, making all carriers at energies below that of g_{max} useless. Therefore, the spectral gain curve has a larger slope A than in 3D. However, the price to

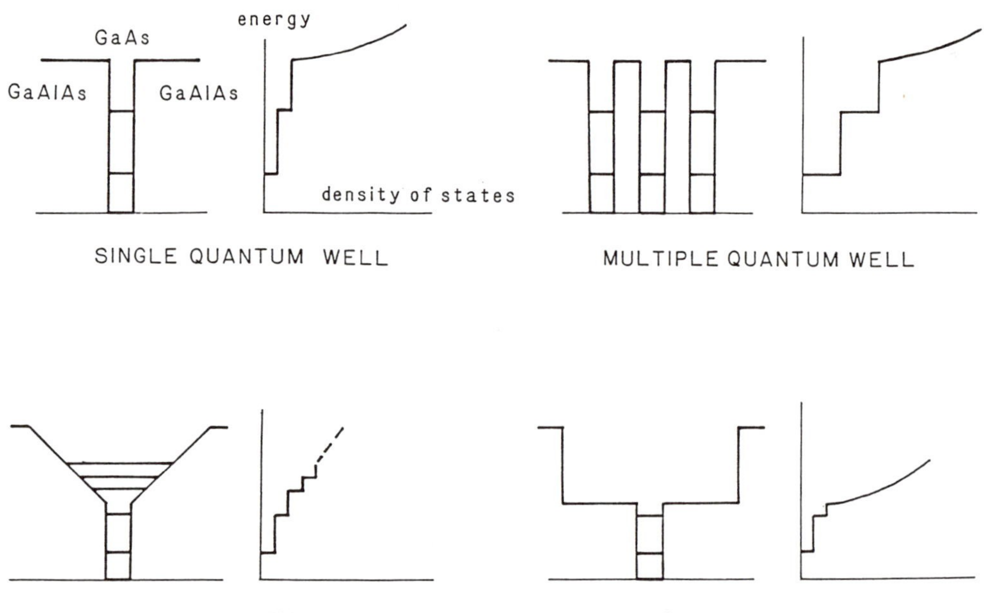

FIG. 93. The various QW structures used as active layers in lasers and the associated density of states.

pay is that gain will saturate at a given finite value when the electron and hole states are fully inverted, whereas g_{max} never saturates in 3D due to the filling of an ever-increasing density of states. Finally, the transparency density or current is much smaller than that for a DH laser because the density of states to be inverted is quite smaller (a GaAs quantum well has $\sim 10^{12}$ states per cm^2 in the minimal energy range of kT to be inverted instead of $\sim 10^{13}$ states in a 1000 Å DH laser).

(3) The confinement factor Γ has to be kept to an optimized value by using a separate optical confinement scheme as seen in Fig. 89. The confinement factor per unit length of active material is then independent of layer thickness. It can be calculated that optimized GRIN–SCH structures yield very comparable values for Γ as straight SCH's.

The detailed calculation yields the results shown in Fig. 94. It also determines an optimal quantum well thickness, shown in Fig. 95:[462,468] the 400 Å laser displays a quasi-3D gain curve, i.e., a linear dependence on injection current. Reducing the thickness reduces the transparency current, i.e., the onset of the gain curve on the current abscissa line. The volume gain increases with diminishing d as the volume diminishes and as

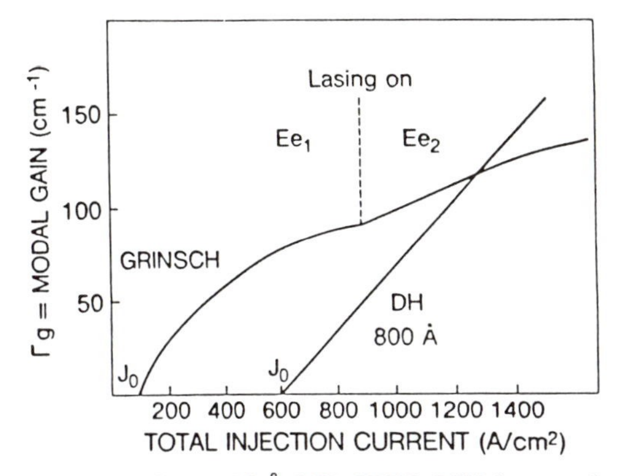

FIG. 94. Gain-current curve for a 120 Å QW GRIN–SCH laser and a 800 Å DH laser, evidencing the changes in transparency currents J_o, and differential gains (slopes of the gain-current curves). Note that additional gain is obtained from $n = 2$ quantum level inversion at high currents.[465]

similar injected numbers of carriers above the transparency density have roughly the same efficiency to amplify an optical wave. [This latter point is also true in 1D and 0D: Since the optical matrix element is the same in all dimensions (neglecting correlation effects), the transition rate *per allowed electron-hole pair* (i.e., $f_e = 1$, $f_h = 1$) is the same, yielding the same total gain. What changes with dimensionality is transparency density and the spectral repartition of gain (see below, Section 27)].

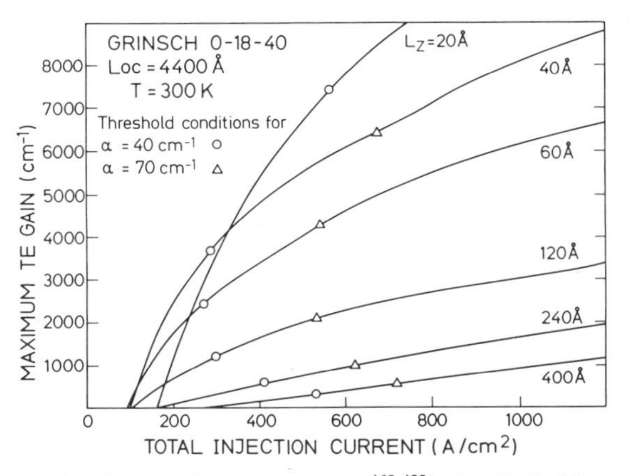

FIG. 95. Calculated volume gain-current curves[468,432a] for $GaAs/Ga_{0.82}Al_{0.18}As/Ga_{0.6}Al_{0.4}As$ GRIN–SCH structures, for various quantum wells. Circles and triangles indicate the threshold current for total loss of 40 and 70 cm^{-1}, respectively.

However, the curves do not scale exactly, owing to second-order effects: Population of higher-lying levels plays a role, in particular for wide wells. Also, for very thin wells, the quasi-Fermi level is so high in the well (due to a large confining energy) that some carriers spill over into the optical confinement cavity, yielding a large J_o. These effects explain why one observes an optimum in the layer thickness for minimizing the current threshold. Also shown on Fig. 95 are the points at which Γg is equal to given losses, i.e., 40 cm^{-1} or 70 cm^{-1}. The occupancy of optical cavity states plays a major role in the difference between SCH and GRIN–SCH structures: Even in optimized structures, the quasi-Fermi level is so high in the conduction band (in the valence band it is quite lower due to the larger density of states; see below, Fig. 100) that population of the cavity states occurs:[469] At threshold, it can be calculated that the number of electrons in the SCH optical cavity is roughly equal to that of the electrons in the active layer, whereas this number is only about 20% in the GRIN–SCH structure due to the smaller density of states in the triangular optical cavity (Fig. 96). The comparison of theory and experiment for both structures is remarkable (Fig. 97), the more so because no extrinsic adjustable mechanism (such as $\eta < 1$) is used in the calculation. This points out the perfect

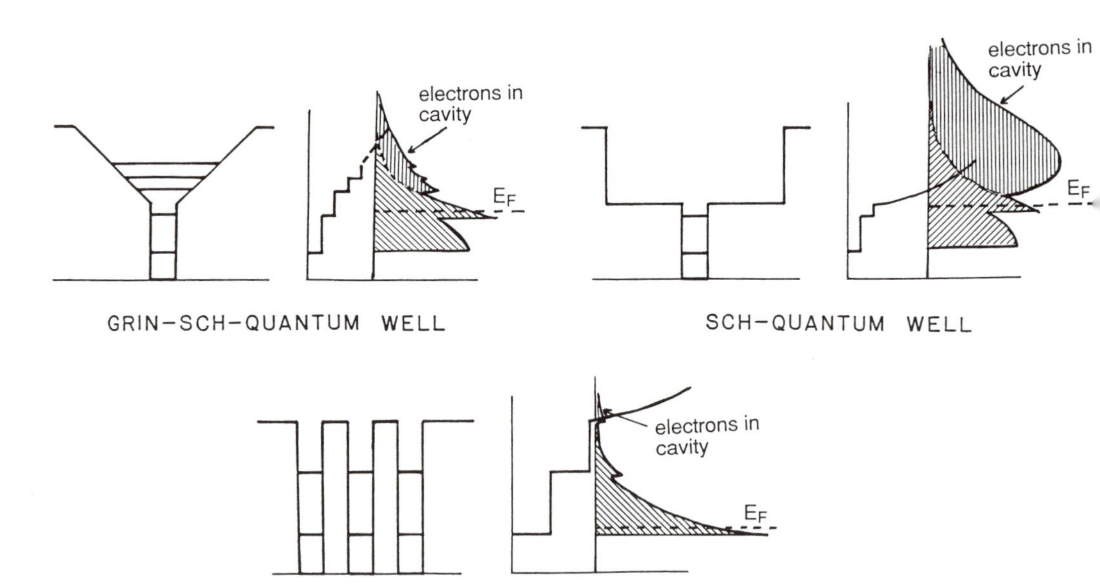

FIG. 96. Schematics of GRIN–SCH and SCH structures, and corresponding density-of-states and occupied states.

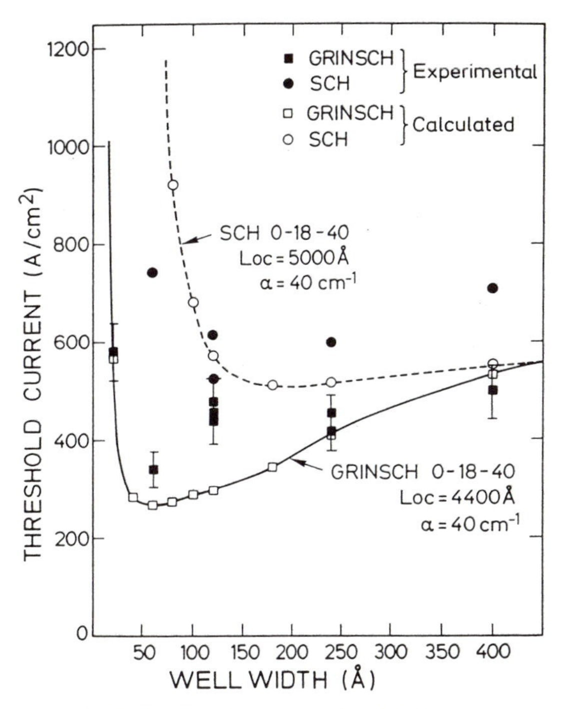

FIG. 97. Comparison of calculated and measured threshold currents for SCH and GRIN–SCH GaAs/GaAlAs structures.[462]

material properties obtained in the GaAs/GaAlAs system. The improvement brought by GRIN–SCH structures was sometimes traced to a "funnel" effect, by which electrons would be more easily captured in the active material QW. Although the correct quantum calculation of relaxation through the quantum states of GRIN and straight SCH structures indeed evidences a shortening of the cascading relaxation time in GRIN structures, the capture times should not play a role in cw operation.[469] However, at high modulation speeds, both the longer trapping times and the reservoir of electrons in the confining layer should put the straight SCH at a disadvantage.

Some "second-order" effects should be taken into account, such as broadening of levels (by collisions of various types), modification of matrix elements due to kinetic energy[470] and valence band mixing,[471] Coulomb interaction between carriers,[471a] but most effects have little influence,[471,472] which explains the good agreement of simplified calculations with experiment.

c. *Multiple Quantum Well Lasers*

Properties of multiple quantum well lasers can easily be deduced from those of a single quantum well if one assumes homogeneous injection of carriers in the various wells: In the approximation where Γ can be considered exactly proportional to the total active layer width, the modal gain Γg-current relation can be deduced from that of a single quantum well by simple scaling transformation, i.e., by multiplying units on both graph-axes by the number of wells (Fig. 98). The advantages and disadvantages of MQW structures are obvious; whether the SQW or MQW is the better one depends on the loss level: At low loss, the SQW laser is always better because of its lower J_o (only the states of one QW have to be inverted) and lower internal losses ($\Gamma\alpha_i$ scales with the number of wells). At high loss, the MQW is always better because the gain stems from a high-slope part of the gain-current curve instead of the saturated part of the SQW gain-curve. In some cases, the saturated gain of SQW is not large enough, and one would absolutely require several QWs to reach the threshold gain. An evident advantage of MQWs is the higher-differential gain, which leads as discussed below to higher modulation frequencies and narrower linewidth (Fig. 99).

The way to ultralow threshold lasers is also clear:[473–5] any diminution of the required threshold gain by using reflection coatings (Eq. (94)) will have

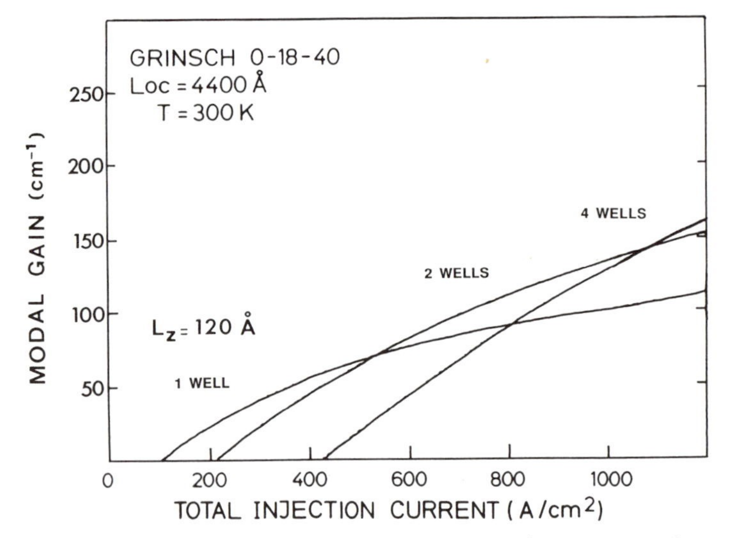

FIG. 98. Modal gain (Γg)-current curve for 1, 2, 4 well GaAs/$Ga_{0.82}Al_{0.18}As$/$Ga_{0.6}Al_{0.4}As$ GRIN–SCHs.[462]

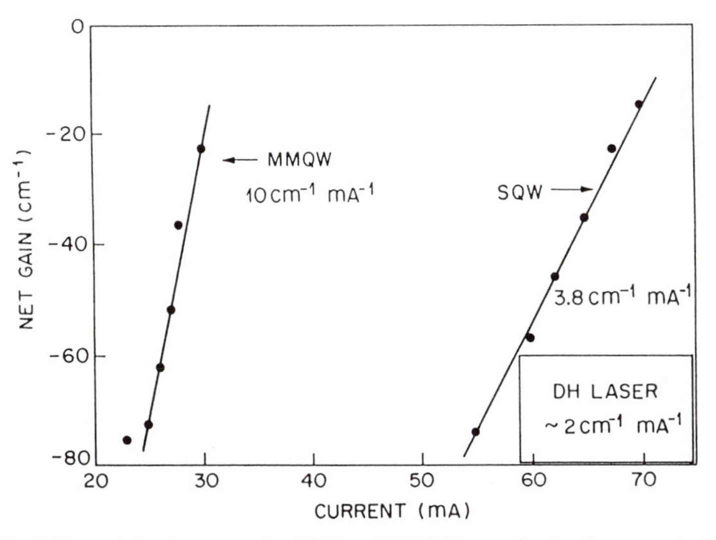

FIG. 99. Differential gain curves for SQW and MQW lasers (in the linear part). (Adapted from W. T. Tsang, *IEEE J. Quantum Electron.* **QE-20**, 1119, © 1984 IEEE.)

a major effect in SQW lasers: It will be translated into lower threshold current, which can be made very near its limit J_o.[473] The effect of such coatings would be much smaller in DH or MQW lasers, since the major part of the current would be the useless J_o. The next step to threshold current decrease is the use of narrower stripe lasers: So far, the progress in that direction has been limited by the heavy optical losses occurring for narrow stripes due to imperfect interfaces and/or lateral carrier confinement. Recent advances in index-guided structures with high lateral perfection such as allowed by patterned growth of V-groved substrates[475a] should yield lasers with much smaller active volumes, i.e., lower current thresholds, in the 100 μA range.

Another path to improvement of the threshold current is the modification of the valence band structure:[476–7] Under usual pumping conditions, the inversion is much smaller in the valence band than in the conduction band as shown in Fig. 100a,b[465] (i.e., $f_v(E) \ll f_c\,(E)$, translating into the fact that E_{fv} is usually quite above the confined valence levels). This arises from the many nearby valence levels (confinement energy for holes is small due to the rather heavy hole mass) with high density of states, which lead to smaller occupancy factors $f_v(E)$ in order to account for the neutrality conditions:

$$N = \int \rho_c(E) f_c(E)\, dE = \sum_i \int \rho_{vi}(E) f_h(E)\, dE \qquad (100)$$

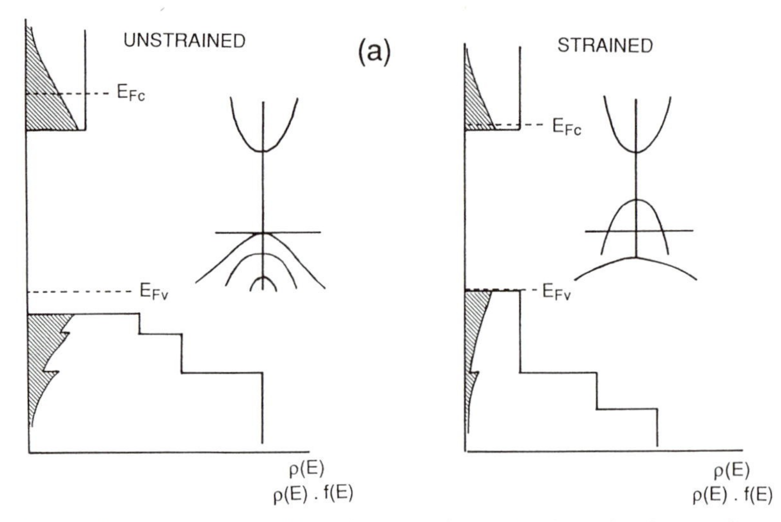

FIG. 100a. Schematics of band-filling for conduction and valence band states for equal numbers of injected electrons and holes in unstrained (left) and strained (right) QW active layers.

where the summation is carried over all valence bands labeled i, considering for the sake of simplicity that only the ground electron level is populated here.

Being able to split apart the various hole states will significantly increase the inversion in the lowest level at a given hole density, which will lead to higher gain according to Eq. (96). This effect has been dramatically evidenced in the excellent results obtained on strained-layer pseudomorphic GaInAs/GaAlAs/GaAs[478–80] and GalnAsP/InP[481] lasers where the strain splits the valence band, raising a valence band with a light transverse hole mass quite above a heavy-hole mass band. Additional useful features of these heterostructures due to the higher refractive index and deeper electron (hole) quantum wells are better optical confinement factor, higher differential gain (and associated superior lasing characteristics[482]), higher catastrophic damage threshold, and transparent substrate for surface emission.

d. The Temperature Dependence of the Threshold Current

The temperature dependence of the current threshold is usually approximated by

$$J_{\text{th}}(T_2) = J_{\text{th}}(T_1)\exp((T_2 - T_1)/T_0) \tag{101}$$

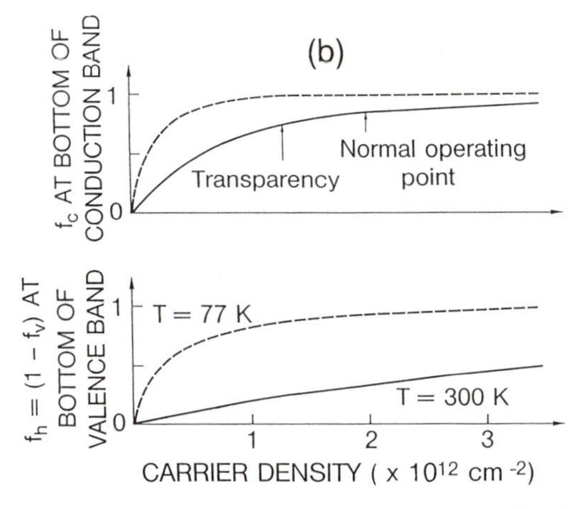

FIG. 100b. Calculated occupancy factors at the bottom of bands of a 60 Å GRIN–SCH unstrained GaAs/GaAlAs laser as a function of carrier densities, at 300 and 77 K. Transparency and normal-operation points are indicated. Note the low occupancy of hole states under 300 K operation, due to the numerous nearby hole states. (From Nagle and Weisbuch.[465])

where T_1 and T_2 are two nearby temperatures. The variation of threshold current is slightest (which is highly desired in most applications) when the temperature coefficient T_o is larger.

Several effects come into play in the calculation of T_o, such as the broadening of the Fermi–Dirac distribution function, the thermal spillover of carriers above confining barriers, etc. A main effect in quantum wells was shown to be the spillover of carriers in the confining layers[469] (Fig. 96). This explains why the structures have better (i.e., larger) T_o's in the order T_o(SCH) < T_o (GRIN–SCH) < T_o (MQW) (Fig. 95).[17,483] The latter is the best because it has the 3D DOS which is the farthest away in energy. It also has a better T_o than the standard DH laser because the high-energy tail of the Fermi–Dirac distribution function only populates a constant 2D DOS instead of populating an energy-increasing 3D DOS with useless carriers.[483a,b] Of course, deeper QW structures such as GaInAs/GaAlAs/GaAs strained QW lasers have improved T_o characteristics, due to both the better confining potential barriers and the lower band filling originating in the more symmetrical electron and hole states described above.

e. *Additional Features of Quantum Well Lasers*

The success of GaAs/GaAlAs QW lasers does not rely only on their low threshold current. Additional properties also play a major role.

Manufacturability and Reliability

QW lasers have been shown to have excellent manufacturability, in particular due to the large performance margin which is available, in particular the very low threshold current density. The thin multilayer active region also allows new manufacturing schemes such as impurity-assisted interdiffusion providing index-guided structures.[483b] Reliability has been shown to be excellent[484–5] as has margin to catastrophic damage failure: The latter, usually due to facet mirror degradation induced by high recombination in DH lasers, is much less efficient in QW lasers. Due to the thin active material compared to the thick optical cavity, the degradation of the facet of the active layer induces only a minor additional loss to the optical cavity and is not a cause for laser action collapse.

Modulation Speed and Spectral Width

Both these parameters, of paramount importance for high-speed telecommunications, are largely improved by the use of QW's as active layers.[486] They both depend on the differential gain $A = dg/dJ$, which is larger in QW lasers due to the square DOS as discussed previously (Fig. 99). It can be shown that the relaxation frequency Ω_R is given by[459,486]

$$\Omega_R = (AP_o/\tau_p)^{1/2}$$

where P_o is the average laser power and τ_p the photon lifetime in the laser cavity. As the diminishing of τ_p increases the cavity losses, the best lasers are MQWs where A is large at large gain, as discussed previously. An additional improvement has been obtained for A by using a p-doped active layer.[487a–c] In that case, the differential gain is larger than for undoped active layers as the quasi-Fermi level for electrons E_{fc} changes very rapidly with injected electron density and inversion is reached as soon as E_{fe} reaches the confined electron level, with the Bernard–Durrafourg relation (97) obeyed (the undoped or n-type doped cases are less favorable because of the large hole DOS to be inverted). Relaxation oscillations as high as 30 GHz have been reported.[487a–c] Higher frequencies could be reached by reducing the damping factor originating in nonlinear gain saturation due to hole burning.[488] Spectral width improvement of QW lasers is by now well-documented,[488a] as is low-chirp of such lasers.[488b]

Power Lasers

The advent of laser-diode pumped solid state lasers opens the way to many applications such as high-power pulses, ultranarrow linewidth, visible-doubled lasers with high efficiency, compact tunable parametric oscillators, etc.[489] At the origin of this explosion is the availability of high-power arrays of laser diodes, which are all based on quantum well structures. The success of QW lasers as generators of high-power is due mainly to two effects:[490]

(1) Quantum efficiency is very high; the transparency current, corresponding to unused excitation, is smaller for QW lasers.
(2) Internal losses are much smaller in QW lasers because of the small confinement factor (contribution $\Gamma\alpha_i$ in Eq. (94)).

Therefore, the electrical-to-optical power efficiency is better, which is essential since laser arrays are often limited by the heat that has to be extracted from the laser die. An additional useful feature is the better catastrophic damage threshold.

Tunability

It has been recognized quite early that quantum-well gain spectra are wider than those of DH lasers, in contradiction to oversimplified pictures (e.g., fig. 91), due to the existence of nearby excited quantum well levels in optimized SQW structures (see the previous discussion of optimal layer thickness for SQW lasers). These excited levels are also mandatory when the gain is not large enough with a single quantum well level and MQWs are not used. Threshold can then be reached using several quantized levels in a single quantum well. This is depicted in Fig. 94 where lasers with losses above 90 cm^{-1} such as due to shorter cavity lengths require significant $n = 2$ gain to reach lasing.[469,491] These wide gain curves have been used to provide wide tuning range up to 15% of the emitting wavelength in external-cavity-tuned lasers.[491–4] It is also used in order to obtain very narrow pulses by mode-locked operation: The constructive interference between the modes of laser yield a light pulse with a time duration given by $\tau\Delta g^{-1}$, where Δg is the width of the gain curve above threshold. QWL's with well-chosen parameters to yield a wide gain curve have been mode-locked with a repetition rate of 108 GHz and a pulse width of 2.4 ps.[494] Others have used structures with controllable internal losses to achieve electronically controlled wavelength jumps between $n = 1$ and $n = 2$ lasing.[495–6] This wavelength jump is a drastic illustration of the fact that maximum gain occurs at the bottom of the 2D-quantized energy band. Wavelength would continuously shift with carrier injection in the case of a 3D continuous DOS.

Vertical Lasers

Vertical-emitting lasers have long been desired because of the various applications they would allow: free-space direct chip-to-chip communication, random placement on a chip, easy access for optical accessories (lenses, fibers), possibility of 2D-arrays for high powers with low divergence or for 2D imaging. Various schemes have been designed, shown in Fig. 101.[496a] The most spectacular realization has been with integral mirrors, which have allowed the simultaneous fabrication of more than one million lasers on a single chip, with fabrication yields in the 95–100% range[496b] (Fig. 102a). Current thresholds in the mA have been obtained.[496c–d] Three main factors determine the performance:

(1) Mirror reflectivity: Due to the limited total thickness allowed by the growth process, the mirror loss factor in a round trip (last term in Eq. (94) becomes prohibitive for simple semiconductor–air interfaces, and reflective coatings are required. Both dielectric and semiconductor multilayer reflectors have been used. The excellent quality of semiconductor multilayer growth has allowed single-step laser fabrication with quarter-wave stack Bragg reflectors having reflectivities in the 99.96% range.

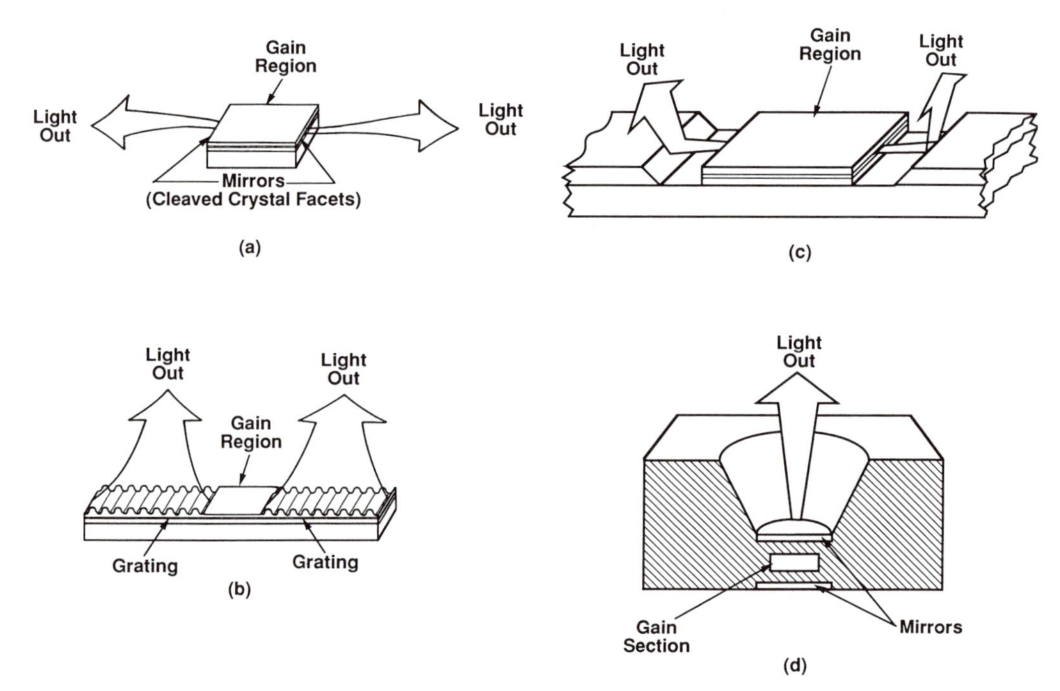

FIG. 101. Various schemes of lasers: standard edge-emitting laser (a), grating couplers (b), integrated mirrors (c), vertical-emitting lasers with integral mirrors (d). (Reprinted with permission from G. A. Evans, N. W. Carlson, J. M. Hammer, and R. A. Bartolini, *Laser Focus World*, Nov. 1989, p. 97).

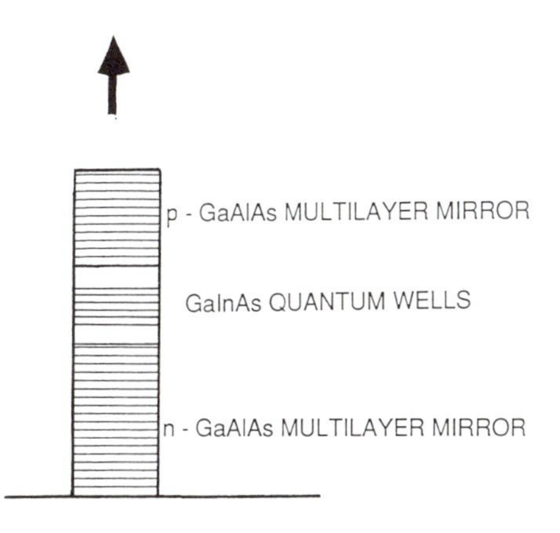

FIG. 102a. (top) Schematics of a mesa, vertical surface-emitting laser; (bottom left) photograph of an etched structure; (bottom right) array of fabricated lasers (from ref. 496b).

When compared with the usual 30% cleaved-facet reflectivity, this allows a ≈ 100-fold decrease in cavity length to reach similar photon-losses, which is the usual range of cavity length in vertical lasers (~1–2 μm). Therefore, such lasers operate with losses (and gain) similar to those of edge-emitting lasers.

(2) Lateral confinement is required to reduce the active volume. For wide lasers, only current confinement is required, whereas below the μm range, optical confinement is also required, as in standard edge-emitting DH lasers. Several schemes have been used: Substrate etching defines a tight optical beam, while current confinement is obtained through lateral *p–n* junction barrier confinement.[496e] Conversely, etching of the mesas through the mirrors and active layers also defines excellent light optical beam and confined-current injection. However, this is at the expense of large surface leakage currents created by the mesa fabrication through reactive ion-etching. Therefore, instead of a decreasing threshold current with mesa-diameter (at least down to 0.1 μm, from which Γ would start to decrease too much), one observes an optimal-threshold current for 5 μm-diameter mesas, below which the diffusion current to the damaged surfaces takes over the standard radiative recombination currents. A planar-type process using tailored ion implantation damage for current confinement has also been demonstrated, with threshold current densities of ~1 kA/cm^2.[496f]

(3) Carrier injection usually occurs through the mirror materials, which have a large series resistance, in the range of several tens of ohms. This leads to severe laser heating under c.w. operation, and is a major limitation where current flows through the mirrors in vertical lasers.

The results achieved so far are impressive, however, and are most often obtained on GaInAs strained-layer QWLs:[496g] submilliamp threshold, 8 GHz operation, 30 ps pulsed output. The losses are in the 10 cm^{-1} range, which shows that dominant losses are cavity losses. The electrical-to-optical power efficiencies are in the 10% range. Power output is limited both by the maximum intensity acceptable on mirrors (~10^7 W.cm^{-2}) and required device cooling.

f. GaInAsP/InP Laser Case

The performance of QW lasers in this materials system has been for a long time rather disappointing. The standard analysis of such lasers, however, does not directly yield such a result. Figure 90 shows the calculation of the confinement factor. The GaInAs/InP QWL Γ is at a disadvantage of two when compared with GaAs/GaAlAs QWLs. The gain-current curve with the only radiative-recombination process more than compensates for this factor, due to the diminished band-gap energy and effective masses, which enter the gain and radiative lifetime expressions.[15,16,459] Experiments, however, directly evidence a very large heating of carriers induced by nonlinear effects, most certainly the Auger effect.[497] It is rather simple to take care of the Auger recombination in the gain calculation when the Auger recombination coefficient C_A is known: One simply calculates for each electron and hole density N a total injected current as the sum of the usual radiative recombination current and an Auger recombination current ($=eLC_AN^3$). The difficult part is to know the Auger coefficient, since it most certainly varies with injection due to the heating effects just described, whereas calculations assume the usual value of 4×10^{-29} cm^{6}s^{-1}. It was even observed in a SQW SCH structure that *increased current* injection leads to *diminished gain* because of that carrier heating.[497] Then, large cavity population builds up and eventually laser emission occurs at the cavity wavelength. That the emission should occur on the cavity quantum states (DH-like) instead of the quantum well might at first seem surprising. One should remark, however, that due to the poor Γ, QWLs operate at high volume gain, i.e., at higher carrier densities than 3D DH lasers. Therefore, as soon as detrimental nonlinear effects are present, the high-equivalent 3D carrier concentration of quantum wells can lead to *poorer performance* than 3D structures. Great care must be devoted to design optimal structures where 2D behavior of quantum wells is conserved while the carrier density is not too high. Another item of concern is the finite barrier height of the SCH structure, which leads to large leakage factors of carriers in the optical cavity for small P concentrations in the GaInAsP intermediate layer. Increasing too much the P concentration of the cavity material, however, translates into a poor Γ, which in turn requires a large volume gain, implying high injection. Figure 102b shows that optimal composition is $y(\mathrm{As}) \approx 0.4$, corresponding to a materials gap ≈ 1.15 μm. Successful SCH operation has indeed evidenced an optimal cavity composition in that range.[498–9]

More recently GRIN–SCH structures have been fabricated in GaInAsP/InP system.[500] They operate with significantly better results due to the better ratio of carrier densities between the QW and the optical cavity. It remains, however, to be shown that theoretical expectations can be reached. These point to ultimate threshold currents of the order of 150 A/cm^{-2}. The fact that best results are of the order of $\approx$500 A/cm^{-2} can either mean that some important deleterious intrinsic effect is left out (such as carrier heating) or that material properties are still far from perfect (η significantly below 1 in Eq (99)).

The best results at the 1.5 μm range have been obtained by using strained-layer GaInAsP/InP active layers.[482] In that case, the symmetrization of conduction and valence bands yields significantly lower thresholds

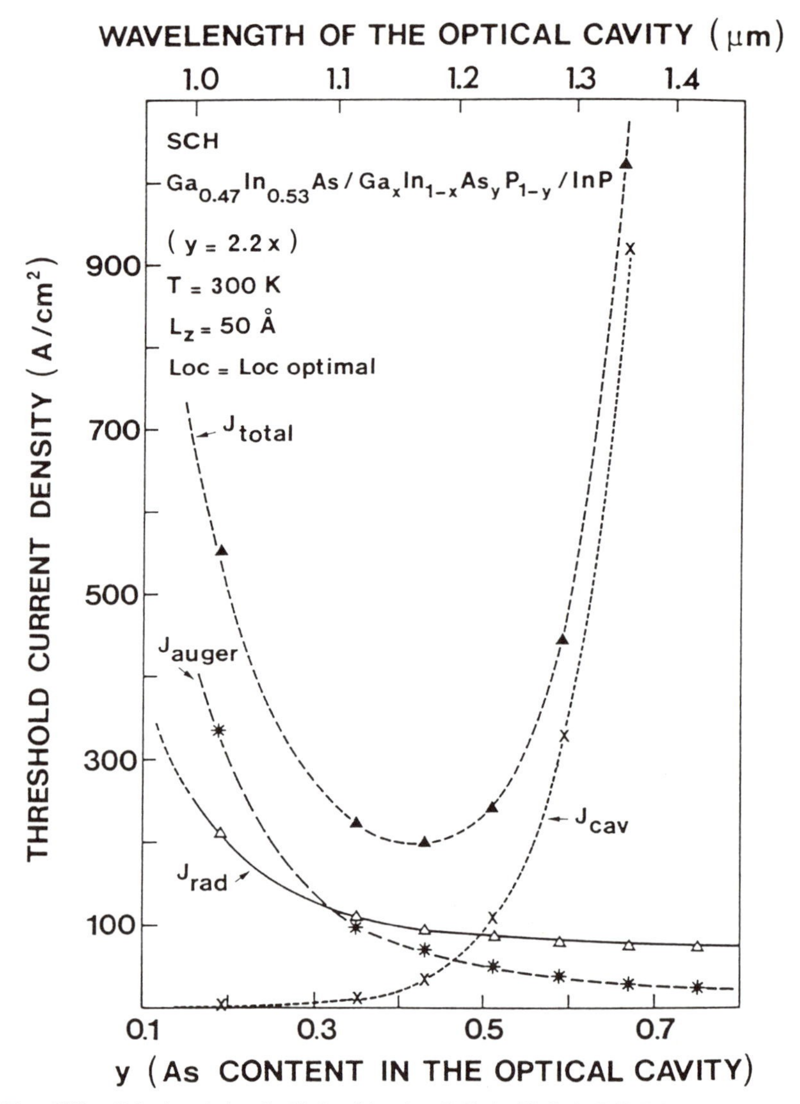

FIG. 102b. Calculated threshold densities for GaInAs/GaInAsP/InP lattice-matched single 50 Å QW SCH lasers as a function of As content in the optical cavity material: At low As content, the cavity material has an index very near InP. Γ is small, which leads to high required volume gain, thus high carrier density, and therefore large Auger recombination. At high As content, Γ is good but energy barriers are small, which leads to carrier spillover in the cavity.[472]

of $\approx$200 A/cm^2, in good agreement with predictions (see, e.g., the "standard" inversion factors in Fig. 100b).

g. Other Materials Systems

The use of QW structure also leads to clear improvements in other material systems. The first room-temperature CW laser in the GaSb/AlGaSb system was obtained using 4 × 120 Å quantum wells inside a graded separate confinement optical cavity.[501] The highest CW operation temperature inn PbTe/PbEuSeTe lasers has been achieved with a single quantum well placed in an asymmetrical separate confinement heterostructure (SCH).[502] In the GaInP/AlGaInP system (MOCVD grown on GaAs substrates), the lowest threshold current density up to now has been obtained for 4 × 100 Å wells in a SCH laser.[503]

It can be concluded from all the good operational features of quantum wells that "quantum well semiconductor lasers are taking over" as formulated by Yariv in a recent tutorial paper on the subject.[504]

CHAPTER VI

Towards 1D and 0D Physics and Devices

Considering the huge success of physics systems and devices based on 2D heterostructures, it was a natural trend to continue to diminish systems' dimensionality and indeed the recent years have witnessed a huge effort toward reducing the dimensionality of systems to 1D and 0D. As well as the whole 2D physics field, the 1D and 0D field is undergoing an exponentially growing phase, with exciting new results appearing on a weekly basis! Due to the size of the field and its rapid progress, we will outline here only its main features, leaving detailed descriptions to review articles or conference proceedings.[504a–510]

24. One- and Zero-Dimensional Systems

If one restricts particles to a narrow line or to a narrow box, further quantization of the particle motion can occur as compared with the 2D situation of quantum wells or heterostructures. In the general case it is impossible to obtain analytical solutions to the Schrödinger equation, but to illustrate the general physics we briefly study the very simplest models.

In the case of a restriction to a one-dimensional line, we assume translational symmetry in the direction y, so that the Schrödinger equation can be written

$$\left(\frac{p_x^2 + p_y^2 + p_z^2}{2m^*} + V(x, z)\right) \psi(x, y, z) = E\psi(x, y, z) \tag{103}$$

Since the potential does not depend on y, the wavefunction is necessarily a plane wave in the y-direction, so that we have

$$\psi(x, y, z) = \frac{1}{\sqrt{L_y}} \zeta(x, z) e^{ik_y y} \tag{104}$$

where $\zeta(x, z)$ satisfies a two-dimensional Schrôdinger equation

$$\left(\frac{p_x^2 + p_z^2}{2m^*} + V(x, z) \right) \zeta_i(x, z) = E_i \zeta_i(x, z) \tag{105}$$

In general these equations describe 1D subbands with $\zeta_i(x, z)$ a state bound in two dimensions and free motion in the y-direction. For a given subband, the energy of the state (i, k_y) is given by

$$E_{i,k_y} = E_i + \frac{\hbar^2 k_y^2}{2m^*} \tag{106}$$

from which we find the 1D density of states by imposing periodic boundary conditions in the y-direction:

$$D_i(E) = D_i(k_y) 2 \left(\frac{dE}{dk_y} \right)^{-1} = \frac{g_s L_y}{\pi \hbar} \sqrt{\frac{m^*}{2(E - E_i)}} \tag{107}$$

with $g_s = 2$ being the spin degeneracy; the factor 2 is for counting both k_y and $-k_y$. The density of states is therefore highly peaked and even has a singularity near the bottom of a subband. With the Fermi level at E_F, the number of electrons in subband i per unit length at temperature $T = 0$ is thus $N_i(E_F) = g_s (2m^*(E_F - E_i))^{1/2} / \pi \hbar$. In 1D structures subbands are more often referred to as channels. The contribution of one state to the current in the channel is $e\hbar k_y / m^* L_y$; in thermal equilibrium there is no net current, since the contribution from one state is exactly cancelled by the contribution from the state of opposite k_y. When one applies a week potential difference V between the two ends of the channel, the chemical potential for the states of $k_y > 0$ lies eV above that of states of $k_y < 0$, so that a net current flows:

$$I_i = \frac{e\hbar k_F}{m^*} \frac{D_i(E_F)}{2} eV \tag{108}$$

which by Eq. (107) gives the conductance of channel i:

$$G_i = \frac{I_i}{V} = \frac{g_s e^2}{h} \tag{109}$$

so that as one adds electrons to a quantum wire its conductance is quantized, each populated channel contributing the same amount, the conductance changing rapidly every time a new channel gets populated. The quantization is not as strict as in the quantum Hall effect; the observed accuracy of the quantization is of the order of 1%.

In order to solve Eq. (105) it is normally necessary to use approximate or numerical methods. In the case of an infinitely deep rectangular wire of dimensions $L_x \times L_z$ the Schrödinger equation can be solved by separation and we have

$$\zeta_i(x, z) = \zeta_{n_x n_z}(x, z) = \frac{1}{2\sqrt{L_x L_z}} \sin\frac{n_x \pi x}{L_x} \sin\frac{n_z \pi z}{L_z} \tag{110}$$

where $n_x = 0,1,2,3\ldots$, and $n_z = 0,1,2,3\ldots$, $(n_x, n_z) \neq (0,0)$ with the subband energy

$$E_i = E_{n_x n_z} = \frac{\hbar^2 \pi^2}{2m^*}\left(\frac{n_x^2}{L_x^2} + \frac{n_z^2}{L_z^2}\right) \tag{111}$$

It can be seen that if L_x is considerably larger than L_z, the n_x levels form a ladder of small steps within the subband ladder of well separated n_z levels. For $L_x = L_z$, however, the two ladders cannot be clearly separated and many energy levels are degenerate.

The extension to zero dimensions is conceptually simple. If one creates a potential that confines the particles in all three dimensions, it is possible to have completely discrete bound states. In that way a "superatom" or a quantum dot has been created. The density of states per unit volume of an ensemble of identical uncoupled quantum dots consists of δ-functions at the discrete energies of each dot; the strength of each δ-function is the spin degeneracy multiplied possibly by the (accidental) degeneracy of the bound state multiplied by the number of dots per unit volume. As for real atoms, the discreteness of the levels strongly restricts the perturbations that will allow transitions between levels; there is therefore the possibility that the lifetime of an electron excited to a higher-lying level can be very long, almost exclusively determined by the radiative transition rate. Evidently the singularly peaked density of states and the possibly very long lifetime in excited states would make optical devices based on 0D structures potentially more performing; the main challenge is in fabricating such structures, in particular making the dots identical.

25. 1D and 0D Semiconductor Fabrication Techniques

As 2D systems are so easily obtained in the growth direction by MBE or MOCVD techniques, an obvious way to obtain 1D (quantum wires) or 0D (quantum dots) systems is to pattern usual 2D heterostructures with nanoscale lithographic techniques. The limits set by the lithographic process are well in the quantum size region, with present state-of-the-art pattern transfer sizes in the 10 nm range.[511]

However, the damage introduced by the etching procedure used in the pattern transfer onto the semiconductor is such that it dominates all electronic properties at small dimensions. One very well documented effect of the etching damage is the carrier depletion, which extends, in the best cases, for standard carrier concentrations, over ≈50 nm. The observations rely on the dependence of electronic population of quantum wires or dots on geometric lateral size: below ≈100 nm, quantum wires or dots are empty.[512] At present time, the GaInAs-based materials seem less sensitive to the effect.[513] Another probe of the damage is the decrease in luminescence efficiency due to the numerous nonradiative defects.[514] Overgrowth over etched structures seems to repair only partially some of the damage.[515]

To obtain less-damaged structures, softer fabrication techniques have been searched for. A few of these have proven useful, although with lower geometric resolution than the direct heterostructure etching (Fig. 103):

(1) Partial etching of the top confining material provides a lateral confining potential due to the different built-in electrostatic potentials in the depleted zones[516] (see, e.g., Fig. 22).

(2) Gates can also be deposited on such patterned confining layers in order to control the electron density.

(3) Applying metal layers on patterned resist layers also induces a periodic potential on a 2D electron gas as the Schottky surface potential gets patterned.[517]

(4) Split-gate electrodes give a voltage-controlled potential well between the electrodes.[518] These structures are widely used in point-contact emitters for longitudinal transport and as potential-controlled wells for lateral transport.

(5) Strain-induced potential wells:[519] Carriers can be laterally confined within a continuous quantum well by superimposing patterned strain-induced potential wells added to the usual heterostructure potentials. The strain is produced from an initially continuous strained layer. The layer is patterned and etched into stressors, which upon relaxation of the material at the edges of and in between the stressors yield a spatial variation of the volume dilations of the quantum well material, resulting in a modulation of the band edge through the deformation potential.

The possibilities of such structures are numerous. Fig. 104 evidences the various situations that can be studied (isolated quantum dots, connected dot array, isolated "antidots") by simply varying the applied voltage of a patterned gate, the energy potential modulation of which is represented by a 2D sinusoidal modulation.[519b]

Finally, a number of direct-fabrication methods are being investigated.

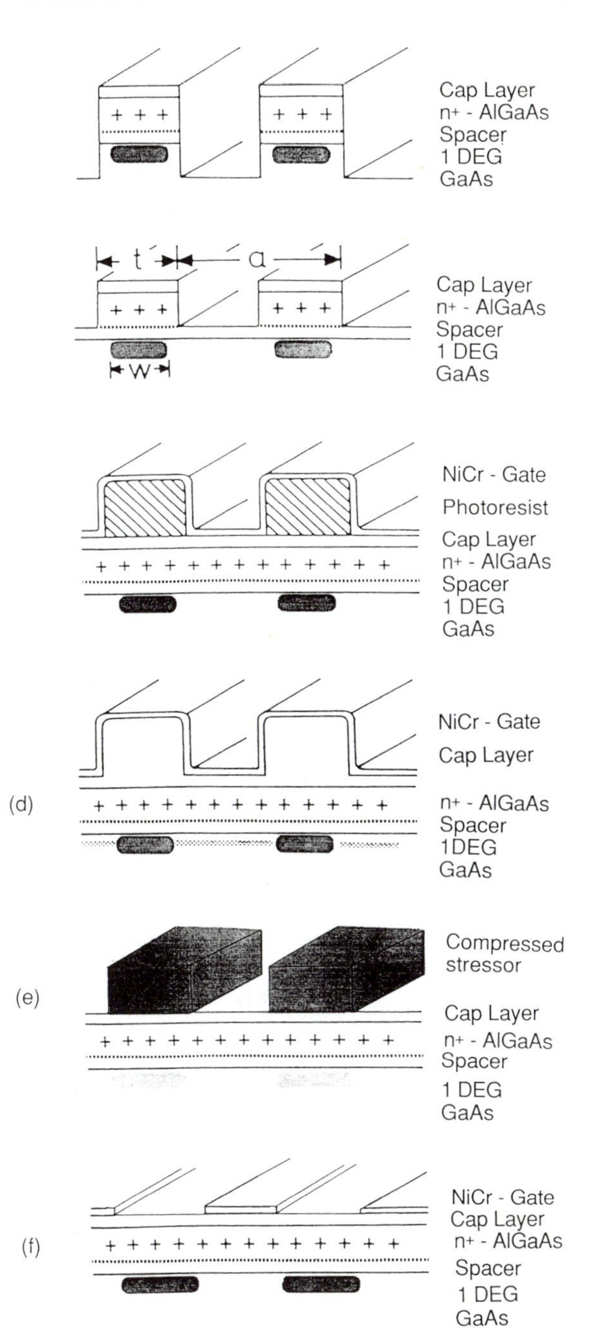

FIG. 103. Various fabrication techniques for laterally nanostructured systems. (After Ref. 519a).

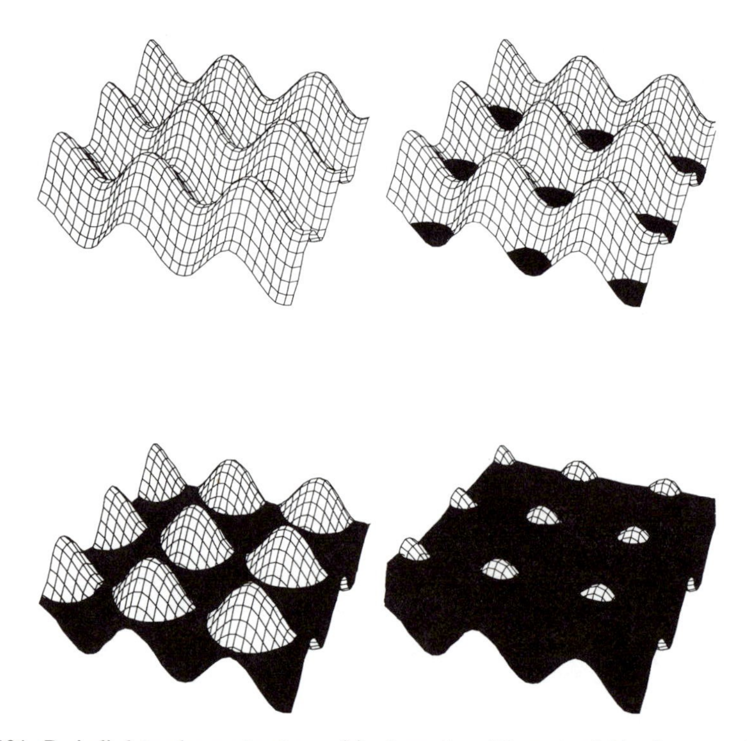

FIG. 104. Periodic lateral nanostructures: (a) schematics of the potential landscape under a 2D-patterned gate; (b, c, d): evolution of occupied states (top of Fermi sea) when varying the applied electric field, changing situation from isolated quantum dots to interconnected quantum dots to "antidots." (A. Lorke and J. P. Kotthaus, private communication).

These appear very promising, for they should yield material with much lower defect densities. These can be classified as follows:

(1) Controlled precipitation in matrix: The oldest method by far, as it is widely used in commercial color filters for optics. From supersaturated glass melts with semiconductor elements such as CdS, CdSe, and their solutions, one fabricates first a supersaturated glass matrix, and then precipitates semiconductor inclusions through annealing.[520] The size distribution of crystallites is controlled by the temperature and duration of the annealing step. Crystallites with radii in the range of tens to hundreds of angstroms have been fabricated with 10–20% size dispersion. Other recent routes, including low-temperature fabrication, can be found in a recent materials report.[521] Some controlled-fabrication techniques consist of cluster formation in cavities of polymer films[522] or of zeolite structures.[522–4] The latter method yields crystallites with less size-dispersion, but with

only very little latitude in size, usually below the (10 Å)^3 range. Zeolites with large atom "cages" will be needed to obtain larger crystallites.

(2) Free-standing nanocrystallites have been fabricated by several methods: Vapor-phase nucleation can deposit nanometer-sized clusters. Controlled reaction from solutions also yield similar crystallites.[525] A major issue is the size-control of clusters, which can be obtained through diffusion control or ligand-controlled size.[525a] A most fascinating fabrication route is provided by bio-organisms: Living cells have DNA-synthesized peptides, which are able selectively to attach heavy metals in order to protect the organism against these killers. By culturing yeasts into Cadmium salt solutions, Dameron *et al.*[526] were able to extract CdS microcrystallites that had been nucleated by the peptides. Particle sizes in the order of 20 ± 3 Å were synthesized, with a predominant rock-salt structure as revealed by X-rays. As can be foreseen, a major drawback of all these directly synthesized crystallites is their free-standing nature, which makes their use in devices somewhat dependent on specific implementations.

(3) Direct semiconductor growth methods seem to be very promising because they yield the usual growth control of MBE and MOCVD with low defect densities (Fig. 105). A first method is the layer-by-layer growth on vicinal surfaces, which yield quantum wire structures through nucleation on well-organized steps.[527] More recent methods rely on step-induced directional growth, which can easily be controlled by the use of a patterned substrate.[528] Further control can be obtained by using selective growth directly on the pattern written on the substrate, through a selection of the orientation of growth-exposed surfaces.[529] Finally, growth on sidewalls of cleaved and etched multiquantum well structures provides excellent 1D channels, although through a delicate two-step growth techniques.[529a,530]

(4) Nucleation of quantum dots during growth can also be contemplated. Already quite controlled in the case of Pt on W,[530a] the technique has been demonstrated for InAs on GaAs, where the growth beyond a strained monolayer of InAs leads to 3D nucleation of well-organized InAs clusters.[530b] The control of clustering growth of semiconductors on semiconductors seems actually quite possible,[530c] with the remaining problem of overgrowth of high-quality imbedding material over a layer of clusters.

(5) Lateral confinement can be obtained from impurity-induced effects: One of these is the carefully controlled, low-energy, ion-implantation-induced damage, which yields carrier confinement due to Fermi-level pinning by defects.[531–2] The other, already widely used as an index-guided laser manufacturing technique, is the impurity-induced interdiffusion disordering (IID).[532a] It has been widely observed that several atom species greatly enhance quantum well interdiffusion processes. By selectively depositing such impurity atoms by masked selective diffusion,[533]

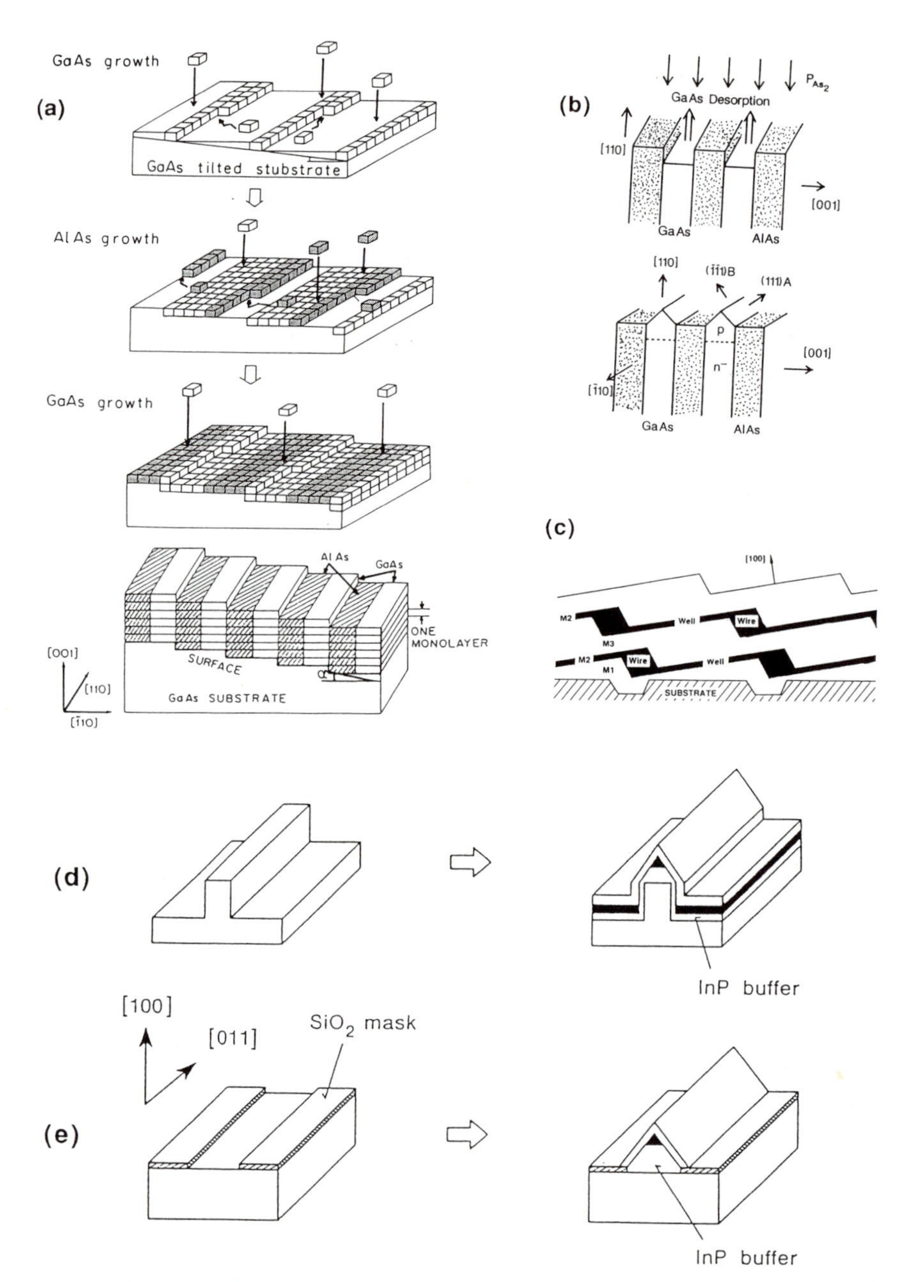

FIG. 105. Various schemes for direct growth of nanostructures: (a) tilted surfaces,[527b] (b) side-growth on quantum wells,[529a] (c) growth on grooved substrates,[529b] (d, e) growth on patterned substrates.[529]

by ion-beam masked implantation,[534] or by direct ion-beam writing, one can define the sample areas that are selectively interdiffused during an annealing phase. Because the interdiffused material has an intermediate gap between those of the layers which interdiffuse, these layers act as laterally confining potential barriers for those carriers in the smaller-carrier-energy materials.

26. Electrical Applications of 1D and 0D Structures

The low-field conductance of an ideal quantum wire was discussed in Section 24 and was shown to be quantized, with the conductance equal to e^2/h times the number of channels occupied. A real quantum wire as realized in, e.g., a quantum point contact is not ideal in several respects: (i) it is not infinitely long and does not have perfect translational symmetry; (ii) the translational symmetry is also broken by impurity potentials in and near the channel; and (iii) whereas scattering events by a small number of defects can give rise to observable, *reproducible,* sample-dependent interference effects such as in the universal conductance fluctuations (UCF) and weak localization, which will be described below, at nonzero temperatures the coupling to phonons creates scattering processes that are random in time and therefore lead to averaging of interference effects. They change the phase of an electron in a *random* way, and the electron loses memory of its phase. The relative importance of these deviations from ideality gives rise to different regimes of conductance.[535,536] The regimes are determined by the length scales of the different processes: the length L and width W of the wire or constriction, the elastic mean free path l_e between impurity scattering processes, and the inelastic or phase-breaking mean free path l_ϕ between phonon scattering events (Fig. 106).

(1) If $l_\phi \gg l_e \gg L, W,$ the only potential felt by the electrons is the boundaries of the wire. Then quantum states exist that extend from one end of the wire to the other. Any such occupied state carries current from one end to the other and if no voltage is applied between the two ends, the left- and right-going currents exactly cancel. If a small voltage V is applied, however, states up to the chemical potential μ_L in the left contact carrying current from left to right are occupied, whereas only states up to the chemical potential of the right contact μ_R are occupied. This imbalance creates the net current from left to right proportional (to first order) to the potential difference $\mu_L - \mu_R = eV$. The conductance is then determined by the quantum-mechanical transmission probability of the states between μ_R and μ_L; i.e., the wire represents a well-defined barrier between the two reservoirs of electrons in either contact.[537–540]

This quantum-mechanical picture is analogous to the coupling of electromagnetic radiation into a waveguide: At a fixed frequency (corresponding

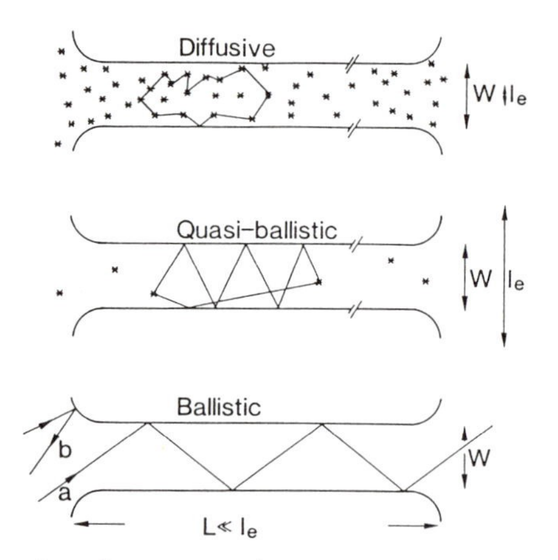

FIG. 106. Various regimes for transport in constricted regions depending on the relative length scales;[536] l_e is the elastic mean free path.

to the chemical potential $\mu_L \approx \mu_R$) electromagnetic radiation is reflected back from the entrance of the waveguide or transmitted through via a waveguide mode. The complementary picture of a waveguide mode is that of optical rays being reflected on the sidewalls of the guide. For the electrons in the quantum wire, this corresponds to looking at the electrons as particles whose trajectory is determined by specular scattering on the potential walls of the wire in a perfectly ballistic way. If the width W is small compared with the Fermi wavelength, then only one or a few channels can be occupied and one talks about quantum ballistic transport.

(2) If $l_\phi \gg L \gg l_e \gg W$, we have a few impurities in the wire. In that case transport is still via channels but the scattering introduced by the impurities mixes the modes, thereby increasing the reflection probability of electrons entering the wire. It is also possible for electrons via multiple scattering on the walls and on a few impurities to be trapped in states that are localized on the scale of l_e. Since those states have no contact to the reservoirs they do not contribute to the transport. In this regime the reflection probability depends strongly on the precise positions of the impurities and the potential that defines wire, so if one changes the latter (or changes the sample) the conductance changes rapidly. Since the changes are usually of the order of e^2/h, this regime is referred to as the universal conductance fluctuation regime.[535]

(3) $l_\phi \gg L > W \gg l_e$. In this situation multiple scattering on the impurities dominates, so wire modes no longer have meaning. Instead, electrons are localized both longitudinally and transversely on the length scale l_e. The electrons no longer see the one-dimensionality of the wire. It can

be shown that no states exist that extend from one end of the wire to the other; the regime is the weakly localized, 2D one, which has no conductivity at low temperatures.

(4) The only possibility then of transporting electrons through the wire is by scattering between the localized states. This requires inelastic scattering, however, in which case we end in the classical 2D Boltzmann regime $L, W \gg l_\phi$ in which electrons diffuse through the wire, effectively averaging over impurity positions. Then the mobility is determined by the average density of impurities. If one extends this regime to high temperatures, the inelastic length may become smaller than the elastic one, so that phonon scattering determines the mobility, as under usual conditions at room temperature.

The ballistic transport in quantum wires has profound effects, since the electrons truly show their wavelike nature. The wires act as waveguides for the electrons so that the resistance of a wire loses its local meaning. The boundary conditions on the contacts become of extreme importance and a system of connected wires shows a nonlocal behavior. Thus, a current flowing between two contacts may influence the voltage between two other contacts, even though classically no current should flow between the two voltage probes. Also, the length and termination of a piece of wire not connected to a contact will change the behavior of the rest of the structure, as any stub circuit in a microwave waveguide setup. Even more spectacular results can be obtained if one terminates a wire in a loop:[541] since the reflection coefficient of the loop oscillates through the Aharonov–Bohm effect with the magnetic flux enclosed in the loop, such oscillations may be observed anywhere in the rest of a connected system such as the one shown in Fig. 107.

Another consequence of the nonlocality of transport is the quenching or even reversal of the Hall effect at low magnetic fields.[542] Usually one measures the Hall voltage between two contacts on opposite sides of a current-carrying wire, but in the ballistic regime the number of electrons reflected into the Hall voltage probe wires depends on the exact geometry of the cross points of the Hall bar.[543,544] The explanations rely on the waveguide properties of narrow constrictions or on a classical billiard ball model.[545,546] Both types of explanations rely on the focusing of electrons occurring whenever a narrow channel opens up at a crossing or at a flared (hornlike) constriction. Then the adiabatic branching in quantum states with larger longitudinal kinetic energy (due to the smaller confining energy of wider channels) induces the focusing effect, which has been directly observed in another setup by using a second point contact as a collector[547] (Fig. 108).

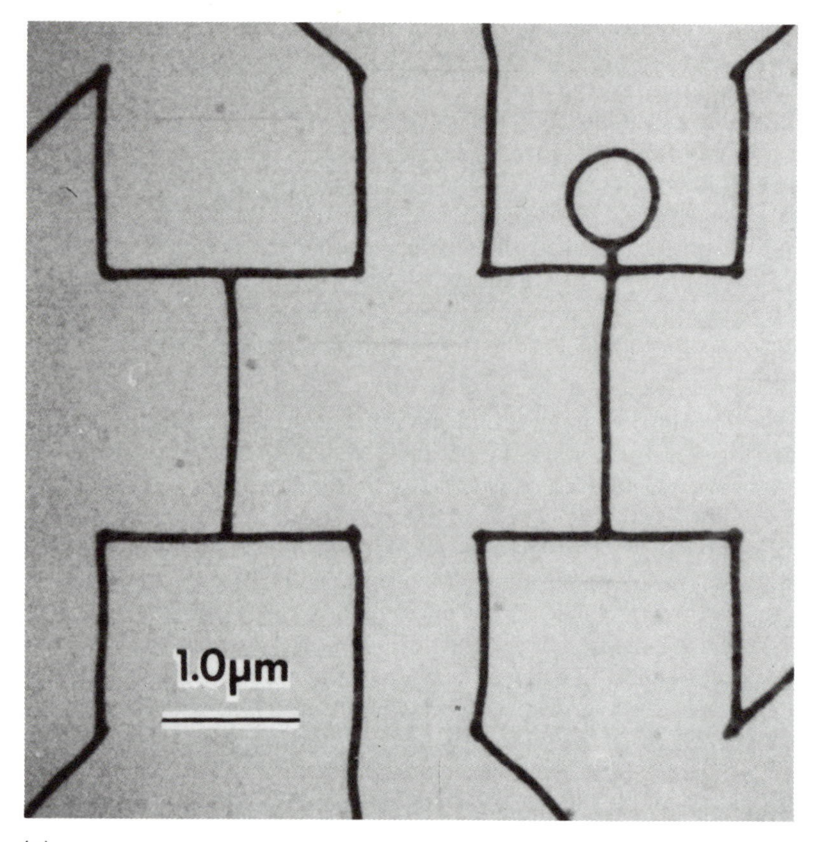

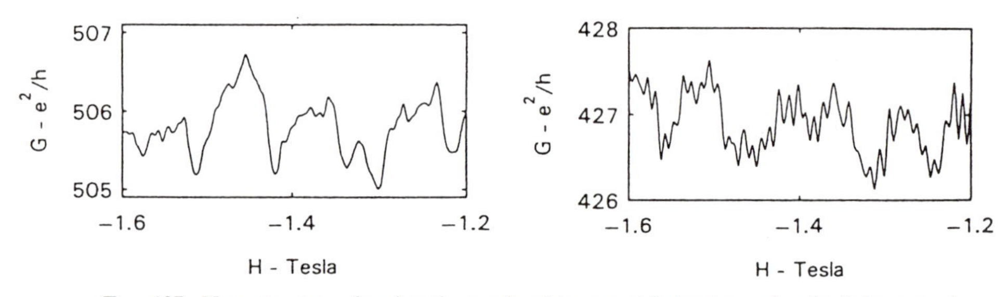

FIG. 107. Nanostructure showing the nonlocal transport behavior under the influence of a magnetic field.[541] The reflection of an electron from the ring structure depends on the magnetic flux enclosed; this creates oscillatory behavior everywhere in the circuit.

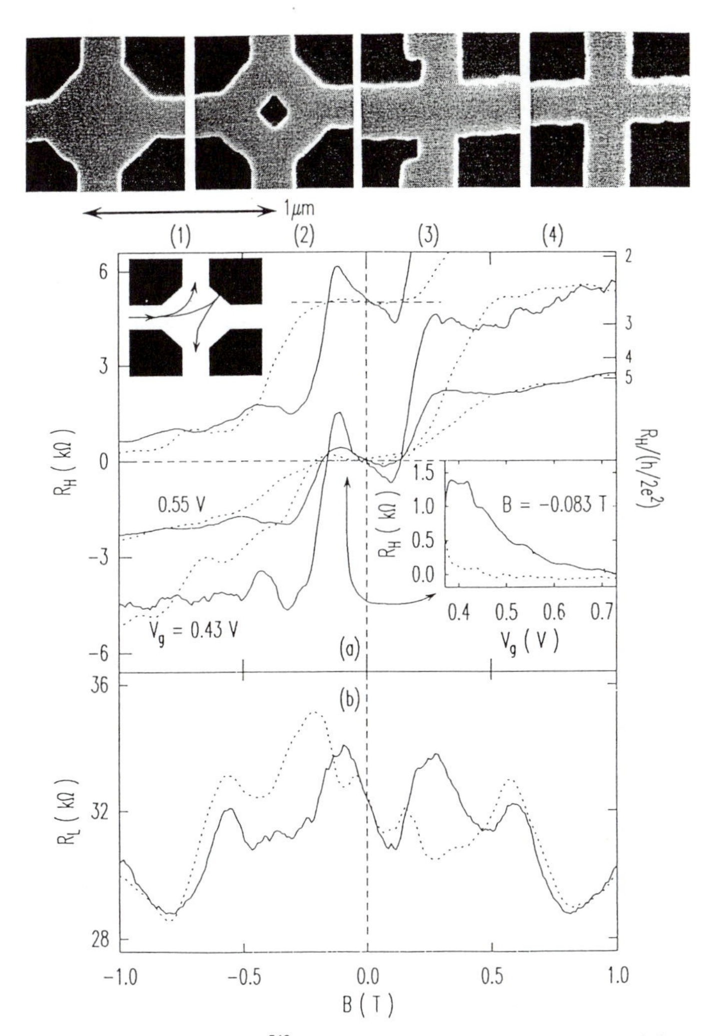

FIG. 108. Anomalous Hall effect:[543] Depending on the exact geometry of the crossing of Hall voltage probes and the central Hall bar, the Hall voltage is quenched or even reversed.

These "mesoscopic" systems have been studied intensively during the last few years. They are quite well understood, as can be seen from several recent reviews.[506,535,548–550]

In the quantum ballistic regime, we have already mentioned the quantized conductance as a novel effect.[551,552] Another important effect of a point contact or lateral constriction is its action as a filter: Only the electrons having an energy corresponding to the channel mode will be transmitted. This may be used to create beams of monoenergetic electrons perfectly analogous to an optical point source. If one introduces a potential by a concave-shaped gate in the path of those electrons (Fig. 109), focusing can be demonstrated in analogy with geometrical optics:[553,554] The refraction is due, as in optics, to the change in forward wavevector when the

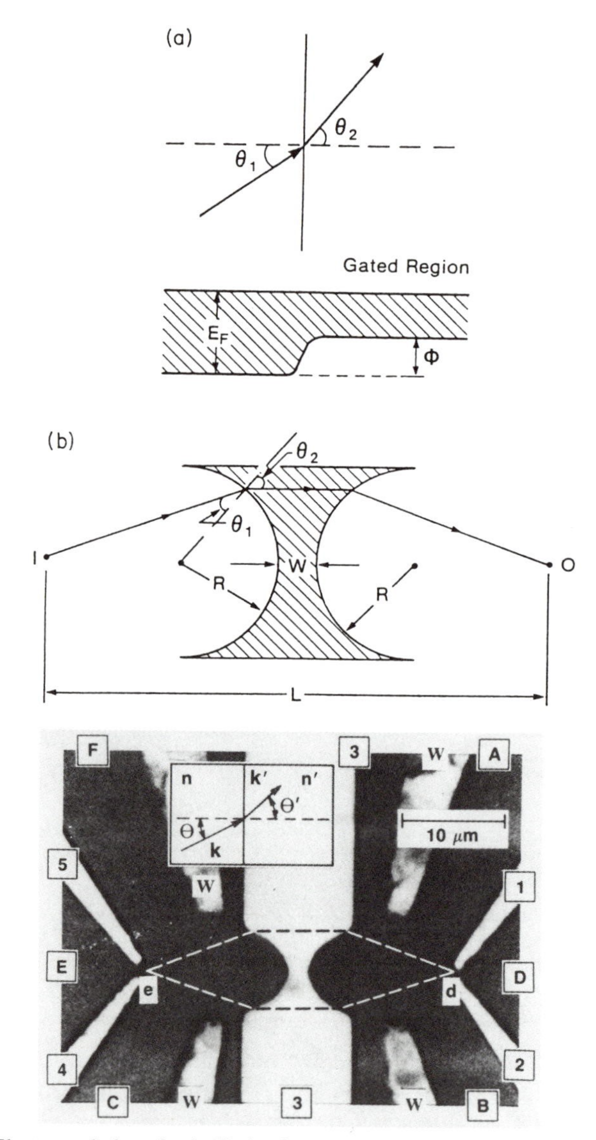

FIG. 109. Electrostatic lens for ballistic electron focusing. (From Refs. 553 and 554).

electron enters a region of different potential. This potential can be changed by variation of the gate voltage.

A corrugated gate (Fig. 110b inset) creates a lateral superlattice.[555] The transmission of such a structure is analogous to the superlattice studied in Section 7c, and for a finite 15 period lattice the transmission should theoretically show 15 peaks inside each miniband, as in Fig. 110a. The experimental result in Fig. 110b proves that the electrons are ballistic or coherent across the whole superlattice.

Enclosed regions between two constrictions as shown in Fig. 111 have also been studied.[556] If one applies a fairly strong magnetic field, the enclosed region acts as a quantum dot weakly coupled via the constrictions to the outside region. Since the dot state energy depends on constriction

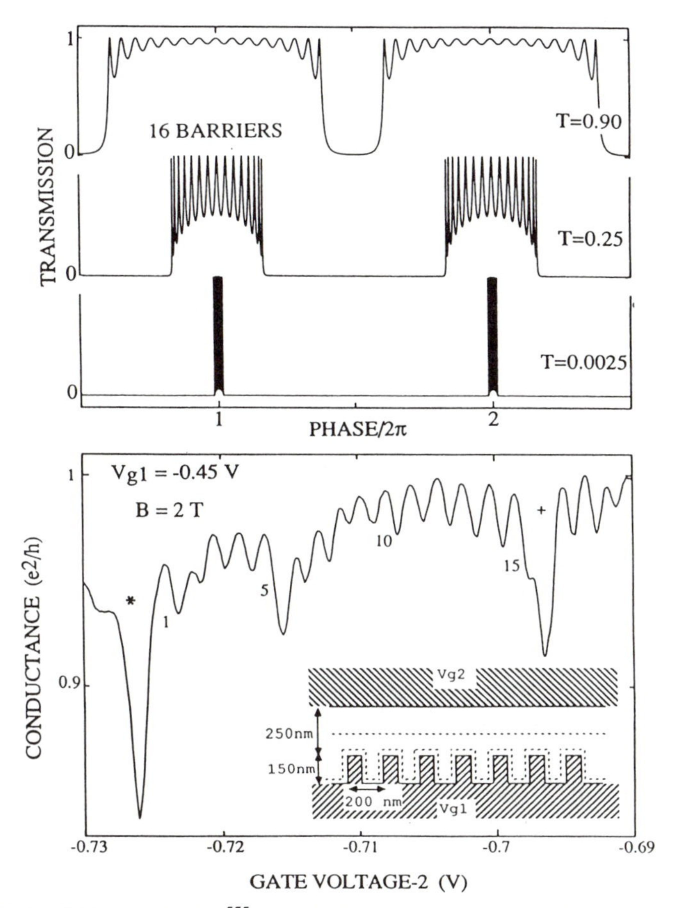

FIG. 110. Artificial superlattice:[555] (a) calculated transmission (b) observed conductance.

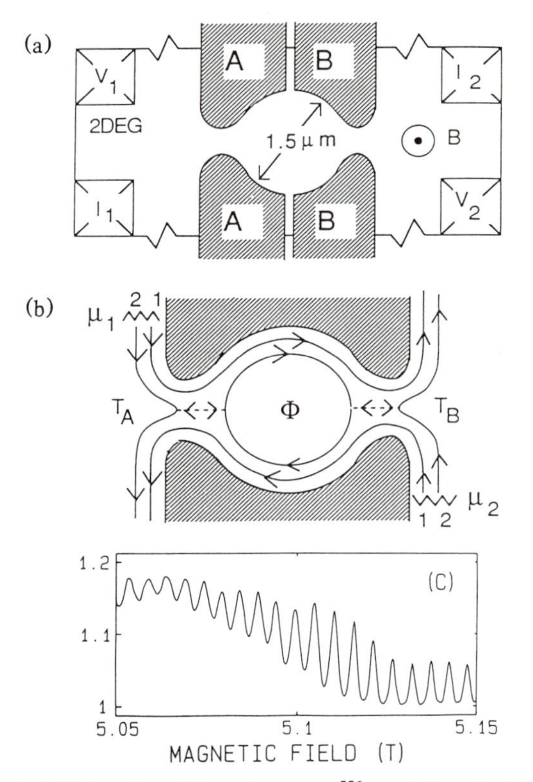

FIG. 111. Electrical "Fabry–Perot" interferometer[556] consisting of a ballistic quantum dot with outside world connections through quantum point contacts. The transmission of the structure is given by the Fabry–Perot formula $T = T_A T_B/(1 - 2\sqrt{R_A R_B}\cos(\theta + \theta_r) + R_A R_B)$ where the phase θ changes with magnetic field, and T_A, T_B, R_A, R_B are transmission and reflection probabilities of the point contacts.

voltages and magnetic field, it is plausible and actually observed that the structure represents the one-dimensional analog of a double-barrier structure or a Fabry–Perot optical interferometer, showing resonant transmission when the quantum dot state is in resonance with the Fermi energies of the outside contacts (see Eq. (33n)). A natural extension of one such dot is to a system of two or more weakly coupled dots or to a periodic two-dimensional lattice of coupled dots. The former system represents an artificial molecule[557] and the latter an artificial 2D crystal (Fig. 104).

It is evident that with mean free paths of tens of microns,[558] rather standard lithography is sufficient to access the ballistic regime of electrons, whose behavior strongly resembles optics with the major difference of the action on the electrons of a magnetic field. This will certainly continue to spawn many highly interesting experiments. Recent reviews of this exploding field are recommended for further study.[508–510]

As usual, the vertical direction provides an easier way to reach a dimension well below the elastic mean free path. Using laterally confined structures through etching techniques, the tunneling properties of quantum dots in double-barrier resonant tunneling diodes have been successfully evidenced[558a] as well as direct measurements of the 1D and 0D electron states by capacitance spectroscopy.[558b]

These breakthroughs now open the way to truly ballistic or quantum-mechanical devices, where one would exploit the wave nature of electrons to produce single-electron, deterministic devices.[560] This is of course at large variance with our present-day well-known transistors operating at best with $\approx 10^4$ electrons, which follow a large number of random, space and time-averaged trajectories.

27. Devices Based on 1D and 0D Effects

The field of 1D and 0D devices is both so new and progressing so fast that it is extremely perilous to make predictions. We can only point out here a few directions and questions which concern things to come.

1D and 0D devices are the results of the natural evolution of technology[559–60]; controlled growth has allowed us to easily obtain 2D materials and devices. Forthcoming lithographic tools will allow systematic, reproducible, "mainstream technique" patterning down to the few-nanometer range sometime in the future (10 years?) (Fig. 111a). More

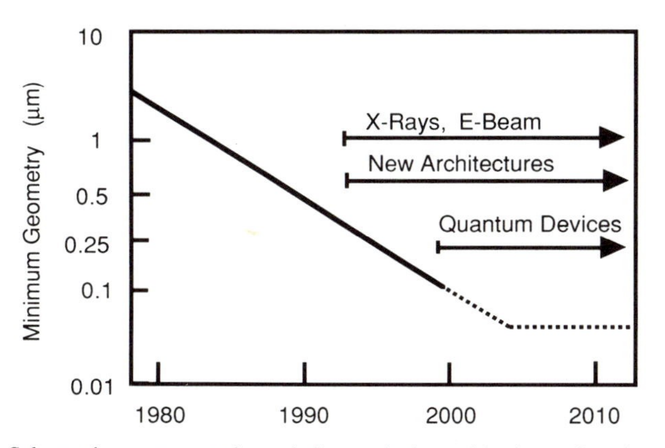

Fig. 111a. Schematic representation of the evolution with time of mainstream, "standard," "available" lithographic minimum feature size. Below 0.25 μm, both new lithographic techniques and chip architectures might be required. Below 0.1 μm, quantum effects will appear.

clever direct fabrication techniques, as described previously, could in fact also produce nanometer-range devices somewhat earlier.

It is therefore time to start thinking about making the best use of the technology which will be at hand tomorrow. In the electron device field, five avenues clearly are open: ultrasmall structures for speed and integration, quantum effects, granular electronics, ballistic deterministic motion, and electron-phase control.

- The *ultrasmall structure* concept for speed and integration is clear-cut and follows the mainstream approach to VLSI and ULSI. Within this natural trend, quantum devices are just the results of the evolution of lithography. It is common knowledge that important issues must be solved in order to capitalize on the forthcoming device densities and speed: As deterministic, zero-defect interconnections seem out of reach in order to conserve reasonable fabrication yields, one will have to leave predetermined interconnection networks and look for new architectures; as error-margins will decrease due to increased fluctuations (both intrinsic, due to diminished carrier numbers for logic operations, and extrinsic, due to fabrication tolerances), one has to allow for high rates of soft errors; synchronization of faster devices on ever-larger dies will become impracticable, etc. Several solutions are being studied, some of them already implemented in leading-edge designs: allowance for redundancy and self-repair, error-correction codes for soft errors etc.
- *Quantum effects* were discussed in Section 22. One uses the existence of internal, fabrication-determined energy levels to yield devices with a higher or complex functionality. This field is just opening, and many specific functions have to be evaluated and implemented to decide about the industrial impact of such an avenue. Clearly, such devices will allow the needed functionality of nodes in cellular automata machines (see later in this section) to be achieved with a single device.
- *Granular electronics*[561] is of course highly desirable, as one would use single-or few-electron events to perform digital operations instead of our present-day, well-known transistors operating at best with $\simeq 10^4$ electrons. Such devices are based on concepts similar to the Coulomb blockade concept[562–4], where the single electron charge changes a "local" potential energy in a structure by $\Delta V \simeq e/C$, where C is a "local" capacitance, determined by the device geometry and materials. The Coulomb blockade effect has been the subject of many investigations in the context of small, superconducting tunnel junctions at low temperatures and can be shown to allow digital electronic functions when several devices are combined[564]. A recent implementation demonstrates a turnstile device activated by an alternative gate voltage[565]. The future issues here are the demonstration of

high-temperature, error-free devices (i.e., device energy e^2/C must be much higher than operating thermal energy k_BT), circuit and systems architecture, including interconnections, self-repair, etc.

• The recent progress in *ballistic-deterministic motion* indicates that many large, interesting effects can be obtained in a non-quantized, classical situation where electrons travel without collisions. Some effects occur in really unquantized modes, such as those due to coherent focussing effects or the lateral hot electron devices[566]. The analog steering of an electron beam into spatially arranged collector electrodes through electrostatic lenses or split-gates might allow A/D conversion in the multi-GHz range. One would use the ballistic electron device as the critical element of the converter, and connections to the outside world would be made through standard devices, achieving an integrated electron-optics component. One can also realize a tunneling-hot-electron transfer amplifier (usually designed in a vertical geometry as described in Section 22) in the 2D electron gas plane thanks to narrow metal gates (~50 nm) (Fig. 112). Due to the excellent transport properties retained from the 2D heterostructure, current gains larger than 100 were obtained (at low temperature).

• Based on the *electron-phase control* of waveguided electrons, new devices relying on electron optics[567] can be designed[568]. One of the most promising could be the Mach-Zehnder electron interferometer, where the electron phase in one arm could be changed by an applied electric field. Another could be the one-dimensional electron Fabry-Perot interferometer where electrons display transmission resonances whenever the 0D-state energy—when moved by a magnetic field—coincides with the Fermi energy[556]. The influence of voltage-controlled remote electrodes placed on "stubs" analogous to their microwave counterparts could also efficiently modulate the waveguide transmission coefficient[569].

For *digital* applications, many difficulties originate both from hardware and software considerations: The low temperature is required for "reasonably" deterministic noiseless operation (but 77K operation might be possible with submicron structures), and also the amplification capability of such devices is still to be demonstrated in order to provide the fan-out required by systems using cascaded architecture (in memory-only types of devices, however, such as EPROMs, the fan-out difficulty disappears and such applications might well be in the range of single- or few-electron devices[570]). Most present devices use gate voltages in the volt range, which would require a large number of electrons to switch conductance from one state to the other (except in the Coulomb-blockade mode where one might require an impedance transformer as most inputs are in the volt range, which for a single-electron current represents a huge impedance).

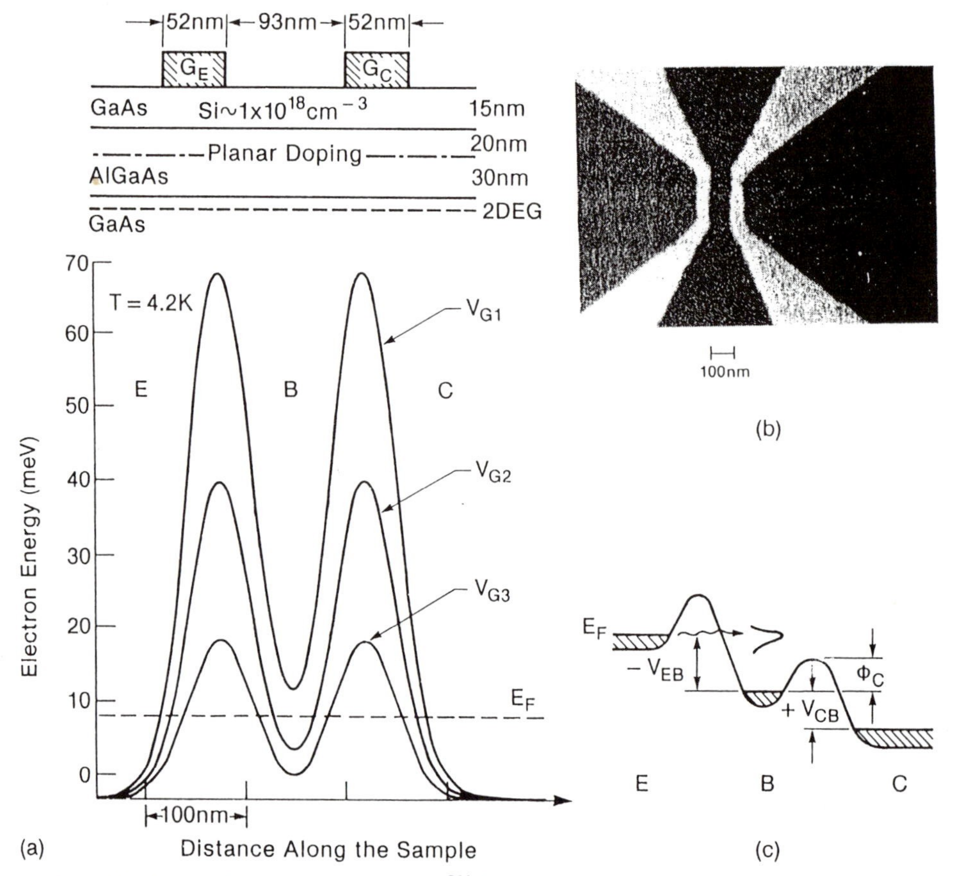

FIG. 112. Lateral hot electron transistor:[566] The two thin gates produce lateral barriers analogous to the vertical hot electron transistor (Fig. 85a).

To alleviate difficulties associated with non-deterministic modes of operation, which would simply add to the unavoidable non-operating devices and connections in ULSI circuits comprising tens or hundreds of millions of transistors, some authors[559,560,571–574b] have evoked the need for revolutionary, fault-tolerant architectures such as *cellular automata.*

Cellular automata machines consist of arrays of elementary digital processors that are located at the nodes of a regular lattice and are connected to a few neighbors. The processors can be as simple as logic gates, but can also be fullscale microprocessors. The main advantage of such machine architecture is the easy synchronization of operations (cells only communicate with nearest neighbors) and modularity. These systems can also be made robust against hard or soft failure by redundancy and verification. Such machines have been shown to be very efficient for implementing specialized functions to solve problems which rely on physical laws which are similar to their architecture, i.e., incorporating high degrees of locality and parallelism[574c]. The debate is still open on whether the universal computer machine can be efficiently designed through a cellular automata architecture. In any event, quantum devices could easily be associated in such architectures to yield standard cellular automata functions such as matrix inner products (Fig. 112a), Fourier transforms, convolution products, etc. An excellent description of processor array operation is given in Chapter 8 of the celebrated Mead and Conway book[574d].

28. 1D and 0D Optical Phenomena

The impact of lower dimensionality in optically active systems is easy to grasp when comparing them with the earlier discussion of optical properties of 2D systems. It relies on several effects, the most important of all being the progressive restriction of allowed states dispersed over $E(k)$ bands toward more concentrated single-energy atomiclike levels of quantum dots.

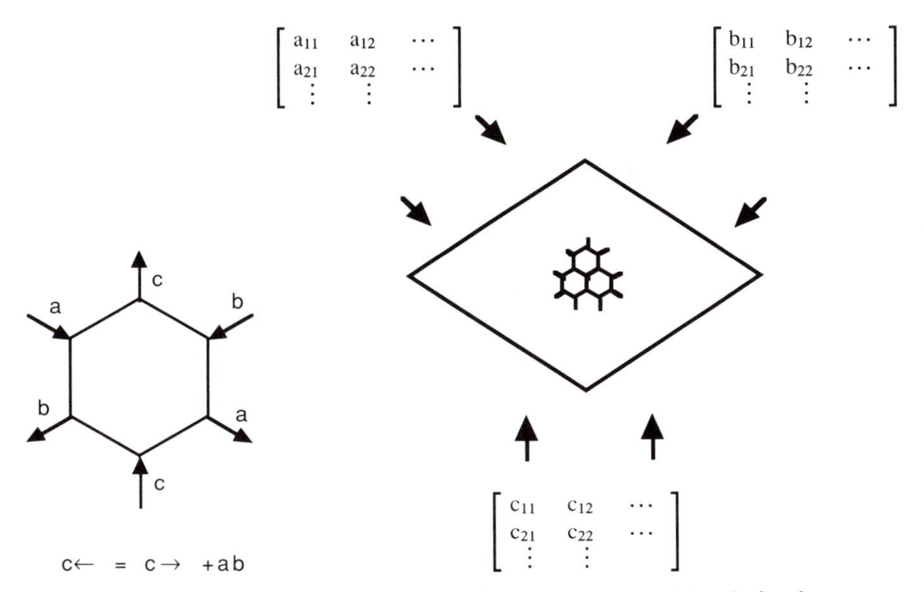

Fig. 112a. Schematic representation of a cellular automata machine. Left: elementary node processor performing the elementary operation for inner product; right: schematic depiction of the motion of data through a cellular machine performing the matrix inner product $c = a \cdot b$ (at the time $t = 0$, all the c's are reset to zero).

(1) The importance of dealing with isolated energy levels: As is obvious from Eq. (54c), dealing with atomlike levels enormously sharpens resonant behavior and therefore energy selectivity. Besides all resonance effects, an important property is the lower dispersion of optical properties over k-states. It is due to the k-selection rule that only vertical transitions are allowed in the $E(k)$ representation of quantum states (k-conserving transitions). Therefore, the occupancy of varied k-states as required by statistics increasingly scatters the properties of injected carriers in 1D, 2D, and 3D. Taking the injection laser as an example (Fig. 113), the occupancy of the same number of electrons of 2D, 1D and 0D states above inversion will of course lead to higher gains due to the concentration of electrons and holes over fewer k-states (as long as only one or very few quantized states are within the relevant energy range from the ground state, usually $\sim kT$). Another way to see this is to remark that the saturated gain per carrier (for fully inverted conduction and valence bands) is independent of dimensionality (because the interband matrix element is such). Therefore, concentrating carriers in a phase-space that is less extended in energy (in 0D all quantum boxes exhibit transitions at the same energy level) will lead to a larger maximum in the spectral gain curve, although the integrated gain, of no use except in switched lasers, is of course constant.

(2) Considering exciton effects, three scales are to be considered in quantum dots:[575–577] When $L \ll a_B$, confining kinetic energy is much larger than the Coulomb interaction between electron and hole. In that case, the latter is a perturbation (shift in energy) and the wavefunctions are the exact quantum box wavefunctions. The oscillator strength per transition in the quantum box is just the usual interband oscillator strength

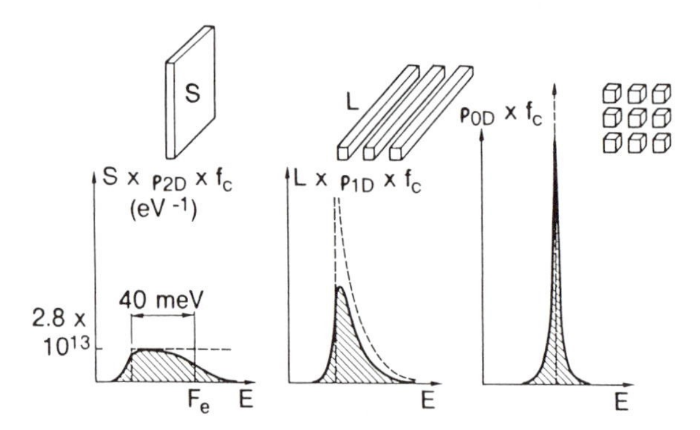

FIG. 113. Schematics of gain curves in 2D, 1D, and 0D structures. Similar numbers of electrons and holes are being injected above transparency, yielding equal integrated gain. (From Nagle and Weisbuch.[465])

because for instance in the case of the rectangular QB, the transition matrix element can be factorized in the three directions along the method of Eq. (54a). Therefore, a planar array of quantum boxes has an oscillator strength $f(D_xD_y)^{-1}$ per unit surface, where D_x and D_y are the center-to-center distances in the x- and y-directions. It should be remarked that in that case the quantum box quantization is not significantly better than the 2D QW with room-temperature excitonic effects. In that case, the oscillator strength per unit surface of quantum well is $(8/\pi a_B^2) f_{at}$, (Eq. 54h), which can be beaten by paving surfaces with quantum boxes only if their center-to-center distance is smaller than $a_B(\pi/8)^{1/2}$—certainly a difficult manufacturing challenge!

When $L \gg a_B$, (but still well-separated, confined energy levels), a "giant" oscillator strength situation develops, which yields a transition matrix element[575–7]

$$f \approx f_{at} \cdot (V_{box}/V_{exc}) \tag{112}$$

where V_{box} and V_{exc} are the QB and exciton volume, respectively. The enhancement of the oscillator strength originates in the coherent excitation of the QB volume (somewhat analogous to the intersubband giant dipole matrix element of Eq. (54e)), which yields an increased dipole moment.

In the intermediate case $L \approx a_B$, the situation can be more complicated as the exciton binding energy can lie in between the confinement energies of electrons and holes. However, the oscillator strength can still be somewhat increased compared with the uncorrelated QB case.

For the quantum wires, it is obvious that exciton properties will lie somewhere in between those of QWs and QDs.

(3) Nonlinear effects will exhibit lower thresholds due to the smaller number of states to fill to reach saturation. In addition, multiple-particle interactions are strongly increased in 1D and 0D: Increasing numbers of electron-hole pairs in a confined volume leads to large coulomb interactions, which can be described as biexciton or multiple-exciton effects.[575–80] These will lead to new spectral nonlinear features. An alternate explanation of the new spectral nonlinear features is the trapping of the excited e-h pair and the resulting Stark effect on the quantum dot energy level.[581]

(4) Electro-optic effects should have very large resonances, due to the sharpening of the 1D and 0D DOSs as compared with QWs.[582]

The fabrication of 1D and 0D optical structures and devices is much more challenging than those of quantized electronic structures based on soft fabrication techniques: The latter only allow smooth potential variations, which usually lead to quantum confinement energies in the meV range sufficient for low-temperature operation of electronic devices. Optical devices require substantially larger quantizing energies in the range of tens of meV, since band filling or heating effects are important, owing to effective carrier temperatures quite higher than the lattice temperature, the more so at low temperatures. The optical structure therefore requires abrupt variations of composition or potential energy with dimensions in the 10 nm range, and therefore one must resort to fabrication techniques other than those of electronic devices.

As a general rule, whereas the linear optical effects were supposed to be more efficient than in QWs, the few measurements in quantum wires and dots have been strikingly disappointing so far, in particular in photoluminescence. This is the exact opposite from QWs, which, as mentioned at the beginning of Section III, have always been much brighter in photoluminescence than 3D structures grown under similar conditions. Quantum wires and dots have at best been as good as QWs, and often much worse, with a systematic worsening when decreasing size.[505,513,583] This might be due to systematically bad interfaces, size fluctuations or to a more profound and intrinsic cause, such as the relaxation bottleneck introduced by the lower dimensionality. The increase in exciton binding energy has been observed.[583a] The influence of periodic structuring of the active medium on luminescence properties of wires and dots, such as grating coupling of polariton modes, has been demonstrated by Heitmann and his group.[583b]

It was predicted that QD lasers should exhibit threshold currents $\approx 10\text{–}10^2$ smaller than those of QW lasers (Fig. 114), as well as increased modulation speed and lower spectral width, both of which are due to the much larger differential gain (as can be seen from Fig. 113, band filling is much more efficient for gain as dimensionality decreases).[584–7] It should be emphasized that conceptually the quantum dot laser is the last step in the long story of the semiconductor laser, where progress was always directed at recapturing the excellent optical properties of isolated, localized atoms: The DH structures optimized the light–matter coupling by electrons and light confinement; the SCH quantum well laser further diminished the number of states to be inverted, while keeping the modal light–matter interaction constant, and started to optimize carrier distribution by reducing allowed k-states, therefore concentrating electrons in those radiative states. The strained QWL continued to optimize both the number of states to be inverted and the efficiency of injected carriers by symmetrizing the conduction and valence bands. The quantum dot laser should finally use carriers in atomlike energy-narrowed states, making all electron-hole pairs efficient for gain at the same energy.

The present attempts of quantum wire lasers only evidence properties somewhat similar to QW lasers with the exception of a better T_o[588] and an enhanced modulation bandwidth,[589] observed on a quantum wire laser obtained through Landau quantization of a 3D DH laser in a strong

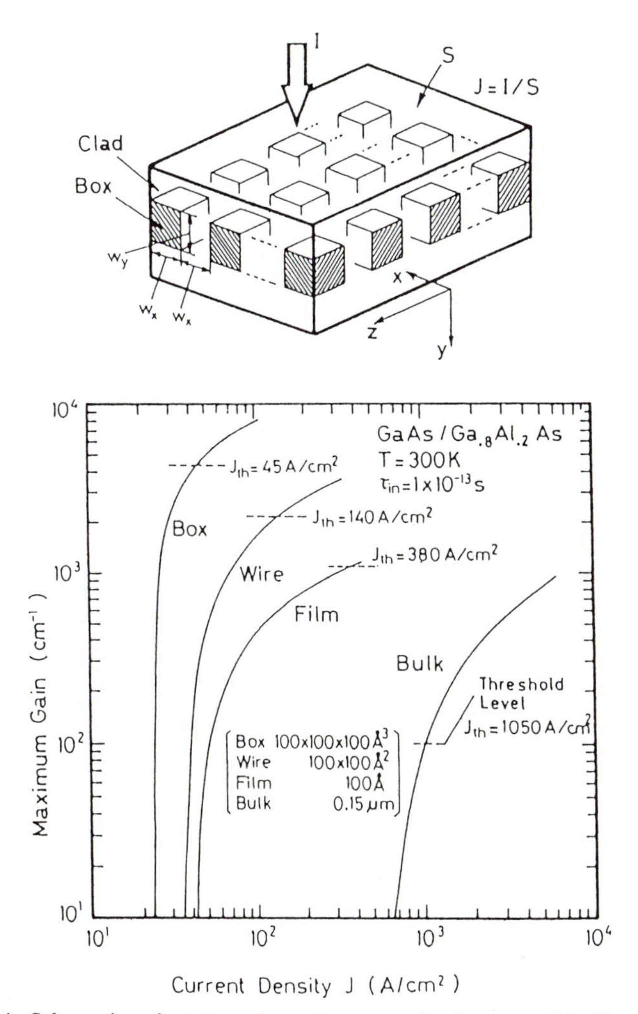

FIG. 114. Schematics of quantum box structure and gain curves 3D, 2D, 1D, 0D lasers, with optimized optical confinement in each case. (Adapted from M. Asada, Y. Miyamoto, and Y. Suematsu, IEEE *J. Quantum Electron.* **QE-22**, 1915, © 1986 1EEE.)

magnetic field. This might be due to the imperfection of the structures and/or the nonoptimal design, as for instance the use of very poor confinement factors Γ:[590,591] The filling factor of quantum wires and dots is usually not optimized from laser physics considerations, but from what is possible with available fabrication techniques. Another severe problem of nanostructured material is the broadening of physical parameters due to size fluctuations:[592–3] It is clear that when the energy fluctuations associated with the size fluctuations are of the order of the energy level separation, all effects associated with lower dimensionality are fully washed out.

An elegant way to obtain quantum-dot–like lasers is to use the magnetic-field quantization of 2D QW lasers. The attempts[594–5] have so far been disappointing, with actually a degradation of current threshold in at least one instance.[595]

The short lifetime predicted from the "giant" dipole effect has been observed in CuCl microcrystals[596–7] (Fig. 115), since the Bohr radius is so small ($a_B \approx 6.8$ Å) that the condition $L \gg a_B$ is easily achieved. In that materials system, the increase of the nonlinear coefficient χ^3 with increasing microcrystallite size has also been observed,[598] with an even faster dependence than predicted by the giant-oscillator-strength theory. Electro-optical effects have been measured in CdS_xSe_{1-x} microcrystallite-doped glasses.[599] No "giant" dipole effect is observed in that case because the system is intermediate, with $L \approx a_B$ in CdS_xSe_{1-x} crystallites. Whenever large crystallites with $L > a_B$ are used, the exciton mass is so heavy that size quantization is not observable. In that case, the transition from QCSE in small crystallites[600] to 3D like the Franz–Keldysh effect has been demonstrated and is in good agreement with theoretical expectations.[601] Photoluminescence is mainly due to recombining donor–acceptor pairs as soon as the crystallite size is reduced below 350 Å.[520] This seems to be due to the

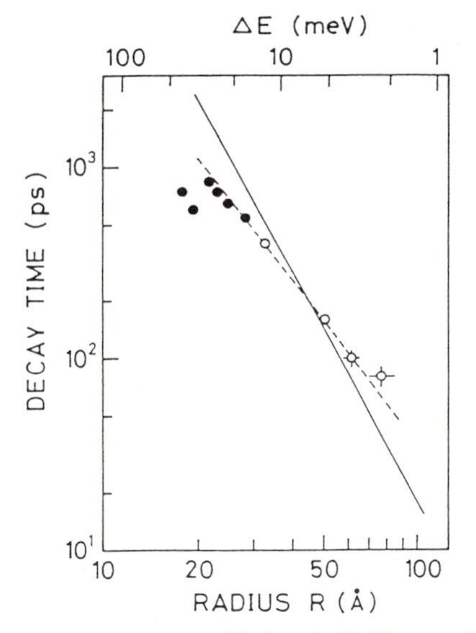

FIG. 115. Variation of photoluminescence lifetime in CuCl nanosphere-doped glasses with diameter, showing the giant oscillator strength increase with sphere size.[596]

very efficient surface traps, which can easily get charged and field-ionize the excitons. The charging of these traps could also explain the frequent photodarkening observed in such semiconductor-doped glasses.[602] However, new unexpected effects can show up, such as the very efficient Auger effect, due to the breakdown of the k-selection rule,[603] allowing very fast response times.

29. 1D and 0D Optical Devices

In order to ascertain the qualities of 1D and 0D optical devices, one has to rely on the separation between devices based on unrelaxed or relaxed excitations of the crystals.

The latter case is the most evident, and is exemplified by the laser (other cases are resonant, absorption-based electro-optical and nonlinear optical devices): The advantage of lowering the dimensionality is to concentrate carriers into k-matched states, with a spectral width determined by size fluctuation, instead of the usual $= kT$ in 2–3D. The gain in laser threshold expected from such lower-dimensionality structures is therefore obvious, and is compounded with the other usual advantages such as those due to higher differential gain (more narrow linewidth, higher modulation speed). However, the challenge is to achieve the large required number of boxes without overly large size fluctuations which tend to destroy the effects of lower dimensionality[587,604]. Good results will clearly depend on progress made on the fabrication of high-quality, low-fluctuation 1D and 0D devices with optimized confinement factors.

Devices based on unrelaxed excitations, such as in off-resonance electro-optical and nonlinear phenomena are much less improved by low dimensionality, since the effects of light coupling and materials modifications are not increased off resonance. Several applications have been demonstrated in CdS Se-doped glasses, such as optical phase conjugation[605], digital optical gates[606], and bistability[607].

It might happen, as usual, that the main advantages of lower dimensionality systems appear because of other effects than those expected from the simplest analysis: Quantum wires and boxes are localized states with excellent quantum efficiency. This might prove essential for some unique applications, like III–V lasers on Si, where localization would prevent carriers from encountering the non-radiative centers associated with the numerous dislocations. Similarly, but for other reasons, quantum boxes might prove useful for IR intersubband detectors, by allowing transitions in the vertical, incoming-light geometry and by diminishing noise thanks to increased relaxation lifetimes. There is clearly room for many new effects and surprises to come in optical applications!

Selected Bibliography

The references cited in this text are both overwhelming, if one searches for a detailed overview of a given subfield, and incomplete with regard to much excellent work that has been omitted. For this we ask indulgence of the authors. To enter more deeply into some aspect of the field, the following bibliography provides a recent set of references which are either (i) review papers on some part of the present chapter, (ii) introductory papers at the nonspecialist level, or (iii) recently published specialized papers which bring new light to some of the outstanding issues discussed in the text. The references will be respectively noted as R (review), I (introductory), and S (specialized).

Growth and Interfaces

E. H. C. Parker, ed. (1985). "The Technology and Physics of Molecular Beam Epitaxy." Plenum, New York. (R)

L. L. Chang and K. Ploog, eds. (1985). "Molecular Beam Epitaxy." NATO ASI Series, Martinus Nijhoff, Dordrecht. (R)

W. T. Tsang (1985). MBE for III–V Compound Semiconductors. In "Semiconductors and Semimetals" (R. K. Willardson and A. C. Beer, eds.), Vol. 22A, Lightwave Communications Technology, volume editor W. T. Tsang. Academic Press, Orlando. (R)

G. B. Stringfellow (1985). Organometallic VPE growth of III–V Semiconductors. In "Semiconductors and Semimetals" (R. K. Willardson and A. C. Beer, eds), Vol. 22A, Lightwave Communications Technology, volume editor W. T. Tsang, Academic Press, Orlando. (R)

M. Razeghi (1985). Low-Pressure MOCVD of $Ga_xIn_{1-x}As_yP_{1-y}$Alloys. In "Semiconductors and Semimetals" (R. K. Willardson and A. C. Beer, eds.), Vol. 22A, Lightwave Communications Technology, volume editor W. T. Tsang, Academic Press, Orlando. (R)

D. B. McWhan (1985). Structure of Chemically Modulated Films. In "Synthetic Modulated Structures" (L. L. Chang and B. C. Giessen, eds.). Academic Press, Orlando. (R)

G. B. Stringfellow (1989). "Organometallic Vapor Phase Epitaxy." Academic Press, Boston. (R)

P. K. Larsen and P. J. Dobson, eds. (1988). "Reflection High-Energy Diffraction and Reflection Electron Imaging of Surfaces." NATO ASI Series, Series B, Physics; **188,** Plenum, New York. (S)

Basic Calculations in Heterostructures

A most useful, highly recommended single reference for much of the physics of heterostructures is G. Bastard (1988), "Wave Mechanics Applied to Semiconductor Heterostructures," Edition de Physique, Les Ulis, France. In particular, all major calculations are carried out in detail at graduate student level (I).

E. E. Mendez and K. von Klitzing (1987). "Physics and Applications of Quantum Wells and Superlattices." NATO ASI Series; Series B, Physics: **170,** Plenum, New York.

T. Ando, A. B. Fowler, and F. Stern (1982). "Electronic Properties of Two-Dimensional Systems", *Rev. Mod. Phys.* **54,** 437. (R)

M. Altarelli (1986). In "Heterojunctions and Semiconductor Superlattices" (G. Allan, G. Bastard, N. Boccara, M. Lannoo, and M. Voos, eds.). Springer-Verlag, Berlin and New York. (R)

B. Ricco and M. Ya. Azbel (1984). Physics of resonant tunneling. the one dimensional double-barrier case. *Phys. Rev. B* **29,** 1970. (S)

J. Barker (1986). Quantum Transport Theory for Small-Geometry Structures. *In* "The Physics and Fabrication of Microstructures and Microdevices" (M. Kelly and C. Weisbuch, eds.). Springer-Verlag, Berlin and New York. (S)

J. Hajdu and G. Landwehr (1985). Quantum Transport Phenomena in Semiconductors in High Magnetic Fields. *In* "Strong and Ultrastrong Magnetic Fields" (F. Herlach, ed.). Springer-Verlag, Berlin and New York. (R)

H. L. Stormer (1986). Images of the Fractional Quantum Hall Effect. *In* "Heterojunctions and Semiconductor Superlattices" (G. Allan, G. Bastard, N. Boccara, M. Lannoo, and M. Voos, eds.). Springer-Verlag, Berlin and New York. (I) (R)

R. E. Prange and S. M. Girvin, eds. (1986). "The Quantum Hall Effect." Springer-Verlag, Berlin and New York. (R)

M. D. Sturge and M. H. Meynadier, eds. (1989, 1990). "The Optical Properties of Semiconducting Quantum Wells and Superlattices." *J. Luminesc.* **44,** 199–414, and **45,** 69–154. (S)

Applications

T. Ikoma, ed. (1990). "Very High Speed Integrated Circuits." Semiconductors and Semimetals, **29,** and **30,** R. K. Willardson and A. C. Beer, Academic Press, Boston (S).

P. M. Solomon. (1982). "A Comparison of Semiconductor Devices for High-Speed Logic." Proc. IEEE **70,** 489 (I, S).

D. S. Chemla, D. A. Miller, and S. Schmitt-Rink, (1991). "Non-Linear Optics and Electro-Optics of Quantum Confined Structures." Academic Press, Boston (R).

S. Schmitt-Rink, D. S. Chemla, and D. A. B. Miller (1989). "Linear and Non-Linear Optical Properties of Semiconductor Quantum Wells." Advances in Physics **38,** 89 (S).

P. M. Solomon (1986). Three Part Series on Heterojunction transistors. *In* "The Physics and Fabrication of Microstructures and Microdevices" (M. Kelley and C. Weisbuch, eds.). Springer-Verlag, Berlin and New York. (R)

B. de Cremoux (1986). Quantum Well Laser Diodes. *In* "Solid State Devices '85" (P. Balk and O. G. Folberth, eds.). Elsevier, Amsterdam. (I)

C. Weisbuch (1987). The Physics of the Quantum Well Laser. Proceedings of NATO ARW, "Optical Properties of Narrow-Gap Low-Dimensional Structures" (C. Sotomayor-Torres and R. A. Stradling, eds.). Plenum, New York. (R)

S. Luryi and A. Kastalsky (1985). Hot electron transport in heterostructure devices. *Physica B* and *C* **134,** 453. (R)

S. Luryi (1987). Hot-Electron-Injection and Resonant-Tunneling Heterojunction Devices. *In* "Heterojunctions: A Modern View of Band Discontinuities and Device Applications" (F. Capasso and G. Margaritondo, eds.). North-Holland Publ., Amsterdam. (I) (R)

F. Capasso, K. Mohammed, and A. Y. Cho (1986). Resonant tunneling through double-barriers, perpendicular quantum phenomena in superlattices, and their device applications. *IEEE J. Quantum Electron.* **QE-22,** 1853. (R)

M. Jaros (1990). "Physics and Applications of Semiconductor Microstructures." Clarendon Press, Oxford. (I)

WHOLE FIELD

The special issue "Semiconductor Quantum Wells and Superlattices: Physics and Applications" of the *IEEE Journal of Electronics* (Vol. **QE-22,** September 1986) contains an excellent set of review articles on various aspects of quantum wells and superlattices. (R)

Proceedings from the Maunterdorf Winterschool (Springer, Berlin) and from the International Conference on the Electronic Properties of 2 Dimensional Systems (Surface Science, vols. **58, 73, 98, 113, 142, 170, 196, 229**). Contain very useful information, providing a good overview of the field at various times.

REFERENCES

1. For a review of early work up to 1975, see R. Dingle, *Festkoerperprobleme* **15,** 21 (1975).
2. Early Russian work is reviewed by B. A. Tavger and V. Ya. Demishovskii, *Usp. Fiz. Nauk* **96,** 61 (1968) [*Sov. Phys.—Usp.* (*Engl. Transl.*) **11,** 644 (1969)]; A. Ya. Shik, *Fiz. Tekh. Poluprovodn.* **8,** 1841 (1974) [*Sov. Phys.—Semicond.* (*Engl. Transl.*) **8,** 1195 (1975)].
3. L. Esaki and R. Tsu, *IBM J. Dev.* **14,** 61 (1970).
4. The subsequent development of the field is reviewed by L. Esaki, *in* "Recent Topics in Semiconductor Physics" (H. Kamimura and Y. Toyozawa, eds.). World Scientific, Singapore, 1983.
5. A. Y. Cho and J. R. Arthur, *Prog. Solid State Chem.* **10,** 157 (1975).
6. K. Ploog, *in* "Crystals: Growth, Properties and Applications" (H. C. Freyhardt, ed.), Vol. 3, p. 75. Springer-Verlag, Berlin and New York, 1980
7. L. L. Chang, *in* "Handbook of Semiconductors" (T. S. Moss, ed.), Vol. 3, p. 563. North-Holland Publ., Amsterdam, 1980.
8. C. T. Foxon and B. A. Joyce, *Curr. Top. Mater. Sci.* 7, 1 (1981).
9. A. C. Gossard, *Treatise Mater. Sci. Technol.* **24,** 13 (1982).
10. A. Y. Cho, *Thin Solid Films* **100,** 291 (1983).
11. L. L. Chang and K. Ploog, eds., "Molecular Beam Epitaxy and Heterostructures," Proc. Erice 1983 Summer School. Martinus Nijhoff, The Hague, 1985.
12. For invaluable information, see the proceedings of the various MBE workshops and conferences, i.e., *J. Vac. Sci. Technol. B* [**1**], 119–205 (1983); *B*[**2**], 162–297 (1984); *B*[**3**], 511–807 (1985).
13. L. Esaki and L. L. Chang, *Phys. Rev. Lett.* **33,** 495 (1974).
14. R. Dingle, W. Wiegmann, and C. H. Henry, *Phys. Rev. Lett.* **33,** 827 (1974).

14a. See, e.g., D. Bohm, "Quantum Mechanics." Prentice-Hall, Englewood Cliffs, New Jersey, 1951; E. Merzbacher, "Quantum Mechanics." Wiley, New York, 1961.

15. H. C. Casey, Jr. and M. B. Panish, "Heterostructure Lasers," Parts A and B. Academic Press, New York, 1978.
16. G. H. B. Thompson, "Physics of Semiconductor Lasers," Wiley, New York, 1980.
17. W. T. Tsang, *in* "Semiconductors and Semimetals" (R. K. Willardson and A. C. Beer, eds.), **24,** p. 397, Academic Press, New York, 1987.
18. R. D. Burnham, W. Streifer, and T. L. Paoli, *J. Cryst. Growth* **68,** 370 (1984).
19. H. Morkoç, *in* "Semiconductors and Semimetals" (R. K. Willardson and A. C. Beer, eds.), **24,** p. 135, Academic Press, New York, 1987.
20. N. T. Linh, *in* "Semiconductors and Semimetals" (R. K. Willardson and A. C. Beer, eds.), **24,** p. 203, Academic Press, New York, 1987.

21. M. Abe, T. Mimura, K. Nishiushi, A. Shibatomi, M. Kobayashi, and T. Misugi, *in* "Semiconductors and Semimetals" (R. K. Willardson and A. C. Beer, eds.), **24,** p. 249, Academic Press, New York, 1987.
22. K. Ploog and G. H. Döhler, *Adv. Phys.* **32,** 285 (1983).
23. H. L. Störmer, *Festkoerperprobleme* **24,** 25 (1984).
24. N. Yokoyama, K. Imamura, T Ohshima, H. Nishi, S. Muto, K. Kondo, and S. Hiyamizu, *Jpn. J. Appl. Phys.* **23,** L311 (1984).
25. T. C. L. G. Sollner, W. D. Goodhue, P. E. Tannenwald, C. D. Parker, and D. D. Peck, *Appl. Phys. Lett.* **43,** 588 (1983).
26. P. M. Petroff, A. C. Gossard, R. A. Logan, and W. Wiegmann, *Appl. Phys. Lett.* **41,** 635 (1982).
27. A. B. Fowler, A. Hartstein, and R. A. Webb, *Phys. Rev. Lett.* **48,** 196 (1982).
28. A. Kastalsky and S. Luryi, *IEEE Electron. Device Lett.* **EDL-4,** 334 (1983).
29. A. Kastalsky, S. Luryi, A. C. Gossard, and R. Hendel, *IEEE Electron. Device Lett.* **EDL-5,** 57 (1984).
30. A. Kastalsky, R. A. Kiehl, S. Luryi, A. C. Gossard, and R. Hendel, *IEEE Electron. Device Lett.* **EDL-5,** 34 (1984).
31. F. Capasso, *in* "Semiconductors and Semimetals" (R. K. Willardson and A. C. Beer, eds.), **24,** p. 319, Academic Press, New York, 1987.
32. G. A. Sai-Halasz, *Conf. Ser.—Inst. Phys.* **43,** 21 (1979).
33. L. L. Chang, *in* "Molecular Beam Epitaxy and Heterostructures" (L. L. Chang and K. Ploog, eds.), Martinus Nijhoff, The Hague. p. 461.
34. M. Voos, *J. Vac. Sci. Technol. B* [**1**], 404 (1983).
35. L. Esaki, *J. Cryst. Growth* **52,** 227 (1981).
36. P. Voisin, *Springer Ser. Solid-State Sci.* **59,** 192 (1984).
37. G. C. Osbourn, *J. Vac. Sci. Technol. B* [**1**], 379 (1983).
38. J.-Y. Marzin, "Heterojunctions and Semiconductor Superlattices," (G. Allan *et al.*, eds), p. 161. Springer-Verlag. Berlin and New York, 1986.
39. G. C. Osbourn, P. L. Gourley, I. J. Fritz, R. M. Biefeld, L. R. Dawson, and T. E. Zipperian, *in* "Semiconductors and Semimetals" (R. K. Willardson and A. C. Beer, eds.), **24,** p. 459, Academic Press, New York, 1987.
40. B. Abeles and T. Tiedje, *Phys. Rev. Lett.* **51,** 2003 (1983).
41. H. Munekata and H. Kubimoto, *Jpn. J. Appl. Phys.* **22,** L544 (1983).
42. M. Hundhausen, L. Ley, and R. Carius, *Phys. Rev. Lett.* **53,** 1598 (1984).
43. J. Kabalios and H. Fritzsche, *Phys. Rev. Lett.* **53,** 1602 (1984).
44. T. P. Pearsall, ed., "Semiconductors and Semimetals" (R. K. Willardson and A. C. Beer, eds.), **33,** Academic Press, New York, 1991.
45. P. D. Dapkus, *Annu. Rev. Mater. Sci.* **12,** 243 (1982).
46. S. Hersee and J. P. Duchemin, *Annu. Rev. Mater. Sci.* **12,** 65 (1982).
47. G. B. Stringfellow, *Annu. Rev. Mater. Sci.* **8,** 73 (1978).
48. M. A. Digiuseppe, H. Temkin, L. Peticolas, and W. A. Bonner, *Appl. Phys. Lett.* **43,** 906 (1983).
49. H. Kinoshita and H. Fujiyasa, *J. Appl. Phys.* **51,** 5845 (1981).
50. D. C. Tsui and R. A. Logan, *Appl. Phys. Lett.* **35,** 99 (1979).
51. J. H. Neave, B. A. Joyce, P. J. Dobson, and N. Norton, *Appl. Phys.* **A31,** 1 (1983).
52. J. H. Neave, B. A. Joyce, and P. J. Dobson, *Appl. Phys.* **A34,** 179 (1984).
53. B. A. Joyce, *in* "Molecular Beam Epitaxy and Heterostructures" (L. L. Chang and K. Ploog, eds.), Martinus Nijhoff, The Hague. Chapter 2.
54. C. T. Foxon, *in* "Heterojunctions and Semiconductor Superlattices" (G. Allan *et al.*, eds.), p. 216. Springer-Verlag, Berlin and New York, 1986.

55. T. Sakamoto, H. Funabashi, K. Ohta, T. Nakagawa, N. J. Kawai, and T. Kojima, *Supplattices Microstruct.* **1,** 347 (1985).
56. N. Sano, H. Kato, M. Nakayama, S. Chika, and H. Terauchi, *Jpn. J. Appl. Phys.* **23,** L640 (1984).
57. P. M. Petroff, R. C. Miller, A. C. Gossard, and W. Wiegmann, *Appl. Phys. Lett.* **44,** 217 (1984).
58. P. M. Frijlink, J. P. Andre, and J. L. Gentner, *J. Cryst. Growth* **70,** 435 (1985).
59. P. M. Frijlink and J. Maluenda, *Jpn. J. Appl. Phys.* **21,** L574 (1982).
60. R. C. Miller, R. D. Dupuis, and P. M. Petroff, *Appl. Phys. Lett.* **44,** 508 (1984).
61. M. R. Leys, C. Van Opdorp, M. P. A. Viegers, and H. J. Talen-Van Der Mheen, *J. Cryst. Growth* **68,** 431 (1985).
61a. J. Nagle, M. Razeghi, and C. Weisbuch, *Conf. Ser.—Inst. Phys.* **74,** 379 (1985).
62. R. Bisaro, G. Laurencin, A. Friederich, and M. Razeghi, *Appl. Phys. Lett.* **40,** 978 (1982).
63. See, e.g., M. P. Seah, *Proc. Int. Vac. Cong., 9th, 1983; Proc. Int. Conf. Solid Surf., 5th, Madrid,* 1983, p. 63 (1983).
64. J. Cazaux, G. Laurencin, and J. Olivier, *J. Phys. Lett.* **45,** L999 (1984).
65. P. M. Petroff, A. C. Gossard, W. Wiegmann, and A. Savage, *J. Cryst. Growth* **44,** 5 (1978).
66. R. M. Fleming, D. B. McWhan, A. C. Gossard, W. Wiegmann, and R. A. Logan, *J. Appl. Phys.* **51,** 357 (1980).
67. D. B. McWhan, *in* "Synthetic Modulated Structures" (L. L. Chang and B. C. Giessen, eds.), p. 43. Academic Press, New York, 1985.
68. G. B. Stringfellow, *Int. Workshop Furture Electron Devices. 1st. Tokyo, 1984* (unpublished).
69. J. Black, P. Norris, E. Koteles, and S. Zemon, *Conf. Ser.—Inst. Phys.* **74,** 683 (1985).
70. N. Kobayashi and T. Fukui, *Electron. Lett.* **20,** 888 (1984).
71. M. Heiblum, E. E. Mendez, and L. Osterling, *J. Appl. Phys.* **54,** 6982 (1983).
72. M. A. Di Forte-Poisson, C. Brylinski, G. Colomer, D. Osselin, J. P. Duchemin, F. Hazan, D. Lechevalier, and J. Lacombe, *Conf. Ser.—Inst. Phys.* **74,** 677 (1985).
73. F. C. Frank and J. H. Van Der Merwe, *Proc. R. Soc. London, Ser. A* 198, 216 (1949).
74. J. W. Matthews and A. E. Blakeslee, *J. Cryst. Growth* **27,** 118 (1974).
75. G. C. Osbourn, *J. Appl. Phys.* **53,** 1586 (1982).
75a. See the review by J. P. Faurie, *IEEE J. Quantum Electron.* **QE-22,** 1656 (1986).
75b. J. M. Berroir, Y. Guldner, and M. Voss, *IEEE J. Quantum Electron.* **QE-22,** 1793 (1986).
76. D. L. Smith, T. C. McGill, and J. N. Schulman, *Appl. Phys. Lett.* **43,** 180 (1983).
77. G. C. Osbourn, *J. Vac. Sci. Technol. B* [2], 176 (1984).
78. R. N. Bicknell, R. W. Yanka, N. C. Giles-Taylor, D. K. Blanks, E. L. Buckland, and J. F. Schetzina, *Appl. Phys. Lett.* **45,** 92 (1984).
79. L. A. Kolodziejski, T. C. Bonsett, R. L. Gunshor, S. Datta, R. B. Bylsma, W. M. Becker, and N. Otsuka, *Appl. Phys. Lett.* **45,** 440 (1984).
79a. See the review by A. V. Nurmikko, R. L. Gunshor, and L. A. Kolodziejski, *IEEE J. Quantum Electron.* **QE-22,** 1785 (1986).
80. J. C. Bean, L. C. Feldman, A. T. Fiory, S. Nakahara, and J. K. Robinson, *J. Vac. Sci. Technol. A* [2], 436 (1984).
80a. See the review by R. People, *IEEE J. Quantum Electron.* **QE-22,** 1696 (1986).
81. We shall not attempt to discuss the formidable challenge created by the theoretical or experimental determinations of band offsets, still a matter of controversy [see, e.g., H. Kroemer, *IEEE Electron Device Lett.* **EDL-4,** 365 (1983) and A. Nussbaum, *ibid.*

EDL-5, 499 (1984), and references therein]. An elementary LCAO theory was proposed by W. Harrison, *J. Vac. Sci. Technol.* **14,** 1016 (1977) and *B* [3], 1231 (1985); *Direct* measurements by XPS of the GaAs-GaAlAs band discontinuities yield values which depend on the growth sequence (!): J. Waldrop, S. P. Kowalczyk, R. W. Grant, E. A. Kraut, and D. L. Miller, *ibid.* **19,** 573 (1981); R. S. Bauer and H. W. Sang, *Surf. Sci.* **132,** 479 (1983). For detailed reviews of recent experiments and theories, see, e.g., G. Margaritondo, *ibid.*, p. 469; A. D. Katnani and G. Margaritondo, *J. Appl. Phys.* **54,** 2522 (1983); H. Kroemer, *Surf. Sci.* **132,** 543 (1983); *J. Vac. Sci. Technol. B* [2], 433 (1984); E. A. Kraut, *ibid.*, p. 486; J. Tersoff, *Phys. Rev. Lett.* **52,** 465 (1984); **56,** 2755 (1986).

82. G. Bastard, *Phys. Rev. B* **24,** 5693 (1981); *B* **25,** 7594 (1982).

82a. For a very detailed account of the application of the envelope wave-function approximation to quantum wells and heterostructures, see the lecture notes by M. Altarelli, *in* "Heterojunctions and Semiconductor Superlattices" (G. Allan *et al.*, eds.), p. 12. Springer-Verlag, Berlin and New York, 1985.

82b. G. Bastard, *in* "Molecular Beam Epitaxy and Heterostructures," p. 381. Proc. Erice 1985 Summer School, Martinus Nijhoff, The Hague (1985).

82c. M. Altarelli, *J. Luminescence* **30,** 472 (1985).

82d. G. Bastard and J. A. Brum, *IEEE J. Quantum Electron.* **QE-22,** 1625 (1986).

82e. L. J. Sham and Y. T. Lu, *J. Lumin.* **44,** 207 (1989).

82f. D. L. Smith and C. Mailhiot, *Rev. Mod. Phys.* **62,** 173 (1990).

83. See, e.g., E. O. Kane, *J. Phys. Chem. Solids* **1,** 249 (1957); *in* "Semiconductors and Semimetals" (R. K. Willardson and A. C. Beer, eds), Vol. 1. Academic Press, New York, 1966.

84. S. R. White and L. J. Sham, *Phys. Rev. Lett.* **47,** 879 (1981).

84a. J. M. Luttinger, *Phys. Rev.* **102,** 1030 (1956).

85. D. S. Chemla, *Helv, Phys. Acta* **56,** 607 (1983).

86. J. C. Hensel and G. Feher, *Phys. Rev.* **129,** 1041 (1963).

87. S. S. Nedozerov, *Fiz. Tverd. Tela* **12,** 2269 (1970) [*Sov. Phys.–Solid State (Engl. Transl.)* **12,** 1815 (1971)].

88. M. I. Dyakonov and A. V. Khaetskii, *Zh. Eksp. Teor. Fiz.* **82,** 1584 (1982) [*Sov. Phys.—JETP* (*Engl. Transl.*) **55,** 917 (1982)].

89. A. Fasolino and M. Altarelli, *Springer Ser. Solid-State Sci.* **59,** 176 (1984).

90. J. C. Maan, A. Fasolino, G. Belle, M. Altarelli, and K. Ploog, *Physica* (*Amsterdam*) **127B,** 426 (1984).

91. R. Sooryakumar, D. S. Chemla, A. Pinczuk, A. Gossard, W. Wiegmann, and L. J. Sham, *J. Vac. Sci. Technol. B* [2], 349 (1984).

92. R. Sooryakumar, A. Pinczuk, A. Gossard, D. S. Chemla, and L. J. Sham, *Phys. Rev. Lett.*, **58,** 1150 (1987).

93. H. L. Stőrmer, Z. Schlesinger, A. Chang, D. C. Tsui, A. C. Gossard, and W. Wiegmann, *Phys. Rev. Lett.* **51,** 126 (1983).

94. U. Ekenberg and M. Altarelli, *Phys. Rev. B* **30,** 3569 (1984).

95. D. A. Broido and L. J. Sham, *Phys. Rev. B***31,** 888 (1985).

96. Y. C. Chang and J. N. Schulman, *Appl. Phys. lett.* **43,** 536 (1983).

97. Y. C. Chang and J. N. Schulman, *Phys. Rev. B* **31,** 2069 (1985).

98. J. P. Eisenstein, H. L. Stőrmer, V. Narayanamurti, A. C. Gossard, and W. Wiegmann, *Phys. Rev. Lett.* **53,** 2579 (1984).

99. The electronic properties of two-dimensional systems are well-documented in the biennial proceedings of the International Conference on the subject: *Surf. Sci.* **58,** (1976); **73** (1978); **98** (1980); **113** (1982); **142** (1984); **170** (1986).

100. See the extensive review on the electronic properties of 2-D systems by T. Ando, A. B. Fowler, and F. Stern, *Rev. Mod. Phys.* **54,** 437 (1982).
101. Useful reviews on excitons in 3-D can be found in R. S. Knox, "Theory of Excitons," Solid State Phys., Suppl. 5. Academic Press, New York, 1963; J. O. Dimmock, *in* "Semiconductors and Semimetals" (R. K. Willardson and A. C. Beer, eds.), Vol. 3. Academic Press, New York, 1967; *in* "Excitons" (E. I. Rashba and M. D. Sturge, eds.). North-Holland Publ., Amsterdam, 1982.
102. Useful reviews on shallow impurities in 3D can be found in W. Kohn, *Solid State Phys.* **5,** 257 (1957); A. K. Ramdas and S. Rodriguez, *Rep. Prog. Phys.* **44,** 1297 (1981).
103. For a review on Coulomb effects in quantum wells, see G. Bastard, *J. Lumin.* **30,** 488 (1985) see also ref. 82d.
104. M. Shinada and S. Sugano, *J. Phys. Soc. Jpn.* **21,** 1936 (1966).
105. F. Stern and W. E. Howard, *Phys. Rev.* **163,** 816 (1967).
106. G. Bastard, E. E. Mendez, L. L. Chang, and L. Esaki, *Phys. Rev. B* **26,** 1974 (1982).
107. R. L. Greene, K. K. Bajaj, and D. E. Phelps, *Phys. Rev. B* **29,** 1807 (1984).
107a. 0. It seems to be also the case of infrared material QW's. See, for the case of GaInAs/AlInAs, J. S. Weiner, D. S. Chemla, D. A. B. Miller, T. H. Wood, D. Sives, and A. Y. Cho, *Appl. Phys. Lett.* **46,** 619 (1985).
107b. See the thorough review by D. S. Chemla and D. A. B. Miller. *J. Opt. Soc. Am.* **B2,** 1155 (1985).
108. J. C. Phillips, "Bonds and Bands in Semiconductors." Academic Press, New York, 1973.
109. G. Bastard, *Phys. Rev. B* **24,** 4714 (1981).
110. R. L. Greene and K. K. Bajaj, *Solid State Commun.* **45,** 825 (1983).
111. C. Mailhot, Y. C. Chang, and T. C. McGill, *Phys. Rev. B* **26,** 4449 (1982).
112. B. V. Shanabrook, J. Comas, T. A. Perry, and R. Merlin, *Phys. Rev. B* **29,** 7096 (1984).
113. N. C. Jarosik, B. D. McCombe, B. V. Shanabrook, J. Comas, J. Ralston, and G. Wicks, *Phys. Rev. Lett.* **54,** 1283 (1985).
114. W. T. Masselink, Y. C. Chang, and H. Morkoc, *Phys. Rev. B* **28,** 7373 (1983); *J. Vac Sci. Technol. B* [2], 376 (1984).
115. S. Chaudhuri, *Phys. Rev. B* **28,** 4480 (1983).
116. J. A. Brum, G. Bastard, and C. Guillemot, *Phys. Rev. B* **31,** 1428 (1985).
117. F. Crowne, T. L. Reinecke, and B. V. Shanabrook, *Solid State Commun.* **50,** 875 (1984).
118. C. Priester, G. Allan, and M. Lannoo, *Phys. Rev. B* **29,** 3408 (1984).
119. T. A. Perry, R. Merlin, B. V. Shanabrook, and J. Comas, *J. Vac. Sci. Technol. B* [3], 636 (1985).
119a. A. A. Reeder, A-M. Mercy, and B. D. McCombe, *IEEE J. Quantum Electron.* **QE-24,** 1690 (1988).
119b. C. Y. Fong, I. P. Batra, and S. Ciraci, eds., (1989). "Properties of Impurity States in Superlattice Semiconductors." NATO ASI Series B; **183,** Plenum, New York.
120. R. Tsu and L. Esaki, *Appl. Phys. Lett.* **22,** 562 (1973).
121. J. N. Schulman and T. C. McGill, *Phys. Rev. B* **19,** 6341 (1979).
122. J. N. Schulman and Y. C. Chang, *Phys. Rev. B* **24,** 4445 (1981).
123. R. Dingle, A. C. Gossard, and W. Wiegmann, *Phys. Rev. Lett.* **34,** 1327 (1975).
125. G. Bastard, *Acta Electron.* **25,** 147 (1983).
126. E. E. Mendez, L. L. Chang, G. Landgren, R. Ludeke, L. Esaki, and F. H. Pollak, *Phys. Rev. Lett.* **46,** 1230 (1981).
127. G. Bastard, *Phys. Rev. B* **30,** 3547 (1984).
128. M. Jaros and K. B. Wong, *J. Phys. C.* **17,** L765 (1984).

129. J. A. Brum and G. Bastard, *Phys. Rev. B* **33,** 1420 (1986).
130. J. E. Zucker, A. Pinczuk, D. S. Chemla, A. C. Gossard, and W. Wiegmann, *Phys. Rev. B* **29,** 7065 (1984).
130a. A detailed account of electrons in heterostructures can be found in F. Stern, *in* "Heterojunctions and Semiconductor Superlattices," p. 38 (G. Allan *et al.*, eds.). Springer-Verlag, Berlin and New York, 1985.
131. H. L. Störmer, R. Dingle, A. C. Gossard, W. Wiegmann, and R. A. Logan, *Conf. Ser.—Inst. Phys.* **43,** 557 (1979).
132. R. Dingle, H. L. Störmer, A. C. Gossard, and W. Wiegmann, *Appl. Phys. Lett.* **33,** 665 (1978).
133. See, e.g., C. Kittel and H. Kroemer, "Thermal Physics," 2nd ed. Freeman, San Francisco, California, 1980.
134. See e.g., I. I. Goldman, V. D. Krivchenkov, V. I. Kogan, and V. M. Galitskii, *in* "Problems in Quantum Mechanics," p. 24. Infosearch, London, 1960.
135. F. Stern, *Phys. Rev. B* **5,** 4891 (1972).
135a. See, e.g., F. Stern, *Appl. Phys. Lett.* **43,** 974 (1983).
136. T. Ando, *J. Phys. Soc. Jpn.* **51,** 3893 (1982).
137. T. Ando, *J. Phys. Soc. Jpn.* **51,** 3900 (1982).
138. L. C. Witkowski, T. J. Drummond, C. M. Stanchak, and H. Morkoc, *Appl. Phys. Lett.* **37,** 1033 (1980).
139. H. L. Störmer, A. Pinczuk, A. C. Gossard, and W. Wiegmann, *Appl. Phys. Lett.* **38,** 69 (1981).
140. F. F. Fang and W. E. Howard, *Phys. Rev. Lett.* **16,** 797 (1966).
141. G. Bastard, *Surf. Sci.* **142,** 284 (1984).
142. F. Stern and S. Das Sarma, *Phys. Rev. B* **30,** 840 (1984).
143. B. Vinter, *Appl. Phys. Lett.* **44,** 307 (1984).
144. B. Vinter, *Surf. Sci.* **142,** 452 (1984).
145. D. Delagebeaudeuf and N. T. Linh, *IEEE Trans. Electron Devices* **ED-29,** 955 (1982).
146. T. Mimura, K. Joshin, and S. Kuroda, *Fujitsu Sci. Tech. J.* **19,** 243 (1983).
147. T. J. Drummond, R. Fischer, S. L. Su, W. G. Lyons, H. Morkoc, K. Lee, and M. Shur, *Appl. Phys. Lett.* **42,** 262 (1983).
148. B. Vinter, *Solid State Commun.* **48,** 151 (1983).
149. K. Hirakawa, H. Sakaki, and J. Yoshino, *Appl. Phys. Lett.* **45,** 253 (1984).
150. H. Morkoc, *in* "Molecular Beam Epitaxy and Heterostructures" (L. L. Chang and K. Ploog, eds.), p. 625. Martinus Nijhoff, The Hague, 1985.
151. F. Stern, *Appl. Phys. Lett.* **43,** 974 (1983).
152. K. Inoue and H. Sakaki, *Jpn. J. Appl. Phys.* **23,** L61 (1984).
153. K. Miyatsuji, H. Hihara, and C. Hamaguchi, *Superlattices Microstruct.* **1,** 43 (1985).
153a. N. H. Sheng, C. P. Lee, R. T. Chen, D. L. Miller, and S. J. Lee, *IEEE Electron Dev. Lett.* **EDL-6,** 307 (1985).
154. N. Shand, T. Henderson, J. Klem, W. T. Masselink, R. Fischer, Y. C. Chang, and H. Morkoc, *Phys. Rev. B* **30,** 4481 (1984), and references therein.
155. E. F. Schubert and K. Ploog, *Phys. Rev. B* **30,** 7021 (1984).
156. T. H. Theis, T. F. Kuech, and L. F. Palmeeter, *Conf. Ser.—Inst. Phys.* **74,** 241 (1985).
157. T. Baba, T. Mizutani, and M. Ogawa, *Jpn. J. Appl. Phys.* **22,** L627 (1983).
158. See e.g., T. J. Drummond, R. J. Fischer, W. F. Kopp, H. Morkoc, K. Lee, and M. Shur, *IEEE Trans. Electron Devices* **ED-30,** 1806 (1983), and references therein.
159. R. Fischer, T. J. Drummond, J. Klem, W. Kopp, T. S. Henderson, D. Perrachione, and H. Morkoc, *IEEE Trans. Electron Devices* **ED-31,** 1028 (1984).
160. M. Heiblum, *J. Vac. Sci. Technol. B* [3], 820 (1985).

160a. K. Inoue, H. Sakaki, and J. Yoshino, *Appl. Phys. Lett.* **46,** 973 (1985).
161. M. Heiblum, E. E. Mendez, and F. Stern, *Appl. Phys. Lett.* **44,** 1064 (1984).
161a. N. C. Cirillo, M. S. Shur, and J. K. Abrokwah, *IEEE Electron Dev. Lett.* **EDL-7,** 63 (1986).
162. J. C. M. Hwang, A. Kastalsky, H. L. Störmer, and V. G. Keramidas, *Appl. Phys. Lett.* **44,** 802 (1984).
163. G. Weimann and W. Schlapp, *Appl. Phys. Lett.* **46,** 411 (1985).
164. For very thorough reviews of *n-i-p-i's,* see K. Ploog and G. H. Döhler, *Adv. Phys.* **32,** 285 (1983); G. H. Döhler, *IEEE J. Quantum Electron.* **QE 22,** 1683 (1986).
165. P. Ruden and G. H. Döhler, *Phys. Rev. B* **27,** 3538 (1983).
166. G. Abstreiter, *Springer Ser. Solid-State Sci.* **59,** 232 (1984).
167. C. Alibert. F. Jiahua, M. Erman, P. Frijlink, P. Jarry, and J. B. Theeten, *Rev. Phys. Appl.* **18,** 709 (1983).
168. C. Weisbuch, R. Dingle, P. M. Petroff, A. C. Gossard, and W. Wiegmann, *Appl. Phys. Lett.* **18,** 840 (1981).
169. D. F. Welch, G. W. Wicks, and L. F. Eastman, *Appl. Phys. Lett.* **43,** 762 (1983).
170. M. Naganuma, Y. Suzuki, and H. Okamoto, *Conf. Ser.—Inst. Phys.* **63,** 125 (1982).
170a. W. T. Tsang, T. H. Chiu, S. N. G. Chu, and J. A. Ditzenberger, *Appl. Phys. Lett.* **46,** 659 (1985).
170b. L. A. Kolodzieski, R. L. Gunshor, T. C. Bonsett, R. Venkatasubramanian, S. Datta, R. B. Bylsma, W. M. Becker, and N. Otsuka, Appl. Phys. Lett. **47,** 169 (1985).
170c. F. Stern, Elementary "Optical Properties of Solids", *in* Solid State Physics **15** p. 299, F. Seitz and D. Turnbull eds, Academic, New York, 1963.
170d. A. Yariv, "Quantum Electronics," third ed., Wiley, New York, 1989.
171. P. Voisin, C. Delalande, M. Voos, L. L. Chang, A. Segmuller, C. A. Chang, and L. Esaki, *Phys. Rev. B* **30,** 2276 (1984).
172. P. Voisin, in "Heterojunctions and Semiconductor Superlattices" (G. Allan *et al.*, eds.), p. 73. Springer-Verlag, Berlin and New York, 1985.
172a. See, e. g., R. S. Knox, Ref. 101, p. 117.
172b. See, e.g., D. E. Aspnes, "Handbook of Semiconductors," (M. Balkanski, vol. ed, T. S. Moss, series ed.,) vol. **2**, p. 109 North-Holland, Amsterdam, 1980; M. Cardona, "Modulation Spectroscopy," *Solid State Physics*, suppl. **11,** Academic, New York (1969); "Modulation Techniques" "Semiconductors and Semimetals" (R. K. Willardson and A. C. Beer, eds.), **9,** Academic Press, New York, 1972; F. Wooten, "Optical Properties of Solids." Academic Press, New York, 1970.
172c. L. C. West and S. J. Eglash, *Appl. Phys. Lett.* **46,** 1156 (1985).
172d. B. Levine, K. K. Choi, C. G. Bethea, J. Walker, and R. J. Malik, *Appl. Phys. Lett.* **50,** 1092 (1987).
172e. B. F. Levine, C. G. Bethea, K. K. Choi, J. Walker, and R. J. Malik, *J. Appl. Phys.* **64,** 1591 (1988).
172f. M. A. Kinch and A. Yariv, *Appl. Phys. Lett.* **55,** 2093 (1989).
172g. S. Lyon, *Surf. Science,* **228,** 508 (1990).
172h. E. Rosencher, P. Bois, J. Nagle, E. Costard, and S. Delaître, *Appl. Phys. Lett.* **55,** 1150 (1989).
172i. M. M. Fejer, S. J. B. Yoo, R. L. Byer, A. Harwit, and J. S. Harris, Jr., *Phys. Rev. Lett.* **62,** 1041 (1989).
172j. E. Rosencher, P. Bois, J. Nagle, S. Delaître, *Electron. Lett.* **25,** 1063 (1989).
172k. E. Rosencher, P. Bois, B. Vinter, J. Nagle, and D. Kaplan, *Appl. Phys. Lett.* **56,** 1822 (1990).
172l. D. S. Chemla, *Helvetica Physica Acta* **56,** 607 (1983).
172m. D. S. Chemla, D. A. B. Miller, and P. W. Smith, *Optical Engineering* **24,** 556 (1985).

173. R. C. Miller, D. A. Kleinman, O. Munteanu, and W. T. Tsang, *Appl. Phys. Lett.* **39,** 1 (1981).
174. A. Pinczuk, J. Shah, R. C. Miller, A. C., Gossard, and W. Wiegmann, *Solid State Commun.* **50,** 735 (1984).
175. See, e.g., M. Born, "Atomic Physics," 8th ed. Blackie, London, 1969.
175a. B. P. Zakharchenya, D. N. Mirlin, V. I. Perel, and I. I. Reshina, *Usp. Fiz. Nauk.* **136,** 459 (1982) *Sov. Phys.—Usp. (Engl. Transl.)* **25,** 143 (1982).
176. For reviews on the so-called optical spin orientation technique in semiconductors, see G. Lampel, *Proc. Int. Conf. Phys. Semicond., 12th 1974*, p. 743 (1974); C. R. Pidgeon, in "Handbook on Semiconductors" (T. S. Moss, ed.), Vol. 2, p. 223, North-Holland, Amsterdam, 1980; "Spin Orientation" (B. P. Zakharchenya and F. Meier, eds.), North-Holland, Amsterdam, 1984.
177. S. Alvarado, F. Ciccacci, and M. Campagna, *Appl. Phys. Lett.* **39,** 615 (1981).
178. C. K. Sinclair, in "High Energy Physics with Polarized Beams and Polarized Targets" (C. Joseph and J. Soffer, eds.). Birkhaeuser, Basel, 1981.
179. M. Campagna, S. F. Alvarado, and F. Ciccacci, *AIP Conf. Proc.* **95,** 566 (1983).
180. R. C. Miller, D. A. Kleinman, and A. C. Gossard, *Conf. Ser.—Inst. Phys.* **43,** 1043 (1979).
181. R. Dingle, *Proc. Int. Conf. Semicond. 13th, Roma, 1976*, p. 65 (1977).
182. H. Kobayashi, H. Iwamura, T. Saku, and K. Otsuka, *Electron. Lett.* **19,** 166 (1983).
183. H. Okamoto, Y. Hirokoshi, and H. Iwamura, *Ext Abstr., Solid State Devices Mater. Conf., Tokyo 15th* (1983).
184. H. Okamoto, *J. Vac. Sci. Technol. B* [**B**], 687 (1985).
184a. For detailed review of exciton effect on luminescence and excitation spectroscopy can be found in R. C. Miller and D. A. Kleinman, *J. Lumin.* **30,** 520 (1985).
185. M. Erman and P. M. Frijlink, *Appl. Phys. Lett.* **43,** 1185 (1983)
185a. M. Erman, J. B. Theeten, P. Frijlink, S. Gaillard, Fan Jia Hia, and C. Alibert, *J. Appl. Phys.* **56,** 3241 (1984).
185b. O. J. Glembocki, B. V. Shanabrook, N. Bottka, W. T. Beard, and J. Colmas, *Appl. Phys. Lett.* **46,** 970 (1985).
185c. B. V. Shanabrook, O. J. Glembocki, and W. T. Beard, *Phys. Rev. B* **35,** 2540 (1987).
186. Y. Suzuki and H. Okamoto, *J. Electron. Mater.* **12,** 397 (1983).
186a. See, e.g., P. W. Yu, D. C. Reynolds, K. K. Bajaj, C. W. Litton, J. Klem, D. Huang, and H. Morkoc, *Solid State Commun.* **62,** 41 (1987).
187. R. C. Miller, D. A. Kleinman, and A. C. Gossard, *Phys. Rev. B* **29,** 7085 (1984).
188. R. C. Miller, A. C. Gossard, D. A. Kleinman, and O. Munteanu, *Phys. Rev. B* **29,** 3740 (1984).
188a. M. H. Meynadier, C. Delalande, G. Bastard, M. Voos, F. Alexandre, and J. L. Lievin, *Phys. Rev. B* **31,** 5539 (1985).
189. R. C. Miller, D. A. Kleinman, W. A. Nordland, Jr., and A. C. Gossard, *Phys. Rev. B* **22,** 863 (1980).
190. H. Kroemer, W. Y. Chien, J. S. Harris, Jr., and D. D. Edwall, *Appl. Phys. Lett.* **36,** 295 (1980).
191. W. I. Wang, E. E. Mendez, and F. Stern, *Appl. Phys. Lett.* **45,** 639 (1984).
192. R. C. Miller, A. C. Gossard, D. A. Kleinman, and O. Munteanu, *Phys. Rev. B* **24,** 1134 (1981).
193. J. C. Maan, G. Belle, A. Fasolino, M. Altarelli, and K. Ploog, *Phys. Rev. B* **30,** 2253 (1984).

194. J. N. Schulman, *J. Vac. Sci. Technol. B* **[1]**, 644 (1983).
195. F. Stern and J. N. Schulman, *Superlattices Microstruc.* **1,** 303 (1985).
196. C. Weisbuch, R. C. Miller, R. Dingle, A. C. Gossard, and W. Wiegmann, *Solid State Commun.* **37,** 219 (1981).
197. The excitation spectra (ES) method has actually been very widely used to study carrier energy relaxation in semiconductors, see, e.g., S. Permogorov, *Phys. Status Solidi B* **68,** 9 (1975), and the relevant chapters in "Excitons" (E. I. Rashba and M. D. Sturge, eds.). North-Holland Publ., Amsterdam, 1982.
198. C. Weisbuch, R. Dingle, A. C. Gossard, and W. Wiegmann, *Conf. Ser.—Inst. Phys.* **56,** 711 (1981).
199. C. Weisbuch, R. Dingle, A. C. Gossard, and W. Wiegmann, *Solid State Commun.* **38,** 709 (1981).
200. G. Bastard, C. Delalande, M. H. Meynadier, P. M. Frijlink, and M. Voos, *Phys. Rev. B* **29,** 7042 (1984).
201. B. Deveaud, J. Y. Emery, A. Chomette, B. Lambert, and M. Baudet, *Appl. Phys. Lett.* **45,** 1078 (1984).
202. D. C. Reynolds, K. K. Bajaj, C. W. Litton, P. W. Yu, J. Singh, W. T. Masselink, R. Fischer, and H. Morkoc, *Appl. Phys. Lett.* **46,** 51 (1985).
202a. H. Sakaki, M. Tanaka, and J. Yoshino, *Jpn. J. Appl. Phys.* **24,** L417 (1985).
202b. T. Fukunaga, K. L. T. Kobayashi, and H. Nakashina, *Jpn. J. Appl. Phys.* **24,** L510 (1985).
202c. T. Hayakawa, T. Suyama, K. Takahashi, M. Kondo, S. Yamamoto, S. Yano, and T. Hijikata, *Surf. Sci.* **174,** 76 (1986).
203. J. C. Phillips, *J. Vac. Sci. Technol.* **19,** 545 (1981).
204. J. Singh and A. Madhukar, *J. Vac. Sci. Technol. B* **[1]**, 305 (1983).
205. J. Singh and K. K. Bajaj, *J. Vac. Sci. Technol. B* **[3]**, 520 (1985).
206. S. V. Ghaisas and A. Madhukar, *J. Vac. Sci. Technol. B*[**3**], 540 (1985).
206a. N. Watanabe and Y. Mori, *Surf. Sci.* **174,** 10 (1986).
207. R. Dingle, *J. Vac. Sci. Technol.* **14,** 1006 (1977).
208. R. M. Fleming, D. B. McWhan, A. C. Gossard, W. Wiegmann, and R. A. Logan, *J. Appl. Phys.* **51,** 357 (1980).
209. P. M. Petroff, *J. Vac. Sci. Technol.* **14,** 973 (1977).
210. W. Laidig, N. Holonyak, Jr., M. D. Camras, K. Hess, J. J. Coleman, P. D. Dapkus, and J. Bardeen, *Appl. Phys. Lett.* **38,** 776 (1981).
211. Y. Suzuki, Y. Horikoshi, M. Kobayashi, and H. Okamoto, *Electron. Lett.* **20,** 383 (1984).
212. J. J. Coleman, P. D. Dapkus, C. G. Kirkpatrick, M. D. Camras, N. Holonyak, Jr., *Appl. Phys. Lett.* **40,** 904 (1982).
213. S. Tarucha, Y. Horikoshi, and H. Okamoto, *Jpn. J. Appl. Phys.* **22,** L482 (1983).
214. See, e.g., D. D. Sell, S. F. Stokowski, R. Dingle, and J. V. Di Lorenzo, *Phys. Rev. B* **7,** 4568 (1973).
215. For reviews, see, e.g., H. Barry Bebb and E. W. Williams, *in* "Semiconductors and Semimetals" (R. K. Willardson and A. C. Beer, eds.), Vol. 8, p. 239. Academic Press, New York, 1972; P. J. Dean, *in* "Collective Excitations in Solids " (B. Dibartolo, ed.), p. 247. Plenum, New York, 1983.
216. J. J. Hopfield, *Phys. Rev.* **112,** 1555 (1958); *Proc. Int. Conf. Phys. Semicond., 8th, Kyoto, 1966* (1966); *J. Phys. Soc. Jpn.* **21,** Suppl., 77 (1966).
217. C. Weisbuch, *J. Lumin.* **24/25,** 373 (1981), and references therein.

218. P. M. Petroff, C. Weisbuch, R. Dingle, A. C. Gossard, and W. Wiegmann, *Appl. Phys. Lett.* **38,** 965 (1981).
219. H. Jung, A. Fischer, and K. Ploog, *Appl. Phys.* **A33,** 97 (1984); *A* **35,** 130 (1984).
220. K. Ploog, private communication.
220a. D. C. Herbert and J. M. Rorison, *Solid State Commun.* **54,** 343 (1985).
221. R. C. Miller, D. A. Kleinman, A. C. Gossard, and O. Munteanu, *Phys. Rev. B* **25,** 6545 (1982).
222. R. C. Miller, A. C. Gossard, W. T. Tsang, and O. Munteanu, *Phys. Rev. B* **25,** 3871 (1982).
223. R. C. Miller, A. C. Gossard, W. T. Tsang, and O. Munteanu, *Solid State Commun.* **43,** 519 (1982).
224. B. V. Shanabrook and J. Comas, *Surf. Sci.* **142,** 504 (1984).
224a. See, e.g., the review by J. Shah, *IEEE J. Quantum Electron.* **QE-22,** 1728 (1986).
225. C. V. Shank, R. L. Fork, R. Yen, J. Shah, B. I. Greene, A. C. Gossard, and C. Weisbuch, *Solid State Commun.* **47,** 981 (1983).
226. W. H. Knox, R. L. Fork, M. C. Downer, D. A. B. Miller, D. S. Chemla, C. V. Shank, A. C. Gossard, and W. Wiegmann, *Phys. Rev. Lett.* **54,** 1306 (1985).
227. Z. Y. Xu and C. L. Tang, *Appl. Phys. Lett.* **44,** 692 (1984).
228. D. J. Erskine, A. J. Taylor, and C. L. Tang, *Appl. Phys. Lett.* **45,** 54 (1984).
229. E. O. Göbel, H. Jung, J. Kuhl, and K. Ploog, *Phys. Rev. Lett.* **51,** 1588 (1983).
230. J. Christen, D. Bimberg, A. Steckenborn, and G. Weimann, *Appl. Phys. Lett.* **44,** 84 (1984).
231. Y. Masumoto, S. Shionoya, and H. Kawaguchi, *Phys. Rev. B* **29,** 2324 (1984).
232. T. Takagahara, *Proc. Int. Conf. Phys. Semicond., San Francisco, 1984* (D. Chadi, ed.). Springer-Verlag, Berlin and New York (1985); *Phys. Rev. B* **31,** 6552 (1985).
232a. D. L. Dexter, *in* "Solid State Physics vol. 6" (F. Seitz and D. Turnbull, eds.), p. 353. Academic Press, New York, 1958.
232b. J. J. Hopfield, *Phys. Rev.* **112,** 1555 (1958).
232c. See, e.g., J. J. Hopfield, *J. Phys. Soc. Japan* **21,** suppl. p77 (1966); D. D. Sell, S. E. Stokowski, R. Dingle, and J. V. Di Lorenzo, *Phys. Rev.* **B7,** 4568 (1973); C. Weisbuch and R. G. Ulbrich, *J. Lumin.* **18/19,** 27 (1978); C. Weisbuch, *J. Lumin.* **24/25,** 373 (1981) and references therein.
232d. L. Schultheis, J. Kuhl, A. Honold, and C. W. Tu, *Phys. Rev. Lett.* **57,** 1797 (1986); L. Schultheis, A. Honold, J. Kuhl, K. Köhler, and C. W. Tu, *Phys. Rev.* **B34,** 9027 (1986).
232e. T. Katsumaya and K. Ogawa, *CLEO 90, Tech. Dig. Ser.* **7** (OSA, Washington D.C., 1990) p. 60; K. Ogawa, T. Katsumaya, and H. Nakamura, *Appl. Phys. Lett.* **53,** 1077 (1988).
232f. J. Feldmann, G. Peter, E. O. Göbel, P. Dawson, K. Moore, C. Foxon, and R. J. Elliott, *Phys. Rev. Lett.* **59,** 2337 (1987).
233. J. Hegarty, *Phys. Rev. B* **25,** 4324 (1982).
234. J. Hegarty, M. D. Sturge, A. C. Gossard, and W. Wiegmann, *Appl. Phys. Lett.* **40,** 132 (1982).
235. J. Hegarty, M. D. Sturge, C. Weisburch, A. C. Gossard, and W. Wiegmann, *Phys. Rev. Lett.* **49,** 930 (1982).
236. J. Hegarty, L. Goldner and M. D. Sturge, *Phys. Rev. B* **30,** 7346 (1984).
237. A detailed review of exciton localization in QW's is given by J. Hegarty and M. D. Sturge, *J. Opt. Soc. Am. B* **B2,** 143 (1985).
238. G. Abstreiter, M. Cardona, and A. Pinczuk, *Top. Appl. Phys.* **54,** 5 (1984).

239. G. Abstreiter, R. Merlin and A. Pinczuk, *IEEE J. Quantum Electron.* **QE-22,** 1771 (1986).
240. E. Burstein, A. Pinczuk, and S. Buchner, *Conf. Ser.—Inst. Phys.* **43,** 585 (1979).
241. P. M. Platzman and P. A. Wolff, "Waves and Interactions in Solid State Plasmas," Academic Press, New York, 1973.
242. E. Burstein, A. Pinczuk, and D. L. Mills, *Surf. Sci.* **98,** 451 (1980).
243. G. Abstreiter and K. Ploog, *Phys. Rev. Lett.* **42,** 1308 (1979).
244. A. Pinczuk, H. L. Stormer, R. Dingle, J. M. Worlock, W. Wiegmann, and A. C. Gossard, *Solid State Commun.* **32,** 1001 (1979).
245. G. H. Döhler, H. Künzel, D. Olego, K. Ploog, P. Ruden, H. J. Stolz, and G. Abstreiter, *Phys. Rev. Lett.* **47,** 864 (1981).
246. A. Pinczuk, H. L. Störmer, A. C. Gossard, and W. Wiegmann, *Proc. Int. Conf. Phys. Semicond., San Francisco, 1984*, p. 329 (D. Chadi, ed.). Springer-Verlag, Berlin and New York, 1985.
247. A. Pinczuk, J. M. Worlock, H. L Störmer, A. C. Gossard, and W. Wiegmann, *J. Vac. Sci. Technol.* **19,** 566 (1981).
248. A. Pinczuk, J. M. Worlock, H. L. Störmer, R. Dingle, W. Wiegmann, and A. C. Gossard, *Solid State Commun.* **36,** 43 (1980).
249. D. Olego, A. Pinczuk, A. C. Gossard, and W. Wiegmann, *Phys. Rev. B* **25,** 7867 (1982).
250. A. Pinczuk, J. Shah, A. C. Gossard, and W. Wiegmann, *Phys. Rev. Lett.* **46,** 1307 (1981).
251. Ch. Zeller, G. Abstreiter, and K. Ploog, *Surf. Sci.* **113,** 85 (1982).
252. Ch. Zeller, B. Vinter, G. Abstreiter, and K. Ploog, *Phys. Rev. B* **26,** 2124 (1982); *Pysica (Amsterdam)* **117, 118B & C,** 729 (1983).
253. J. M. Worlock, A. Pinczuk, Z. J. Tien, C. H. Perry, H. L. Störmer, R. Dingle, A. C. Gossard, W. Wiegmann, and R. Aggarwal, *Solid State Commun.* **40,** 867 (1981).
253a. D. S. Chemla, D. A. B. Miller, and S. Schmitt-Rink, "Non-linear Optics and Electro-Optics of Quantum Confined Structures," Academic Press, New York, to be published.
254. See the excellent, self-contained review by S. Schmitt-Rink, D. S. Chemla, and D. A. B. Miller, *Advances in Physics* **38,** 89 (1989) and references therein. See also "Optical Non-Linearities and Instabilities in Semiconductors" (H. Haug, ed.), Academic Press, Boston (1988).
255. D. S. Chemla and D. A. B. Miller, *J. Opt. Soc. Am.* **B2,** 1155 (1985).
256. D. A. B. Miller, D. S. Chemla, and S. Schmitt-Rink, *in* "Optical Non-Linearities and Instabilities in Semiconductors" (H. Haug, ed.), p. 325. Academic Press, Boston, 1988.
257. J. D. Dow and Redfield, *Phys. Rev.* **B1,** 3358 (1970).
258. D. A. B. Miller, D. S. Chemla, T. C. Damen, A. C. Gossard, W. Wiegmann, T. H. Wood and C. A. Burrus, *Phys. Rev. Lett.* **53,** 2173 (1984).
259. H.-J. Polland, L. Schultheis, J. Kuhl, E. O. Göbel, and C. W. Tu, *Phys. Rev. Lett.* **55,** 2610 (1985).
260. See, e.g., J. E. Zucker, T. Y. Chang, M. Wegener, N. J. Sauer, K. L. Jones, and D. S. Chemla, *IEEE Photonics Tech. Lett.* **2,** 29 (1990) and references therein.
261. H. Sakaki and H. Yoshimura, "Optical Switching in Low-Dimensional Systems" (H. Haug and L. Bányai, eds.), NATO ASI Series B : Physics **194,** p. 25. Plenum, New York, 1989.
262. K. Nishi and T. Hiroshima, *Appl. Phys. Lett.* **51,** 320 (1987).
263. See, e.g., P. F. Yuh and K. L. Wang, *IEEE J. Quant. El.* **QE-25,** 1671 (1989); Deyeol

Ahn, ibid. **QE-25,** 2260 (1989) ; K. W. Steijn, R. P. Leavitt, and J. W. Little, *Appl. Phys. Lett.* **55,** 383 (1989).

264. See, e.g., T. B. Norris, N. Vodjdani, B. Vinter, C. Weisbuch, and G. A. Mourou, *Phys. Rev.* **B40,** 1392 (1989); H. W. Liu, R. Ferreira, G. Bastard, C. Delalande, J. F. Palmier and B. Etienne, *Appl. Phys. Lett.* **54,** 2082 (1989); R. Sauer, T. D. Harris and T. W. Tsang, *Phys. Rev.* **B39,** 12929 (1989) and references therein.
265. B. Deveaud, A. Chomette, A. Regreny, J. L. Oudar, D. Hulin, and A. Antonetti, *in* "High-Speed Electronics" (B. Källbäck and H. Beneking, eds.) Springer Series Electronics and Photonics, **22,** p. 101. Springer, Berlin, (1986).
266. J. W. Little, J. K. Whisnant, R. P. Leavitt, and R. A., Wilson, *Appl. Phys. Lett.* **51,** 1786 (1987).
267. A. Tackenchi, S. Muto, T. Inata, and T. Fujii, *Japan J. Appl. Phys.* **28,** L 1098 (1989).
267a. I. Bar-Joseph, K. W. Gossen, J. M. Kuo, R. F. Kopf, D. A. B. Miller, and D. S. Chemla, *Appl. Phys. Lett.* **55,** 340 (1989).
267b. K. K. Law, R. H. Yan, J. L. Merz, and L. A. Coldren, *Appl. Phys. Lett.* **56,** 1886 (1990).
268. J. Bleuse, G. Bastard, and P. Voisin, *Appl. Phys. Lett.* **60,** 220 (1988); E. E. Mendez, F. Agullo-Rueda and J. M. Hong, ibid., 2624 (1988); P. Voisin, J. Bleuse, C. Bouche, S. Gaillard, C. Alibert, and A. Regreny, ibid. **61,** 1639 (1988).
269 A. Chavez-Pirson, H. M. Gibbs and S. Koch, in "Non-Linear Optics of Organics and Semiconductors" (T. Kobayashi, ed.), p. 44. Springer, Berlin, 1989.
269a. H-C. Lee, A. Kost, M. Kawase, A. Hariz, P. D. Dapkus, and E. M. Garmire, *IEEE J. Quantum Electron.* **QE-24,** 1581 (1988).
270. For a tutorial review, see, e.g., D. A. B. Miller, *Optics and Photonics News*, pp. 7–15, Feb. 1990.
271. J. E. Midwinter, *IEE Proc.* (London), **132J.** 371 (1985).
272. See, e.g., the review by M. Erman, "GaAs and related Compounds 1987," *Inst. Phys. Conf. Ser. n° 91*, p. 33 (1988).
273. S. Nojima, *Appl. Phys. Lett.* **55,** 1868 (1989).
274. J. E. Zucker, K. L. Jones, B. I. Miller and V. Koren, *IEEE Photonics Tech. Lett.* **2,** 32 (1990).
275. See, e.g., the complicated electrode structure required to operate at high-frequencies through velocity-matching electrodes in V. L. Veselka and S. K. Korotky *in* Tech. Dig. Conf. on Photonic Switching, Salt Lake City, 1989, paper ThA2-1; see also R. G. Walker, I. Bennion, and A. C. Carter, *Electron. Lett.* **23,** 1549 (1989).
276. See, e.g., the special issues of *IEEE Communications Magazine*, May 1988; *IEEE Journal of Lightwave Technology*, August 1988.
277. See, e.g., "Optical Signal Processing" (J. L. Horner, ed.), Academic Press, San Diego, 1987; *Proc IEEE*, January 1990.
278. Ran-Hong Yan, R. J. Simes, and L. A., Coldren, *IEEE J. Quantum Electr.* **QE-25,** 2272 (1989) and references therein.
279. M. Whitehead, A. Rivers, G. Parry, J. S. Roberts, and C. Button, *Electron. Letters,* **25,** 985 (1989).
280. E. Rosencher, P. Bois, J. Nagle, and S. Delaître, *Electron Letters,* **25,** 1063 (1989).
281. E. Rosencher, P. Bois, J. Nagle, E. Costard, and S. Delaître, *Appl. Phys. Lett.* **55,** 1597 (1989).
282. D. A. Miller, D. S. Chemla, T. C. Damen, A. C. Gossard, W. Wiegmann, T. Wood, and C. A. Burrus, *Appl. Phys. Lett.* **45,** 13 (1984).
283. D. Jäger and F. Forsmann, *in* "Optical Non-Linearities and Instabilities in Semiconductors" (H. Haug, ed.) p. 361. Academic Press, Boston, 1988.

284. D. A. B. Miller, D. S. Chemla, T. C. Damen, T. H. Wood, C. A. Burrus, A. C. Gossard, and W. Wiegmann, *IEEE J. Quantum Electron.* **QE-21,** 1462 (1985).
285. A. L. Lentine, D. A. B. Miller, J. E. Henry, J. E. Cunningham, and L. M. F. Chirovsky, *IEEE J. Quantum Electron.* **QE-25,** 1921 (1989).
286. C. R. Giles, T. Li, T. Wood, C. A. Burrus, and D. A. B. Miller, *Electronics Lett.* **4,** 848 (1988).
286a. D. A. B. Miller, *Optical and Quantum Electronics,* Nov. 1990.
287. A. L. Lentine, H. S. Hinton, D. A. B. Miller, J. E. Henry, J. E. Cunningham, and L. M.F. Chirovsky, *Appl. Phys, Lett.* **52,** 1419 (1988).
288. A. L. Lentine, F. B. MacHormick, R. A. Novotny, L. M. F. Chirovsky, L. A. D'Asaro, R. F. Kopf, J. M. Kuo, and G. D. Boyd, *IEEE Photonics Tech. Lett.* **2,** 51 (1990).
289. H. L. Störmer, *Surf. Sci.* **132,** 519 (1983).
290. For a review on electron mobility in III-V compounds, see D. L. Rode, in "Semiconductors and Semimetals" (R. K. Willardson and A. C. Beer, eds.), Vol. 10, Chapter 1. Academic Press, New York, 1975.
291. For a review on hole mobility in III-V compounds, see Wiley, in "Semiconductors and Semimetals" (R L Willardson and A. C. Beers, eds.), Vol 10, Chapter 2. Academic Press, New York, 1975.
292. G. E. Stillman and C. M. Wolfe, *Thin Solid Films* **31,** 69 (1976).
293. D. Chattopadhyay and H. J. Queisser, *Rev. Mod. Phys.* **53,** 745 (1981), and references therein.
294. W. Bludau, E. Wagner, and H. J. Queisser, *Solid State Commun.* **18,** 861 (1976).
295. B. Clerjaud, A. Gelineau, D. Galland, and K. Saminadayar, *Phys. Rev. B* **19,** 2056 (1979).
296. R. Romestain and C. Weisbuch, *Phys. Rev. Lett.* **45,** 2067 (1980).
297. K. Hess, *Appl. Phys. Lett.* **35,** 484 (1979).
298. P. Price, *Ann. Phys. (N.Y)* **133,** 217 (1981).
299. T. Ando, *J. Phys. Soc. Jpn.* **51,** 3900 (1982).
300. P. Price, *Surf. Sci.* **143,** 145 (1984).
301. P. Price and F. Stern, *Surf. Sci.* **132,** 577 (1983).
302. F. Stern, *Appl. Phys. Lett.* **43,** 974 (1983).
303. G. Fishman and D. Calecki, *Physica B (Amsterdam)* **117B–118B,** 744 (1983).
304. B. Vinter, *Appl. Phys. Lett.* **45,** 581 (1984).
305. W. Walukiewicz, H. E. Ruda, J. Lagowski, and H. C. Gatos, *Phys. Rev. B* **30,** 4571 (1984).
306. S. Hiyamizu, K. Nanbu, T. Mimura, T. Fujii, and H. Hashimoto, *Jpn. J. Appl. Phys.* L378 (1981).
307. See, e.g., the original papers on modulation-doping by R. Dingle and H. L. Störmer.[131,132]
308. T. J. Drummond, W. Kopp, H. Morkoc, K. Hess, A. Y. Cho, and B. G. Streetman, *J. Appl. Phys.* **52,** 5689 (1981).
308a. E. E. Mendez, *IEEE J. Quantum Electron.* **QE-22,** 1720 (1986).
309. E. E. Mendez, P. J. Price, and M. Heiblum, *Appl. Phys. Lett.* **45,** 294 (1984).
310. B. J. F. Lin, D. C. Tsui, M. A. Paalanen, and A. C. Gossard, *Appl. Phys. Lett.* **45,** 695 (1984).
311. B. J. F. Lin, Ph. D. Thesis, Princeton University, Princeton, New Jersey (1984) (unpublished).
312. P. J. *Price, Phys. Rev. B* **32,** 2643 (1985).
313. B. Vinter, *Phys. Rev. B* **33,** 5904 (1986).

314. M. A. Littlejohn, J. R. Hauser, T. H Glisson, D. K. Ferry, and J. W. Harrison, *Solid State Electron.* **21,** 107 (1978).
315. Y. Takeda, *in* "GaInAsP Alloy Semiconductors" (T. P. Pearsall, ed.), p. 213. Wiley, Chichester, England, 1982.
316. G. Bastard, *Appl. Phys. Lett.* **43,** 591 (1983).
317. J. A. Brum and G. Bastard, *Solid State Commun.* **53,** 727 (1985).
318. H. Morkoc, T. J. Drummond, and R. Fischer, *J. Appl. Phys.* **53,** 1030 (1982).
319. H. L. Störmer, A. C. Gossard, and W. Wiegmann, *Solid State Commun.* **41,** 707 (1982).
320. T. Englert, J. C. Maan, D. C. Tsui, and A. C. Gosard, *Solid State Commun.* **45,** 989 (1983).
320a. L. Pfeiffer, K. W. West, H. L. Störmer, and K. W. Baldwin, *Appl. Phys. Lett.* **55,** 1888 (1989); C. T. Foxon, J. J. Harris, D. Hilton, J. Hewett, and C. Roberts, *Semicond. Sci. Technol.* **4,** 582 (1989).
320b. R. Lassnig, *Solid State Commun.* **65,** 765 (1987).
321. H. L. Störmer and W. T. Tsang, *Appl. Phys. Lett.* **36,** 685 (1980).
322. H. L. Störmer, A. C. Gossard, W. Wiegmann, R. Blondel, and K. Baldwin, *Appl. Phys. Lett.* **44,** 139 (1984).
323. H. L. Störmer, K. Baldwin, A. C. Gosard, and W. Wiegmann, *Appl. Phys. Lett.* **44,** 1062 (1984).
324. S. Tiwari and W. I. Wang, *IEEE Electron Device Lett.* **EDL-5,** 333 (1984).
325. R. A. Kiehl and A. C. Gossard, *IEEE Electron Device Lett.* **EDL-5,** 420 (1984).
326. S. Mori and T. Ando, *Phys. Rev. B* **19,** 6433 (1979).
327. S. Mori and T. Ando, *J. Phys. Soc. Jpn.* **48,** 865 (1980).
328. K. Inoue and H. Sakaki, *Jpn. J. Appl. Phys.* **23,** L61 (1984).
329. R. Fischer, W. T. Masselink, Y. L. Sun, T. J. Drummond, Y. C. Chang, M. V. Klein, and H. Morkoc, *J. Vac. Sci. Technol. B* [**2**], 170 (1984).
330. K. Inoue, H. Sakaki, J. Yoshino, and Y Yoshoka, *Appl. Phys. Lett.* **46,** 973 (1985).
331. H. Sakaki, *Jpn. J. Appl. Phys.* **21,** L381 (1982).
332. K. Hirakawa, H. Sakaki, and J. Yoshino, *Phys. Rev. Lett.* **54,** 1279 (1985).
333. Review on hot electron phenomena can be found in "Physics of Non-linear Transport." (D. K. Ferry, J. R. Barker, and C. Jacoboni, eds.), *NATO Adv. Study Inst. Ser., Ser. B* **52** (1980); The various proceedings of the Hot-Electrons Conferences, i.e., *Solid State Electron.* **43** (1978); *J. Phys. (Paris), Colloq.* **C7** (1981); *Physica (Amsterdam)* **134B & C** (1985); and E. M. Conwell, *Solid State Phys., Suppl.* **9.** (1967).
334. T. L. Drummond, W. Kopp, H. Morkoc and M. Keever, *Appl. Phys. Lett.* **41,** 3 (1982).
335. P. J. Price, *Ann. Phys. (N.Y.)* **133,** 217 (1981).
336. P. J. Price, *J. Appl. Phys.* **53,** 6863 (1982).
337. B. K. Ridley, *J. Phys. C* **15,** 5899 (1982).
338. F. A. Riddoch and B. K. Ridley, *J. Phys. C* **16,** 6971 (1984).
339. B. K. Ridley, *J. Phys. C* **17,** 5357 (1984).
340. K. Hess, *Physica B (Amsterdam)* **117B-118B,** 723 (1983), and references therein.
341. T. J. Drummond, M. Keever, W. Kopp, H. Morkoc, K. Hess, B. G. Streetman, and A. Y. Cho, *Electron Lett.* **17,** 545 (1981).
342. T. J. Drummond, W. Kop, H. Morkoc, and M. Keever, *Appl. Phys. Lett.* **41,** 277 (1982).
343. T. J. Drummond, S. L. Su, W. G. Lyons, R. Fischer, W. Kopp, H. Morkoc, K. Lee, and M. Shur, *Electron Lett.* **18,** 1057 (1982).
344. E. F. Schubert and K. Ploog. *Appl. Phys.* **A33,** 183 (1984).
344a. W. T. Masselinkk, N. Braslau, W. I. Wang, and S. L. Wright, *Appl. Phys. Lett.*

51, 1533 (1987); W. T. Masselink, N. Braslau, D. LaTulipe, W. I. Wang, and S. L. Wright, *Solid State Electron.* **31,** 337 (1988).

344b. M. Tomizawa, K. Yokoyama, and A. Yoshii, *IEEE Electron Device Lett.* **EDL-5,** 464 (1984); K. Yokoyama and K. Hess, *Phys. Rev. B* **33,** 5595 (1986); K. Yokoyama and K. Hess, *J. Appl. Phys.* **59,** 3798 (1986); K. Yokoyama, *J. Appl. Phys.* **63,** 938 (1988).

344c. K. Tsubaki, A. Sugimura, and K. Kumabe, *Appl. Phys. Lett.* **46,** 764 (1985).

344d. K. Hirakawa, and H. Sakaki, *Appl. Phys. Lett.* **49,** 889 (1986).

345. J. Shah, A. Pinczuk, H. L. Störmer, A. C. Gossard, and W. Wiegmann, *Appl. Phys. Lett.* **42,** 55 (1983).

346. J. Shah, A. Pinczuk, A. C. Gossard, and W. Wiegmann, *Phys. Rev. Lett.* **54,** 2045 (1985).

347. H. Sakaki, K. Hirakawa, J. Yoshino, S. P. Svensson, Y. Sekiguchi, T. Hotta, and S. *Sishii, Surf. Sci.* **142,** 306 (1984); K. Inoue, H. Sakaki, and J. Yoshino, *Appl. Phys. Lett.* **47,** 614 (1985).

348. M. Inoue, M. Inayama, S. Hiyamizu, and Y. Insuishi, *Jpn. J. Appl. Phys.* **22,** L213 (1983).

348a. See also the review by F. Capasso, K. Mohammed, and A. Y. Cho, *IEEE J. Quantum Electron.* **QE-22,** 1853 (1986).

349. For a recent discussion on the existence and detection of Bloch oscillations, see R. O. Grondin, W. Porod, J. Ho, D. K. Ferry, and G. J. Iafrate, *Superlattices Microstruct.* **1,** 183 (1985).

350. L. Esaki and R. Tsu, *IBM Res. Note* **RC-2418** (1969).

351. A. B. Pippard, "The Dynamics of Conduction Electrons." Gordon & Breach, New York, 1965.

352. J. Zak, *Phys. Rev. Lett.* **20,** 1477 (1968).

353. J. N. Churchill and F. E. Holmstrom, *Phys. Lett.* **85A,** 453 (1981).

354. R. Tsu, L. L. Chang; G. A. Sai-Halasz, and L. Esaki, *Phys. Rev. Lett.* **34,** 1509 (1975).

355. L. L. Chang, L. Esaki, and R. Tsu, *Appl. Phys. Lett.* **24,** 593 (1974).

355a. R. F. Kazarinov and R. A. Suris, *Fiz, Tekh, Poluprov,* **6,** 148 (1972) [*Sov. Phys.—Semicond. (Engl. Transl.)* **6,** 120 (1972)].

355b. J. F. Palmier and A. Chomette, *J. Phys. C* **17,** 5017 (1984); see also the very detailed account by J. F. Palmier *in* "Heterojunctions and Semiconductor Superlattices" (G. Allen *et al.*, eds.), p. 127. Springer-Verlag, Berlin and New York, 1986.

355c. J. R. Barker, *Springer Proceedings in Physics* **13,** 210 (1986).

356. T. W. Hickmott, P. M. Solomon, R. Fischer, and H. Morkoç, *Appl. Phys. Lett.* **44,** 90 (1984).

356a. M. Tsuchiya, H. Sakaki, and J. Yoshino, *Jap. J. Appl. Phys.* **24,** L466 (1985); M. Tsuchiya and H. Sakaki, *Jap. J. Appl. Phys.* **25,** L185 (1986); M. Tsuchiya and H. Sakaki, *Appl. Phys. Lett.* **49,** 88 (1986); M. Tsuchiya and H. Sakaki, *Appl. Phys. Lett.* **50,** 1503 (1986).

356b. M. J. Paulus, C. A. Bozada, C. I. Huang, S. C. Dudley, K. R. Evans, C. E. Stutz, R. L. Jones, and M. E. Cheney, *Appl. Phys. Lett.* **53,** 204 (1988); C. I. Huang, M. J. Paulus, C. A. Bozada, S. C. Dudley, K. R. Evans, C. E. Stutz, R. L. Jones, and M. E. Cheney, *Appl. Phys. Lett.* **51,** 204 (1987).

356c. S. K. Diamond, E. Ozbay, M. Rodwell, D. Bloom, Y. C. Pao, E. Wolak, and J. J. Harris, *IEEE Electron Device Lett.* **EDL-10,** 104 (1989).

356d. E. E. Mendez, W. I. Wang, B. Ricco, and L. Esaki, *Appl. Phys. Lett.* **47,** 415 (1985).

356e. T. Inata, S. Muto, Y. Nakata, T. Fujii, H. Ohnishi, and S. Hiyamizu, *Jap. J. Appl. Phys.* **25,** L983 (1986).

356f. T. Inata, S. Muto, Y. Nakata, S. Sasa, T. Fujii, and S. Hiyamizu, *Jap. J. Appl. Phys.* **26,** L1332 (1987).
356g. T. P. E. Brokaert, W. Lee, and C. Fonstad, *Appl. Phys. Lett.* **53,** 1545 (1988).
357. R. T. Collins, J. Lambe, T. C. McGill, and R. D. Burnham, *Appl. Phys. Lett.* **44,** 532 (1984).
357a. V. J. Goldman, D. C. Tsui, and J. E. Cunningham, *Phys. Rev. B* **36,** 7635 (1987).
357b. M. L. Leadbeater, E. S. Alves, L. Eaves, M. Henini, O. H. Hughes, A. Celeste, J. C. Portal, G. Hill, and M. A. Pate, *Phys. Rev. B* **39,** 3438 (1989).
357c. N. S. Wingreen, K. W. Jacobsen, and J. W. Wilkins, *Phys. Rev. Lett.* **61,** 1396 (1988).
357d. B. G. R. Rudberg, *Semicon. Sci. Technol.* **5,** 328 (1990).
357e. G. Y. Wu and T. C. McGill, *Phys. Rev. B* **40,** 9969 (1989).
357f. F. Chevoir and B. Vinter, *Appl. Phys. Lett.* **55,** 1859 (1989); F. Chevoir and B. Vinter, *Surf. Sci.* **229,** 158 (1990).
357g. see, e.g., M. Büttiker, *IBM J. Res. Develop.* **32,** 63 (1988).
357h. A. D. Stone and P. A. Lee, *Phys. Rev. Lett.* **54,** 1196 (1985).
357i. T. Weil and B. Vinter, *Appl. Phys. Lett.* **50,** 1281 (1987).
357j. M. Jonson and A. Grincwajg, *Appl. Phys. Lett.* **51,** 1729 (1987).
358. For recent calculations of tunneling transport, see, e.g., C. Mailhiot, T. C. McGill, and J. N. Schulman, *J. Vac. Sci. Technol. B* **1,** 439 (1983) and references therein.
358a. A. Zaslavsky, V. J. Goldman, and D. C. Tsui, *Appl. Phys. Lett.* **53,** 1408 (1988).
358b. E. S. Alves, L. Eaves, M. Henini, O. H. Hughes, M. L. Leadbeater, F. W. Sheard, G. A. Toombs, G. Hill, and M. A. Pate, *Electron. Lett.* **24,** 1190 (1988).
359. W. R. Frensley, *Phys. Rev. B* **36** 1570 (1987); H. Ohnishi, T. Inata, S. Muto, N. Yokoyama, and A. Shibatomi, *Appl. Phys. Lett.* **49,** 1248 (1986); W. Pötz, *J. Appl. Phys.* **66,** 2458 (1989); N. C. Kluksdahl, A. M. Kriman, and D. K. Ferry, *Phys. Rev. B* **39,** 7720 (1989); W. R. Frensley, *Solid-State Electronics* **32,** 1235 (1989).
359a. D. Y. Oberli, J. Shah, T. C. Damen, C. W. Tu, T. Y. Chang, D. A. B. Miller, J. E. Henry, R. F. Kopf, N. Sauer, and A. E. Giovanni, *Phys. Rev. B* **40,** 3028 (1989).
359b. T. Weil and B. Vinter, *J. Appl. Phys.* **60,** 3227 (1986).
359c. T. Norris, N. Vodjdani, B. Vinter, C. Weisbuch, and G. A. Mourou, *Phys. Rev. B* **40,** 1392 (1989).
359d. H. W. Liu, R. Ferreira, G. Bastard, C. Delalande, J. F. Palmier, and B. Etienne, *Appl. Phys. Lett.* **54,** 2082 (1989).
359e. R. Ferreira and G. Bastard, *Phys. Rev. B* **40,** 1074 (1989); G. Bastard, C. Delalande, R. Ferreira, and H. W. Liu, *J. Luminescence* **44,** 207 (1989).
359f. B. Lambert, F. Clerot, B. Deveaud, A. Chomette, G. Talalaeff, A. Regreny, and B. Sermage, *J. Luminescence* **44,** 277 (1989) and references therein.
359g. M. Tsuchiya, T. Matsusue, and H. Sakaki, *Phys. Rev. Lett.* **59,** 2356 (1987).
359h. J. F. Young, B. M. Wood, G. C. Aers, R.L. S. Devine, H. C. Liu, D. Landheer, M. Buchanan, A. J. SpringThorpe, and P. Mandeville, *Phys. Rev. Lett.* **60,** 2085 (1988).
359i. N. Vodjdani, F. Chevoir, D. Thomas, D. Cote, P. Bois, E. Costard, and S. Delaître, *Appl. Phys. Lett.* **55,** 1528 (1989).
359j. M. S. Skolnick, A. W. Higgs, P. E. Simmonds, D. G. Hayes, G. W. Smith, H. J. Huchinson, A. D. Pitt, C. R. Whitehouse, L. Eaves, M. Henini, and O. H. Hughes, *Surf. Sci.* **229,** 185 (1990).
360. B. Deveaud, A. Chomette, J. Y. Emery, A. Regreny, and B. Lambert, *Solid State Commun.* **54,** 75 (1985).
360a. B. Deveaud, A. Chomette, B. Lambert, A. Regreny, R. Romestain, and P. Edel, *Solid State Commun.* **57,** 885 (1986).
361. Good reviews of the dynamics of semiconductor electrons in magnetic fields can be found in L. M. Roth and P. N. Argyres, *in* "Semiconductors and Semimetals" (R. K.

Willardson and A. C. Beer, eds.), Vol. 1, p. 159. Academic Press, New York, 1966; L. M. Roth, in "Handbook on Semiconductors" (T. S. Moss, ed.), Vol. 1, p. 451. North-Holland Publ., Amsterdam, 1982; R. Kubo, S. Miyake, and N. Hashitsume, *Solid State Phys.* **17,** 269 (1965); K. Seeger, "Semiconductor Physics," *Springer Ser. Solid-State Sci.* **40,** (1982).

362. r_c can be calculated from the Bohr-Sommerfeld quantization condition $\int P\, dr = \hbar$ and the dynamic equilibrium condition $m\omega^2 r_c = qvB$ of classical mechanics.
363. W. Zawadski and R. Lassnig, *Surf. Sci.* **142,** 225 (1984).
364. W. Zawadski and R. Lassnig, *Solid State Commun.* **50,** 537 (1984).
365. E. Gornik, R. Lassnig, G. Strasser, H. L. Störmer, A. C. Gossard, and W. Wiegmann, *Phys. Rev. Lett.* **54,** 1820 (1985).
366. H. Obloh, K. von Klitzing, and K. Ploog, *Surf. Sci.* **142,** 236 (1984).
367. H. L. Störmer, R. Dingle, A. C. Gossard, W. Wiegmann, and M. Sturge, *Solid State Commun.* **29,** 705 (1979).
368. M. A. Brummel, R. J. Nicholas, L. C. Brunel, S. Huant, M. Bay, J. C. Portal, M. Razeghi, M. A. Di Forte-Poisson, K. Y. Chang, and A. Y. Cho, *Surf. Sci.* **142,** 380 (1984).
369. H. Bluyssen, J. C. Maan, P. Wyder, L. L. Chang, and L. Esaki, *Solid State Commun.* **31,** 35 (1979).
370. W. Seidenbusch, G. Lindemann, R. Lassnig, J. Edlinger, and E. Gornik, *Surf. Sci.* **142,** 375 (1984).
371. T. Englert, J. C. Maan, C. Uihlein, D. C. Tsui, and A. C. Gossard, *Solid State Commun.* **46,** 545 (1983).
372. Z. Schlesinger,, S. J. Allen, J. C. M. Hwang, P. M. Platzmann, and N. Tzoar, *Phys. Rev. B* **30,** 43 (1984).
373. H. L. Störmer, T. Haavasoja, V. Narayanamurti, A. C. Gossard, and W. Wiegmann, *J. Vac. Sci. Technol. B* [**1**], 423 (1983).
374. J. P. Eisenstein, H. L. Störmer, V. Narayanamurti, and A. C. Gossard, *Proc. Int. Conf. Phys. Semicond., 1984,* p. 329 (D. Chadi, ed.). Springer-Verlag, Berlin and New York, 1985.
375. T. Ando, *J. Phys. Soc. Jpn.* **44,** 475 (1978).
376. J. Yoshino, H. Sakaki, and T. Hotta, *Surf. Sci.* **142,** 326 (1984).
377. A. B. Fowler, F. Fang, W. E. Howard, and P. J. Stiles, *Phys. Rev. Lett.* **16,** 901 (1966).
378. J. P. Harrang, R. J. Higgins, R. K. Goodall, P. R. Jay, M. Laviron, and P. Delescluse, *Phys. Rev. B: Condens. Matter* [3] **31,** (1985).
379. T. Englert, D. C. Tsui, A. C. Gossard, and C. Uihlein, *Surf. Sci.* **113,** 295 (1982).
380. R. J. Nicholas, M. A. Brummell, J. C. Portal, K. Y. Cheng, A. Y. Cho, and T. P. Pearsall, *Solid State Commun.* **45,** 911 (1983).
381. C. Weisbuch and C. Hermann, *Phys. Rev. B* **15,** 816 (1977).
382. C. Hermann and C. Weisbuch, *in* "Spin Orientation" (B. P. Zakharchenya and F. Meier, eds.), p. 463. North-Holland Publ., Amsterdam, 1984.
383. D. Stein, K. von Klitzing, and G. Weimann, *Phys. Rev. Lett.* **51,** 130 (1983).
384. E. F. Schubert, K. Ploog, H. Dambkes, and K. Heime, *Appl. Phys.* **A33,** 63 (1984).
385. A very simple and elegant description of the Quantum Hall effect using-classical and quantum analysis can be found in H. L. Störmer and D. C. Tsui, *Science* **220,** 1241 (1983); another simple description (in French) is given by G. Toulouse, M. Voos, and B. Souillard, *C. R. Acad. Sci. (Paris) Vie Sci.* **1,** 321 (1984).
386. For review, see H. L. Störmer, *Festkoerperprobleme* **24,** 25 (1984); K. von Klitzing and G. Ebert, *Springer Ser. Solid-State Sci.* **59,** 242 (1984); B. I. Halperin, *Helv. Phys. Acta* **56,** 75 (1983).
387. K. von Klitzing, G. Dorda, and M. Pepper, *Phys. Rev. Lett.* **45,** 494 (1980).

388. D. C. Tsui and A. C. Gossard, *Appl. Phys. Lett.* **38,** 552 (1981).
389. D. C. Tsui, H. L. Störmer, J. C. M. Hwang, J. S. Brooks, and M. J. Naughton, *Phys. Rev. B* **28,** 2274 (1983).
390. K. von Klitzing, *Festkoerperprobleme* **21,** (1981).
390a. D. C. Tsui, A. C. Gossard, B. F. Field, M. E. Cage, and R. F. Dziuba, *Phys. Rev. Lett.* **48,** 3 (1982).
390b. T. Quinn, Metrolgia **26,** 69 (1989).
391. H. Aoki and T. Ando, *Solid State Commun.* **38,** 1079 (1981); see also the review by T. Ando, *in* "Recent Topics in Semiconductor Physics" (H. Kamimura and Y. Toyozawa, eds.), p. 72, World Scientific, Singapore, 1983.
392. R. E. Prange, *Phys. Rev. B* **23,** 4802 (1981).
393. R. B. Laughlin, *Phys. Rev. B* **25,** 5632 (1981); see also the illuminating discussion by R. B. Laughlin, *Springer Ser. Solid-State Sci.* **59,** 272, 288 (1984).
394. H. Aoki, *Lect, Notes Phys.* **177,** 11 (1983).
395. J. Hajdu, *Lect. Notes Phys.* **177,** 23 (1983).
396. D. C. S. Tsui, H. L. Störmer, and A. C. Gossard, *Phys. Rev. B* **25,** 1405 (1982).
397. M. A. Paalanen, D. C. Tsui, and A. C. Gossard, *Phys, Rev. B* **25,** 5566 (1982).
398. M. E. Cage, B. F. Field, R. F. Dziuba, S. M. Girvin, A. C. Gossard, and D. C. Tsui, *Phys. Rev. B* **30,** 2286 (1984).
399. H. L. Störmer, A. M. Chang, D. C. Tsui, and J. C. M. Hwang, *Proc. Intl. Conf. Phys. Semicond., San Francisco, 1984,* p. 267 (D. Chadi, ed.). Springer-Verlag, Berlin and New York, 1985.
400. S. Komiyama, T. Takamasu, S. Hiyamizu, and S. Sasa, *Solid State Commun.* **54,** 479 (1985).
401. F. F. Fang and P. J. Stiles, *Phys. Rev. B* **29,** 3749 (1984).
402. See, e.g., the discussions in H. L. Störmer and B. L. Halperin's papers.[386]
403. D. C. Tsui, H. L. Störmer, and A. C. Gossard, *Phys. Rev. Lett.* **48,** 1559 (1982).
404. H. L. Störmer, A. Chang, D. C. Tsui, J. C. M. Hwang, and A. C. Gossard, *Phys. Rev. Lett.* **50,** 1953 (1983).
405. E. E. Mendez, M. Heiblum, L. L. Chang, and L. Esaki, *Phys. Rev. B.* **28,** 4886 (1983).
405a. R. Willett, J. P. Eisenstein, H. L. Störmer, D. C. Tsui, A. C. Gossard, and J. H. English, *Phys. Rev. Lett.* **59,** 1776 (1987).
406. See the discussion in B. I. Halperin.[386]
407. D. Yoshioka, *Phys. Rev.* **B29,** 6833 (1984), and references therein.
408. R. B. Laughlin, *Phys. Rev. Lett.* **50,** 1395 (1983).
409. F. D. M. Haldane, *Phys. Rev. Lett.* **51,** 605 (1983).
410. R. E. Prange and S. M. Girvin, eds., "The Quantum Hall Effect," Springer-Verlag, New York, 1987.
411. Chakraborty and P. Pietiläinen, "The Fractional Quantum Hall Effect: Properties of an Incompressible Quantum Fluid," Springer Series in Solid State Physics **85** (1988).
412. C. T. Van Degrift, M. E. Cage, and S. M. Girvin, *Am. J. Phys.* **58,** 109 (1990).
413. Y. Guldner, J. P. Hirtz, J. P. Vieren, P. Voisin, M. Voos, and M. Razeghi, *J. Phys. Lett. (Orsay, Fr.)* **43,** L613 (1982).
414. E. E. Mendez, L. L. Chang, C. A. Chang, L. F. Alexander, and L. Esaki, *Surf. Sci.* **142,** 215 (1984).
415. M. A. Paalanen, D. C. Tsui, A. C. Gossard, and J. C. M. Hwang, *Solid State Commun.* **50,** 841 (1984).
416. A. P. Long, H. W. Myron, and M. Pepper, *J. Phys. C* **17,** L433 (1984).

417. C. McFadden, A. P. Long, H. W. Myron, M. Pepper, D. Andrews, and G. J. Davies, *J. Phys. C* **17,** L439 (1984).
419. J. A. Cooper, Jr., D. F. Nelson, S. A. Schwarz, and K. K. Thornber, *in* "VLSI Electronics Microstructure Science" (N. G. Einspruch and R. S. Bauer, eds.), vol. 10, p. 324, Academic Press, Orlando, 1985.
420. B. Vinter, *in* "Heterojunctions and Semiconductors Superlattices" (G. Allan *et al.,* eds.), p. 238. Springer-Verlag, Berlin and New York, 1985.
421. See, e.g., S. M. Sze, "Physics of Semiconductor Devices," Wiley, New York, 1981.
421a. K. Yokoyama, *J. Appl. Phys.* **63,** 938 (1988).
422. K. Hikosaka, N. Hidaka, Y. Hirachi, and M. Abe, *IEEE Electron Device Lett.* **EDL-8,** 521 (1987).
422a. M. A. Hollis and R. A. Murphy, *in* "High-Speed Semiconductor Devices" (S. M. Sze, ed.), p. 211, Wiley, New York, 1990.
422b. H. Fukui, *IEEE Trans. Electron. Dev.* **ED-26,** 1032 (1979).
423. P. M. Smith, P. C. Chao, K. H. G. Duh, L. F. Lester, and B. R. Lee, *Electron. Lett.* **22,** 780 (1986).
424. K. H. G. Duh, P. C. Chao, P. M. Smith, L. F. Lester, and B. R. Lee, *Electron. Lett.* **22,** 647 (1986).
425. A. N. Lepore, H. M. Levy, R. C. Tiberio, P. J. Tasker, H. Lee, E. D. Wolf, L. F. Eastman, and E. Kohn, *Electron. Lett.* **24,** 364 (1988).
426. T. P. Pearsall, R. Hendel, P. O'Connor, K. Alavi, and A. Y. Cho, *IEEE Electron Device Lett.* **EDL-4,** 5 (1983).
427. C. Y. Chen, A. Y. Cho, K. Alavi, and P. A. Garbinski, *IEEE Electron Device Lett.* **EDL-3,** 205 (1982).
428. U. K. Mishra, A. S. Brown, S. E. Rosenbaum, C. E. Hooper, M. W. Pierce, M. J. Delaney, S. Vaughn, and K. White, *IEEE Electron Device Lett.* **EDL-9,** 647 (1988).
429. L. D. Nguyen, D. C. Radulescu, M. C. Foisy, P. J. Tasker, and L. F. Eastman, *IEEE Trans. Electron. Dev.* **ED-36,** 833 (1989); L. D. Nguyen, D. C. Radulescu, P. J. Tasker, W. J. Schaff, and L. F. Eastman, *IEEE Electron Device Lett.* **EDL-9,** 374 (1988); G. W. Wang, Y. K. Chen, D. C. Radulescu, and L. F. Eastman, *IEEE Electron Device Lett.* **EDL-9,** 4 (1988); L. D. Nguyen, W. J. Schaff, P. J. Tasker, A. N. Lepore, L. F. Palmateer, M. C. Foisy, and L. F. Eastman, *IEEE Trans. Electron. Dev.* **ED-35,** 139 (1989).
430. P. C. Chao, K. H. G. Duh, P. Ho, P. Smith, J. M. Ballingall, A. A. Jabra, and R. C. Tiberio, *Electron. Lett.* **25,** 504 (1989).
430a. D. Kiefer and J. Heightley, *Technical Digest, GaAs IC Symposium*, p. 3, IEEE, New York, 1987.
430b. M. Abe, T. Mimura, and M. Kobayashi, *Fujitsu Sci. Tech. J.* **24,** 271 (1988); M. Abe, T. Mimura, N. Kobayashi, M. Suzuki, M. Kosugi, M. Nakayama, K. Odani, and I. Hanyu, *IEEE Trans. Electron. Dev.* **ED-36,** 2021 (1989).
430c. Vitesse Semiconductor Corporation, Camarillo, California. Press release, September 1990.
430d. N. J. Shah, S.-S. Pei, C. W. Tu, and R. C. Tiberio, *IEEE Trans. Electron. Dev.* **ED-33,** 543 (1986).
430e. Y. Awano, M. Kosugi, T. Mimura, and M. Abe, *IEEE Electron Device Lett.* **EDL-8,** 451 (1987).
430f. P. M. Solomon, *Proc. IEEE* **70,** 489 (1982).

430g. T. Ikoma, ed., "Very High Speed Integrated Circuits" *in* "Semiconductors and Semimetals" (R. K. Willardson and A. C. Beer, eds.) **29** and **30,** Academic Press, Boston, 1990.
431. P. M. Solomon, C. M. Knoedler, and S. L. Wright, *IEEE Electron Device Lett.* **EDL-5,** 379 (1984).
432. K. Matsumoto, M. Ogura, T. Wada, N. Hashizume, T. Yao, and Y. Hayashi, *Electron. Lett.* **23,** 462 (1984).
432a. C. Weisbuch, *in* "Technologies for Optoelectronics" (F. Potter and J. M. Bulabois, eds.). *Proc. SPIE* **869,** 155 (1987).
433. K. Hess, H. Morkoç, H. Shichijo, and B. G. Streetman, *Appl. Phys. Lett.* **35,** 469 (1979).
434. A. Kastalsky, R. A. Kiehl, S. Luryi, A. C. Gossard, and R. H. Hendel, *IEEE Electron Device Lett.* **EDL-5,** 321 (1984); S. Luryi, A. Kastalsky, A. C. Gossard, and R. H. Hendel, *Appl. Phys. Lett.* **45,** 1294 (1984).
435. S. Luryi, "Hot-Electron-Injection and Resonant-Tunneling Heterojunction Devices" *in* "Heterojunctions: A Modern View of Band Discontinuities and Device Applications" (F. Capasso and G. Magaritondo, eds.), p. 489. North-Holland, Amsterdam, 1987.
435a. C.-T. Sah, *Proc. IEEE* **76,** 1280 (1988).
436. T. C. G. L. Sollner, W. D. Goodhue, P. E. Tannenwald, C. D. Parker, and D. D. Peck, *Appl. Phys. Lett.* **43,** 588 (1983).
437. E. R. Brown, T. C. G. L. Sollner, C. D. Parker, W. D. Goodhue, and C. L. Chen, *Appl. Phys. Lett.* **55,** 1777 (1989); T. C. G. L. Sollner, E. R. Brown, W. D. Goodhue, and H. Q. Le, *Appl. Phys. Lett.* **50,** 332 (1987); for a recent review see T. C. L. G. Sollner, E. R. Brown, and H. Q. Le, *in* "Physics of Quantum Electron Devices" (F. Capasso, ed.) Springer Series in Electronics and Photonics **28,** p. 147. Springer-Verlag, Berlin, Heidelberg, 1990.
437a. P. H. Ladbrooke, *GEC J. Res.* **4,** 114 (1986).
437b. S. I. Long, *IEEE Trans. Electron. Dev.* **ED-36,** 1274 (1989).
437c. D. D. Tang and P. M. Solomon, IEEE J. Solid-State Circ. **SC-14,** 679 (1979).
438. K. Nagata, O. Nakajima, T. Nittono, H. Ito, and T. Ishibashi, *Electron. Lett.* **23,** 566 (1987).
438a. L. P. Ramberg, P. M. Engquist, Y. K. Chen, F. E. Najjar, L. F. Eastman, E. A. Fitzgerald, and K. L. Kavanaugh, *J. Appl. Phys.* **61,** 1234 (1987).
439. T. Ishibashi, *in* "20th International Conference on Solid State Devices and Materials," p. 515, Tokyo, 1988.
440. Y. Yamaguchi, K. Nagata, O. Nakajima, H. Ito, T. Nittino, and T. Ishibashi, *Electron. Lett.* **23,** 881 (1987); Y. Yamauchi, O. Nakajima, K. Nagata, H. Ito, and T. Ishibashi, "Technical Digest, GaAs IC Symposium, San Diego," p. 121, IEEE, New York, 1989.
441. B. Bayraktaroglu, N. Camilleri, and H. Q. Tserng, *Proceedings IEEE/Cornell Conference on Advanced Concepts in High Speed Semiconductor Devices and Circuits,* 265 (1987).
442. H. Fukano, Y. Kawanura, H. Asai, and Y. Takanashi, *Jap. J. Appl. Phys.* **28,** L1737 (1989).
443. B. Jalali, R. N. Nottenburg, W. S. Hobson, Y. K. Chen, T. Fullovan, S. J. Pearton, and A. S. Jordan, *Electron. Lett.* **25,** 1496 (1989).
444. Y. K. Chen, A. F. J. Levi, R. N. Nottenburg, P. H. Beton, and M. B. Panish, *Appl. Phys. Lett.* **55,** 1789 (1989).
445. F. Capasso, "Band-Gap Engineering and Interface Engineering: From Graded-Gap

Structures to Tunable Band Discontinuities," *in* "Heterojunctions: A Modern View of Band Discontinuities and Device Applications" (F. Capasso and G. Magaritondo, eds.), p. 399, North-Holland, Amsterdam, 1987.
446. R. Katoh, M. Kurata, and J. Yoshida, *IEEE Trans. Electron. Dev.* **ED-36,** 846 (1989).
447. M. A. Reed, W. R. Frensley, R. J. Matyi, J. N. Randall, and A. C. Seabaugh, *Appl. Phys. Lett.* **54,** 1034 (1989).
448. A. F. J. Levi, J. R. Hayes, P. M. Platzmann, and W. Wiegmann, *Phys. Rev. Lett.* **55,** 2071 (1985); A. F. J. Levi, J. R. Hayes, and R. Bhat, *Appl. Phys. Lett.* **48,** 1609 (1986).
449. S. Muto, K. Imamura, N. Yokoyama, S. Hiyamizu, and H. Nishi, *Electron. Lett.* **21,** 555 (1985).
450. M. Heiblum, D. I. Thomas, C. M. Knoedler, and M. I. Nathan, *Appl. Phys. Lett.* **47,** 1105 (1985); M. Heiblum, M. I. Nathan, D. I. Thomas, and C. M. Knoedler, *Phys. Rev. Lett.* **55,** 2200 (1985).
451. M. Heiblum, I. M. Anderson, and C. M. Knoedler, *Appl. Phys. Lett.* **49,** 207 (1986); M. Heiblum, E. Calleja, I. M. Anderson, W. P. Dumke, C. M. Knoedler, and L. Osterling, *Phys. Rev. Lett.* **56,** 2854 (1986).
452. K. Seo, M. Heiblum, V. M. Knoedler, J. E. Oh, J. Pamulpati, and P. Bhattacharya, *IEEE Electron Device Lett.* **EDL-10,** 73 (1989).
453. A. F. J. Levi and T. H. Chiu, *Appl. Phys. Lett.* **51,** 984 (1987).
454. A. F. J. Levi, R. N. Nottenburg, Y. K. Chen, and J. E. Cunningham, *Appl. Phys. Lett.* **54,** 2250 (1989).
455. T. Mori, H. Ohnishi, K. Imamura, S. Muto, and N. Yokoyama, *Appl. Phys. Lett.* **49,** 1779 (1986); K. Imamura, S. Muto, H. Ohnishi, T. Fujii, and N. Yokoyama, *Electron. Lett.* **23,** 870 (1987).
456. T. Futatsugi, Y. Yamaguchi, K. Imamura, S. Muto, N. Yokoyama, and A. Shibatomi, *Jap. J. Appl. Phys.* **26,** L131 (1987); T. Futatsugi, Y. Yamaguchi, K. Imamura, S. Muto, N. Yokoyama, and A. Shibatomi, *J. Appl. Phys.* **65,** 1771 (1989).
457. N. Yokoyama, K. Imamura, S. Muto, S. Hiyamizu, and H. Nishi, *Jap. J. Appl. Phys.* **24,** L853 (1985).
458. G. P. Agrawal and N. K. Dutta, "Long Wavelength Semiconductor Lasers," Van Nostrand Reinhold, New York, 1986.
459. A. Yariv, "Quantum Electronic" 3rd ed. chapters 11 and 12. Wiley, New York, 1989.
460. I. Hayashi, *IEEE Trans. Electron. Devices* **ED-31,** 1630 (1984).
461. W. T. Tsang, *Appl. Phys. Lett.* **39,** 134 (1981); ibid. **40,** 217 (1982).
462. J. Nagle and C. Weisbuch, "Technical Digest," *ECOC'87,* Helsinki, part II, p. 25.
463. W. T. Tsang, *Appl. Phys. Lett.* **39,** 786 (1981).
464. M. G. A. Bernard and G. Durrafourg, *Phys. Stat. Sol.* **1,** 699 (1961).
465. J. Nagle and C. Weisbuch, *in* "Science and Engineering of 1 and 0 Dimensional Semiconductor Systems" C. M. Sotomayor-Torres and S. P. Beaumont, eds.), Plenum, New York, 1990.
466. J. P. Noblanc, *Surf. Sci.* **168,** 847 (1986).
467. See, e.g., S. Schmitt-Rink, D. S. Chemla, and D. A. B. Miller, *Adv. Phys.* **38,** 89 (1989).
468. J. Nagle, Thesis, Université Paris VI, 1987 (unpublished).
469. J. Nagle, S. D. Hersee, M. Krakowski, T. Weil, and C. Weisbuch, *Appl. Phys. Lett.* **49,** 1325 (1986).
469a. J. A. Brum, T. Weil, J. Nagle, and B. Vinter, *Phys. Rev.* **B34,** 2381 (1986).

470. M. Asada, A. Kameyama, and Y. Suematsu, *IEEE J. Quantum Electron* **QE-20,** 745 (1984); M. Yamanishi and I. Suemune, *Jpn. J. Appl. Phys.* **23,** 135 (1984).
471. S. Colak, R. Eppenga, and M. F. H. Schurmans, *IEEE J. Quantum Electron.* **QE-19,** 960 (1987).
471a. W. W. Chow, S. W. Koch, and M. Sargent III, *in* "International Conference on Quantum Electronics Technical Digest Series 1990," **8,** p. 208. OSA, Washington, D. C., 1990.
472. J. Nagle and C. Weisbuch, *in* "Quantum Wells and Superlattices Physics II," *SPIE Proc. 943,* (F. Capasso, G. Döhler, and J. N. Schulman, eds), p. 76, 1988.
473. P. L. Derry, A. Yariv, K. Y. Lau, N. Bar-Chaim, K. Lee, and J. Rosenberg, *Appl. Phys. Lett.* **50,** 1773 (1987).
474. K. Y. Lau, P. L. Derry, and A. Yariv, *Appl. Phys. Lett.* **52,** 88 (1988).
475. A. Yariv, *Appl. Phys. Lett.* **53,** 1033 (1988); erratum **55,** 603 (1989).
475a. E. Kapon, S. Simhony, J. P. Harbison, and L. T. Florez, *Appl. Phys. Lett.* **56,** 1825 (1990).
476. A. R. Adams, *Electron. Lett.* **22,** 250 (1986).
477. E. Yablonovitch and E. O. Kane, *IEEE J. Lightwave Technol.* **LT-4,** 504 (1988).
478. D. Fekete, K. T. Chan, J. M. Ballantyne, and L. F. Eastman, *Appl. Phys. Lett.* **49,** 1659 (1986).
479. P. K. York, K. Berenik, G. E. Fernandez, and J. J. Coleman, *Appl. Phys. Lett.* **54,** 499 (1989).
480. L. E. Eng, T. R. Chen, S. Sanders, Y. H. Zhuang, B. Zhao, A. Yariv, and H. Morkoc. *Appl. Phys. Lett.* **55,** 1378 (1989).
481. H. Temkin, T. Tanbun-Ek, and R. A. Logan, *Appl. Phys. Lett.* **56,** 1210 (1990).
482. T. Ohtoshi and N. Chinone, *IEEE Phot. Technol. Lett.* **1,** 117 (1989).
483. See, e.g., W. T. Tsang, *IEEE J. Quantum Electron.* **QE-20,** 1119 (1984).
483a. Y. Arakawa and H. Sakaki, *Appl. Phys. Lett.* **40,** 939 (1982).
483b. See, e.g., M. Sugimoto, N. Hamao, N. Takado, K. Asakawa, and T. Yuasa, *Japan J. Appl. Phys.* **28,** L1013 (1989) and references therein.
484. C. Lindstrom, T. L. Paoli, R. D. Burnham, D. R. Sirfres, and W. Streifer, *Appl. Phys. Lett.* **43,** 278 (1983).
485. R. D. Dupuis, R. L. Hartmann, and F. R. Nash, *IEEE Electron Device Lett.* **EDL-4,** 286 (1983).
486. Y. Arakawa and A. Yariv, *IEEE J. Quantum Electron.* **QE-21,** 1666 (1985).
487a. K. Uomi, T. Mishima, and N. Chinone, *Appl. Phys. Lett.* **51,** 78 (1987).
487b. K. Uomi, *Jpn. J. Appl. Phys.* **29,** 81 (1990).
487c. K. Uomi, T. Mishima, and N. Chinone, *Jpn. Appl. Phys.* **29,** 88 (1990).
488. K. Uomi and N. Chinone, *Japan J. Appl. Phys.* **28,** L1424 (1989).
488a. See, e.g., S. Takamo, T. Sasaki, H. Yamada, M. Kitamura, and I. Mito, *Electron. Lett.* **25,** 357 (1989).
488b. See, e.g., Y. Sakakibara, H. Watanabe, Y. Ohkura, Y. Kawama, N. Yoshida, S. Kakimoto and S. Ibuki, to be published.
489. See, e.g., the tutorial reviews by R. L. Byer, "Diode laser-pumped-Solid state laser," *Science* **239,** 742 (1988); *Laser Focus World,* March 1989.
490. W. Streifer, D. R. Scifres, G. L. Harnagel, D. F. Welch, J. Berger, and M. Sakamoto, *IEEE J. Quantum Elect.* **QE-24,** 883 (1988).
491. M. Mittelstein, Y. Arakawa, A. Larsson, and A. Yariv, *Appl. Phys. Lett.* **49,** 1689 (1986).
492. J. E. Epler, N. Holonyak, Jr., R. D. Burnham, C. Lindström, W. Streifer, and T. L. Paoli, *Appl. Phys. Lett.* **43,** 740 (1983).

493. M. Mittelstein, D. Mehuys, and A. Yariv, *Appl. Phys. Lett.* **54,** 1092 (1989).
494. M. Bagley, R. Wyatt, D. J. Elton, H. J. Wickes, P. C. Spurdens, C. P. Seltze, D. M. Cooper, and W. J. Devlin, *Electron Lett.* **26,** 269 (1990).
494a. S. Sanders, L. Eng, and A. Yariv, *CLEO 90, Tech. Digest Ser.* **7,** p. 480 USA, Washington, D. C., 1990.
495. Y. Tokuda, N. Tsokuda, K. Fujiwara, K. Hamanaka, and T. Nakayama, *Appl. Phys. Lett.* **49,** 1629 (1986).
496. K. Berthold, A. F. J. Levi, S. J. Pearton, R. J. Malik, W. Y. Jan, and J. E. Cunningham, *Appl. Phys. Lett.* **55,** 1382 (1989); K. Berthold, A. F. J. Levi, T. Tanbun-Ek, and R. A. Logan, *Appl. Phys. Lett.* **56,** 122 (1990).
496a. G. A. Evans, N. W. Carlson, J. M. Hammer, and R. A. Bartolini, *Laser Focus World*, p. 97, Nov. 1989.
496b. J. L. Jewell, A. Scherer, S. L. McCall, Y. H. Lee, S. Walker, J. P. Harbison, and L. T. Florez, *Electron. Lett.* **25,** 1124 (1989); A. Scherer, J. L. Jewell, Y. H. Lee, J. P. Harbison, and L. T. Florez, *Appl. Phys. Lett.* **55,** 2724 (1989).
496c. Y. H. Lee, J. L. Jewell, A. Scherer, S. L. McCall, J. P. Harbison, and L. T. Florez, *Electron. Lett.* **25,** 1377 (1989).
496d. R. S. Geels, S. W. Corzine, J. W. Scott, B. B. Young, and L. A. Coldren, *IEEE Photon. Technol. Lett.* **2,** 235 (1990).
496e. F. Koyama, S. Kinoshita, and K. Iga, *Appl. Phys. Lett.* **55,** 221 (1989); see also K. Iga and F. Koyama, *IEEE J. Quantum Electron.* **QE-24,** 1845 (1988) and references therein.
496f. Jemell M. Orenstein, A. C. Von Lehmen, N. G. Stoffel, C. Chang Husmain, J. P. Harbison, L. T. Florez, E. Clausen, J. L. Jewell, *CLEO 90, Tech. Dig. Ser.* **7,** p. 504. USA, Washington, D. C., 1990.
496g. J. L. Jewell, Y. H. Lee, R. S. Tucker, C. A. Burrus, A. Scherer, J. P. Harbison, L. T. Florez, C. J. Sandroff, S. L. McCall, and N. A. Olsson, *CLEO 90, Tech, Dig. Ser.* **7,** p. 500. USA, Washington, D. C., 1990.
497. J. Nagle and C. Weisbuch, *Inst. Phys. Conf. Ser.* **91,** 617 (1988).
498. M. B. Panish *et al.*, *J. Vac. Sci. Technol.* **B3,** 657 (1985).
499. Y. Miyamoto *et al.*, *Jpn. Appl. Phys.* **26,** L176 (1987).
500. A. Kasukawa, I. J. Murgatroyd, Y. Imago, N. Matsumoto, T. Fukushima, H. Okamoto, and S. Kashiwa, *Jpn. J. Appl. Phys.* **28,** L661 (1989); T. Tanbun-Ek, R. A. Logan, H. Temkin, K. Berthold, A. F. J. Levi, and S. N. G. Chu, *Appl. Phys. Lett.* **55,** 2283 (1989).
501. Y. Ohmori, Y. Suzuki, and H. Okamoto, *Jpn. J. Appl. Phys.* **24,** L657 (1985).
502. D. L. Partin, *IEEE J. Quantum Electron.* **QE-24,** 1716 (1988).
503. M. Ikeda, A. Torda, K. Nakano, Y. Mori, and N. Watanabe, *Appl. Phys. Lett.* **50,** 1033 (1987).
504. A. Yariv, *IEEE Circuits and Devices Magazine,* p. 25, Nov. 1989.
504a. For fabrication techniques, see e.g., "The Physics and Fabrication of Microstructures and Microdevices" (M. J. Kelly and C. Weisbuch, eds.), Springer, Berlin, 1986; T. H. P. Chang, *et al.*, *IBM J. Res. Develop.* **32,** 462 (1988).
505. Optical properties of 1D and 0D structures are reviewed by K. Kash, in *J. Lumin.* **46,** 69 (1990).
506. For a review of transport properties up to 1988, see *IBM J. Res. Devel.,* **32,** May and July 1988 special issues.
507. *TI Technical Journal,* July–August 1989.
508. "Nanostructure Physics and Fabrication" (M. A. Reed and W. P. Kirk, eds.), Academic Press, Boston, 1989.

509. "Science and Engineering of 1 and 0-Dimensional Semiconductor Systems" (C. M. Sotomayor-Torres and S. P. Beaumont, eds.), Plenum, New York, 1990.

509a. M. Reed, ed., "Nanostructured Systems," Academic Press, Boston, 1991.

510. C. W. J. Beenakker, and H. van Houten, "Quantum Transport in Semiconductor Nanostructures," *in* "Solid State Physics," (H. Ehrenreich and D. Turnbull, eds.), Academic Press, Boston, 1991.

511. See, e.g., A. N. Broers, *in* "The Physics and Fabrication of Microstructures and Microdevices" (M. J. Kelly and C. Weisbuch, eds.), p. 421 Springer, Berlin, 1986; *IBM J. Res. Devel.* **32,** 502 (1988); B.P. Van der Gaag and A. Scherer, *Appl. Phys. Lett.* **56,** 481 (1990).

512. See, e.g., R. Cheung, Y. H. Lee, K. Y. Lee, T. P. Smith III, D. P. Kern, S. P. Beaumont, and C. D. W. Wilkinson, *J. Vac. Sci. Technol.* **B7,** 1462 (1989) and references therein.

513. B. E. Maile, A. Forchel, R. Germann, D. Grützmacher, H. P. Meier, and J. P. Reithmaier, *J. Vac. Sci. Technol.* **B7,** 2030 (1989).

514. E. M. Clausen, H. G. Craighead, J. P. Harbison, A. Scherer, M. Schiavone, B. Van der Gaag, and L. T. Florez, *J. Vac. Sci. Technol.* **B7,** 2011 (1989).

515. S. P. Beaumont, *in* "Nanostructure Physics and Fabrication" (M. A. Reed and W. P. Kirk, eds.), p. 77. Academic Press, Boston, 1989.

516. H. Van Houten, B. J. Van Wees, M. G. J. Heigman, and J. P. André, *Appl. Phys. Lett,* **49,** 1781 (1986).

517. W. Hansen, M. Horst, J. P. Kotthaus, U. Merkt, Ch. Sikorski and K. Ploog, *Phys. Rev. Lett.* **58,** 2586 (1987).

518. T. J. Thornton, M. Pepper, H. Ahmed, D. Andrews, and G. J. Davies, *Phys. Rev. Lett.* **56,** 1198 (1986).

519. See, e.g., K. Kash, R. Bhat, D. B. Mahoney, P. S. D. Lin, A. Scherer, J. M. Worlock, B. P. Van der Gaag, M. Kosa and P. Grabbe, *Appl. Phys. Lett.* **55,** 681 (1989).

519a. T. Demel, D. Heitmann, and P. Grambow, *in* "Spectroscopy of Semiconductor Microstructures" (G. Fasol, A. Fasolino, and P. Lugli, eds.), NATO ASI Series B: Physics Vol. **206,** p. 75. Plenum, New York, 1990.

519b. A. Lorke and J. P. Kotthaus, private communication.

520. A. I. Ekimov and A. A. Onushenko, *Fiz, Tekh, Poluprovodn,* **16,** 1215 (1982) [*Sov. Phys. Semicond.* **16,** 775 (1982)]; A. I. Ekimov, I. A. Kudryavtsev, M. G. Ivanov and Al. L. Efros, *J. Lumin.* **46,** 83 (1990) and references therein.

521. R. P. Andres, R. S. Averback, W. L. Brown, L. E. Brus, W. A. Goddard III, A. Kaldon, S. G. Louie, M. Moscovits, P. S. Peercy, S. J. Riley, R. W. Siegel, F. Spaepen, and Y. Wang, *J. Mat. Res.* **4,** 704 (1989).

522. E. F. Hilinski, P. A. Lucas, and Y. Wang, *J. Chem. Phys.* **89,** 3435 (1988).

523. J. E. Mac Dougall, H. Eckert, G. D. Stucky, N. Herron, Y. Wang, K. Moller, T. Beon, and D. Cox, *J. Am. Chem. Soc.* **111,** 8006 (1989); G. D. Stucky and J. E. Mac Dougall, *Science* **247,** 669 (1990).

524. S. Hayashi, H. Sanda, M. Agata, and K. Yamamoto, *Phys. Rev.* **B40,** 5544 (1989).

525. L. Brus, *IEEE J. Quantum Electron.* **QE-22,** 1909 (1986).

525a. M. L. Steigerwald and L. E. Brus, *Ann. Rev. Mater. Sci.* **19,** 471 (1989).

526. C. T. Dameron, R. N. Reese, R. K. Mehra, A. R. Kortan, P. J. Caroll, M. L. Steigerwald, L. E. Brus, and D. R. Winge, *Nature,* **338,** 596 (1989).

527. P. M. Petroff, J. Gaines, M. Tsuchiya, R. Simes, L. Coldren, H. Kroemer, J. English and A. C. Gossard, *J. Cryst. Growth,* **95,** 260 (1989).

527a. M. Tanaka and H. Sakaki, *Jpn. J. Appl. Phys.* **27,** L2025 (1988).

527b. Y. Tokura, H. Saito, and T. Fukui, *J. Cryst. Growth* **94,** 46 (1989); T. Fukui, H. Saito, Y. Tokura, K. Tsubaki, and N. Susa, *Surf. Science* **228,** 20 (1990).
528. P. C. Morais, H. M. Cox, P. L. Bastos, D. M. Hwang, J. M. Worlock, E. Yablonovitch and R. E. Nahory, *Appl. Phys. Lett.* **54,** 442 (1989); H. M. Cox, P. S. Lin, A. Yi-Yan, K. Kash, M. Sets and P. Bastos, *Appl. Phys. Lett.* **55,** 472 (1989).
529. Y. D. Galeuchet, P. Roentgen, S. Nilson and V. Graf, *in* "Science and Engineering of 1-and 0-Dimensional Semiconductor Systems" (C. M. Sotomayor-Torres and S. P. Beaumont, eds.), Plenum, New York, 1990, p. 1.
529a. Y. Iimura, S. Shimomura, K. Nagata, S. Den, Y. Aoyagi, and S. Namba, *Jpn. J. Appl. Phys.* **28,** L1083 (1989).
529b. E. Colas, E. Kapon, S. Simhony, H. M. Cox, R. Bhat, K. Kash, and P. S. D. Lin, *Appl. Phys. Lett.* **55,** 867 (1989).
530. A. Usui, H. Sunakawa, F. J. Stützler, and K. Ishidu, *Appl. Phys. Lett.* **56,** 289 (1990).
530a. R. A. Demmin and T. E. Madey, *J. Vac. Sci. Technol.* **A7,** 1954 (1989).
530b. L. Goldstein, F. Glas, J. Y. Marzin, M. N. Charasse, and G. Leroux, *Appl. Phys. Lett.* 1099 (1985).
530c. M. Zinke-Allmang, L. C. Feldman, and S. Nakahara, *J. Vac. Sci. Technol.* **B6,** 1137 (1988).
531. A. Scherer and M. L. Roukes, *Appl. Phys. Lett.* **55,** 377 (1989).
532. T. Hiramoto, K. Hirakawa, Y. Iye and T. Ikoma, *Appl. Phys. Lett.* **54,** 2103 (1989); D. G. Deppe and N. Holonyak, Jr., *J. Appl. Phys.* **64,** R93 (1988).
533. H. A. Zarem, P. C. Sercel, M. E. Hoenk, J. A. Lebens, and K. J. Vahala, *Appl. Phys. Lett.* **54,** 2692 (1989).
534. J. Cibert, P. M. Petroff, G. J. Dolan, S. J. Pearton, A. C. Gossard, and J. H. English, *Appl. Phys. Lett.* **49,** 1275 (1986).
535. See, e.g., the detailed tutorial review by A. D. Stone *in* "Physics and Technology of Submicron Structures "(H. Heinrich, G. Bauer, and F. Kuchar eds.), p. 108, Springer, Berlin, 1988, and references therein.
536. H. van Houten, B. J. van Wees, and C. W. J. Beenakker, *in* "Physics and Technology of Submicron Structures" (H. Heinrich, G. Bauer, and F. Kuchar eds.), p. 198 Springer, Berlin, 1988.
537. R. Landauer, *IBM J. Res. Devel.* **1,** 223 (1970).
538. R. Landauer, *IBM J. Res. Devel.* **32,** 306 (1988).
539. A. D. Stone and A. Szafer, *IBM J. Res. Develop.* **32,** 384 (1988).
540. R. Landauer, *in* "Nanostructure Physics and Fabrication" (M. A. Reed and W. P. Kirk, eds.), p. 17, Academic Press, Boston, 1989.
541. C. P. Umbach, P. Santhanam, C. Van Haesendonck, and R. A. Webb, *Appl. Phys. Lett.* **50,** 1289 (1987).
542. M. L. Roukes, A. Scherer, S. J. Allen Jr., H. G. Craighead, R. M. Ruthen, E. D. Beebe, and J. P. Harbison, *Phys. Rev. Lett.* **59,** 3011 (1987).
543. C. J. B. Ford, S. Washburn, M. Büttiker, C. M. Knoedler, and C. M. Jong *Phys. Rev. Lett.* **62,** 2724 (1989).
544. M. L. Roukes, A. Scherer, and B. P. Van der Gaag, *Phys. Rev. Lett.* **64,** 992 (1990).
545. See, e.g., H. U. Baranger and A. D. Stone, *Phys. Rev. Lett.* **63,** 414 (1989), H. U. Baranger and A. D. Stone, *in* "Science and Engineering of 1 and 0-Dimensional Semiconductors" (S. P. Beaumont and C. M. Sotomayor Torrés, eds.), Plenum, New York, 1990 p. 121 and references therein.
546. C. W. J. Beenaker and H. Van Houten, *Phys. Rev. Lett.* **63,** 1857 (1989).
547. L. W. Molenkamp, A. A. M. Staring, C. W. J. Beenakker, R. Eppenga, C. E. Tim-

mering, J. G. Williamson, C. J. P. M. Harmans, and C. T. Foxon, *Phys. Rev.* **B41,** 1274 (1990).

548. Y. Imry, *in* "Directions in Condensed Matter Physics" (G. Grinstein and G. Mazenko, eds.), p. 101. World Scientific, Singapore, 1986.
549. P. A. Lee, R. A. Webb, and B. L. Al'tshuler, "Mesoscopic Phenomena in Solids." Elsevier, Amsterdam, 1990.
550. R. A. Webb, *in* "Nanostructure Physics and Fabrication" (M. A. Reed and W. P. Kirk, eds.), p. 43. Academic Press, Boston, 1989.
551. B. J. Van Wees, H. Van Houten, C. W. J. Beenakker, J. G. Williamson, L. P. Kouwenhoven, D. Van der Marel, and C. T. Foxon, *Phys. Rev. Lett.* **60,** 848 (1988).
552. D. A. Wharam, T. J. Thornton, R. Newbury, M. Pepper, H. Ahmed, J. E. F. Frost, D. G. Hasko, D. C. Peacock, D. A. Ritchie, and G. A. C. Jones, *J. Phys. C.* **21,** L209 (1988).
553. J. Spector, H. L. Störmer, K. W. Baldwin, L. N. Pfeiffer, and K. W. West, *Appl. Phys. Lett.* **56,** 1290 (1990).
554. U. Sivan, M. Heiblum, C. P. Umbach, and H. Shtrikman, *Phys. Rev.* **B41,** 7937 (1990).
555. L. P. Kouwenhoven, B. J. Van Wees, F. W. J. Hekking, and K. J. P. M. Harmans, *in* "Localization and Confinement of Electrons in Semiconductors" (K. von Klitzing, ed.), Springer, Berlin, 1990.
556. B. J. Van Wees, L. P. Kouwenhoven, J. G. Williamson, C. E. Timmering, M. E. I. Borkaart, C. T. Foxon, and J. J. Harris, *Phys. Rev. Lett.* **62,** 2523 (1989).
557. A. Lorke, J. P. Kotthaus, and K. Ploog, *Phys. Rev. Lett.* **64,** 2559 (1990).
558. J. Spector, H. L. Störmer, K. W. Baldwin, L. N. Pfeiffer, and K. W. West, *Surf. Science,* **228,** 283 (1990).

558a. M. A. Reed, J. N. Randall, R. J. Aggarwal, R. J. Matyi, T. M. Moore, and A. E. Wetzel, *Phys. Rev. Lett.* **60,** 1535 (1988).

558b. T. P. Smith III, K. Y. Lee, C. M. Kinsedler, J. M. Hong, P. Kern, *Phys. Rev.* **B38,** 2172 (1988).

559. J. N. Randall, M. A. Reed and G. A. Frazier, *J. Vac Sci. Technol.* **B7,** 1398 (1989).
560. R. Bate, G. Frazier, W. Frensley, and M. Reed, *TI Technical Journal,* p. 13, July–August 1989.
561. See, e.g., *Proceedings of the NATO ARW "Granular Electronics Il Ciocco (Italy), 1990,* Plenum, New York, 1991.
562. J. H. F. Scott, S. B. Field, M. A. Kastner, H. I. Smith, and D. A. Antoniadis, *Phys. Rev. Lett.* **62,** 583 (1989).
563. H. Van Houten and C. W. J. Benakker, *Phys. Rev. Lett.* **53,** 1893 (1989); M. A. Kastner, S. B. Field, U. Meirav, J. H. F. Scott Thomas, D. A. Antoniadis, and H. I. Smith, *Phys. Rev. Lett.* **63,** 1894 (1989).
564. K. K. Likharev, *IBM J. Res. Devel.* **32,** 144 (1988).
565. L. J. Geerligs, V. F. Anderegg, P. A. M. Holweg, J. E. Mooij, H. Pothier, D. Esteve, C. Urbina, and M. H. Devoret, *Phys. Rev. Lett.* **64,** 2691 (1990).
566. A. Palevski, C. P. Umbach, and M. Heiblum, *Phys. Rev. Lett.* **62,** 1776 (1989) and *Appl. Phys. Lett.* **55,** 1421 (1989).
567. T. F. Gaylord and K. F. Brennan, *J. Appl. Phys.* **65,** 814 (1989).
568. S. Datta, *Superlattices and Microstructures* **6,** 83 (1989) and references therein.
569. F. Sols, M. Macucci, U. Ravaioli, and K. Hess, *Appl. Phys. Lett.* **54,** 350 (1989).
570. S. Beaumont, private communication.
571. G. Frazier, *in* "Concurrent Computations" (S. Tewsburg, B. W. Dickinson, and S. C. Schwartz, eds.), Plenum, New York, 1988.
572. R. T. Bate, *Solid State Technol.* p. 101, November 1989.

573. G. Weisbuch, "Dynamics of Complex Systems," Addison-Wesley, Reading, Massachusetts, 1991 [in French, "Dynamique des Systèmes Complexes," Interéditions, 1989].
574a. N. Margolus and T. Toffoli, *Complex Systems* **1,** 967 (1987).
574b. S. Wolfram, *Rev. Mod. Phys.* **55,** 601 (1984).
574c. P. Manneville *et al.*, eds, "Cellular Automata and Modeling of Complex Physical Systems," Springer, Berlin, 1989.
574d. C. A. Mead and L. Conway, "Introduction to VLSI Systems," Addison-Wesley, Reading, Massachusetts, 1980.
575. E. Hanamura, *Phys, Rev.* **B37,** 1273 (1988).
576. T. Takagahara, *Phys. Rev.* **B36,** 9293 (1987).
577. G. W. Bryant, *Phys. Rev.* **B37,** 8763 (1988).
578. Y. Z. Hu, S. W. Koch, M. Lindberg, N. Peyghambarian, E. L. Pollock, and F. F. Abraham, *Phys. Rev. Lett.* **64,** 1805 (1990).
579. S. Schmitt-Rink, D. A. B. Miller and D. S. Chemla, *Phys. Rev.* **B38,** 8113 (1988).
580. L. Banyai, Y. Z. Hu, M. Lindberg, and S. W. Koch, *Phys. Rev.* **B38,** 8142 (1988).
581. L. Rothberg, T. M. Jedju, W. L. Wilson, M. G. Bawendi, M. L. Steigerwald, and L. E. Brus, *IQEC 90, Tech. Dig. Ser.* p. 34. USA, Washington, D. C., 1990.
582. D. A. B. Miller, D. S. Chemla, and S. Schmitt-Rink, *Phys. Rev.* **B33,** 6976 (1986).
583. See, e.g., K. Kash, ref. 505; B. E. Maile, ref. 513, and A. Izrael, B. Sermage, J. Y. Marzin, A. Oyrazzadan, R. Azoulay, J. Etrillard, V. Thierry-Mieg, and L. Henry, *Appl. Phys. Lett.* **56,** 830 (1990).
583a. M. Kohl, D. Heitmann, P. Grambow, and K. Ploog, *Phys. Rev. Lett.* **63,** 2124 (1989).
583b. M. Kohl, D. Heitmann, P. Grambow, and K. Ploog *in* "Condensed Systems of Low Dimensionality" (J. L. Beeby, ed.). NATO ARW, to be published. Plenum, London.
584. Y. Arakawa and H. Sakaki, *Appl. Phys. Lett.* **24,** 195 (1982).
585. M. Asada, Y. Miyamoto, and Y. Suematsu, *IEEE J. Quantum Electron.* **QE-22,** 1915 (1986).
586. Y. Arakawa, K. Vahala, and A. Yariv, *Surf. Sci.* **174,** 155 (1986).
587. T. Takahashi and Y. Arakawa, *Optoelectronics* **3,** 155 (1988).
588. Y. Arakawa and H. Sakaki, *Appl. Phys. Lett.* **40,** 939 (1982).
589. Y. Arakawa, K. Vahala, A. Yariv, and K. Lau, *Appl. Phys. Lett.* **47,** 1142 (1985).
590. E. Kapon, S. Simhony, R. Bhat and D. M. Hwang, *Appl. Phys. Lett.* **55,** 2715 (1989).
591. Y. Miyamoto, M. Cao, Y. Shingai, K. Furuya, Y. Suematsu, K. G. Ravikemar, and S. Arai, *Jpn. J. Appl. Phys.* **26,** L225 (1987).
592. K. J. Vahala, *IEEE J. Quantum Electron.* **QE-24,** 523 (1988).
593. H. Sakaki, *Inst. Phys. Conf. Ser.*, **96,** 13, (1989).
594. K. Vahala, Y. Arakawa and A. Yariv, *Appl. Phys. Lett.* **51,** 365 (1987).
595. T. T. J. M. Berendschet, H. A. J. M. Reinen, H. A. Bluyssen, C. Harder, and H. P. Meier, *Appl. Phys. Lett.* **54,** 1827 (1989).
596. A. Nakamura, H. Yamada, and T. Tokizaki, *Phys. Rev.* **B40,** 8585 (1989).
597. T. Itoh, M. Furumiya, T. Ikehara, and C. Gourdon, *Solid State Commun.* **73,** 271 (1990).
598. A. Nakamura, T. Tokizaki, T. Kataoka, N. Sugimoto, and T. Manabe, *CLEO 90, Tech. Dig. Ser.* **7** p. 178. USA, Washington, D. C., 1990.
599. F. Hache, D. Ricard and C. Flytzannis, *Appl. Phys. Lett.* **55,** 1504 (1989).
600. F. Hache, D. Ricard, and C. Flytzannis, *Appl. Phys. Lett.* **55,** 1584 (1989).
601. D. Cotter and H. P. Guidlestone, *IQEC 90, Tech. Dig. Ser.* p. 34. USA, Washington, D. C., 1990; *Electron. Lett.* **26,** 183 (1990).
602. P. Horan and W. Blau, *J. Opt. Soc. Am.* **B7,** 304 (1990).

603. F. de Rougemont, R. Frey, P. Roussignol, D. Ricard, and C. Flytzannis, *Appl. Phys. Lett.* **50,** 1619 (1987).
604. Y. Arakawa, *in* "Waveguide Optoelectronics," Plenum, New York, 1991.
605. J. T. Remillard *et al.*, *IEEE J. Quant. Electr.* **QE-25,** 408 (1989).
606. Z. Jakubcyk *et al.*, *Opt. Lett.* **12,** 750 (1987).
607. J. Yumoto *et al.*, *Opt. Lett.* **12,** 832 (1987).

Index

E

F

G

H

I

L

M

N